SEASONAL PATTERNS OF STRESS, IMMUNE FUNCTION, AND DISEASE

As the seasons change, animals face alterations in environmental stressors. In particular, the prevalence and intensity of pathogenic infection are often seasonal. This book presents evidence that infection is cyclical with the seasons and that this phenomenon is mirrored in cycles of immune function.

The goal of this book is to identify the mechanisms by which the immune system is bolstered to counteract seasonally recurrent stressors, such as extreme temperature reductions and food shortages. The authors consider how such environmental changes create energetically demanding conditions that can compromise host immunity and lead to illness and death. Specifically, stress, infectious diseases, autoimmune diseases, and human cancers are examined, and the role of hormones such as melatonin and glucocorticoids is considered. The book begins with an overview of seasonality, biological rhythms and photoperiodism, and basic immunology, and continues with the characterization of seasonal fluctuations in disease prevalence, energetics, and endocrinology as they relate to immune function. Finally, the clinical significance of seasonal patterns in immune function is addressed to emphasize the role that seasonal changes in host immunity and hormones may play in the development and treatment of infections.

This is the first monograph to examine seasonal immune function from an interdisciplinary perspective. Practitioners as well as advanced undergraduates and graduate students in biology, immunology, human and veterinary medicine, neuroscience, endocrinology, and zoology will find its approach both insightful and relevant.

Randy J. Nelson is Distinguished Professor of Social and Behavioral Sciences in the Departments of Psychology and Neuroscience at The Ohio State University.

Gregory E. Demas is Assistant Professor in the Department of Biology at Indiana University.

Sabra L. Klein is a Postdoctoral Fellow in the Department of Molecular Microbiology and Immunology at The Johns Hopkins Bloomberg School of Public Health.

Lance J. Kriegsfeld is a Postdoctoral Fellow in the Department of Psychology at Columbia University.

SEASONAL PATTERNS OF STRESS, IMMUNE FUNCTION, AND DISEASE

RANDY J. NELSON
The Ohio State University

GREGORY E. DEMAS
Indiana University

SABRA L. KLEIN
*Johns Hopkins School of
Public Health*

LANCE J. KRIEGSFELD
Columbia University

CAMBRIDGE
UNIVERSITY PRESS

CAMBRIDGE UNIVERSITY PRESS
Cambridge, New York, Melbourne, Madrid, Cape Town, Singapore, São Paulo

Cambridge University Press
The Edinburgh Building, Cambridge CB2 2RU, UK

Published in the United States of America by Cambridge University Press, New York

www.cambridge.org
Information on this title: www.cambridge.org/9780521590686

First published 2002
This digitally printed first paperback version 2005

A catalogue record for this publication is available from the British Library

Library of Congress Cataloguing in Publication data
Seasonal patterns of stress, immune function, and disease / Randy J. Nelson . . . [et al.].
p. ; cm.
Includes bibliographical references and index.
ISBN 0-521-59068-X (hb)
1. Diseases – Seasonal variations.   2. Immunity.   I. Nelson, Randy Joe.
[DNLM: 1. Periodicity.   2. Communicable Diseases – epidemiology.   3. Immunity –
physiology.   4. Seasons. QT 167 S439 2002]
RA793.S43   2002
616.07′.9 – dc21   2001037415

ISBN-13 978-0-521-59068-6 hardback
ISBN-10 0-521-59068-X hardback

ISBN-13 978-0-521-02117-3 paperback
ISBN-10 0-521-02117-0 paperback

For our Families . . .

Contents

Acknowledgments

We benefited from the assistance of many individuals during the preparation of this book. We thank Amanda Byrne, Noelle Shearer, Courtney DeVries, Jim Power, and Brian Prendergast for assistance with figures. We are also grateful to Noelle Shearer for bibliographic assistance.

We also thank the individuals at Cambridge University Press, including Robin Smith, who initially encouraged us to write this book, and Ellen Carlin, who inherited this project and served as editor. Others who deserve special mention are Karin Pearl Horler, Cambridge University Press associate production controller; Michie Shaw, project manager at TechBooks, Inc., compositors; and Vivian Mason, copyeditor. This book is much better than our original manuscript because of their hard work, dedication, and uncompromising efforts.

Foreword

Most animals exhibit seasonal variation in their reproduction and survival simply because most of them live in environments in which food availability varies seasonally. This is true in the tropics as well as at the higher latitudes, but at the higher latitudes the combined challenge of food shortage and low temperature makes winter a particularly difficult time to reproduce and survive.

Traditionally, vertebrate physiologists interested in seasonal phenomena have focused on reproduction and largely ignored survival. In part, this is because of the discovery 70-odd years ago that some temperate zone animals use variation in day length to synchronize their reproduction with seasonally changing environmental conditions. This discovery meant that researchers interested in seasonality could bring the power of the scientific method into play simply by installing a light timer on the wall of an animal room. The result of several decades of experimentation using this technique is a robust body of knowledge about how photoperiod regulates reproduction. Indeed, this is an epic, albeit an as yet unfinished epic, in biological research.

Randy Nelson and his colleagues have added a whole new dimension to the study of seasonality by shifting the focus from reproduction to survival. This book is a pioneering effort to define this field of study, and it will prove to be a milestone in research on seasonality. The specific hypothesis under consideration is that "individuals have evolved mechanisms to bolster immune function in order to counteract seasonally recurrent stressors that may otherwise compromise immune function." Stated differently, the authors visualize seasonal variation in survival as an interaction between two factors: the suppression of immune response attributable to changing energetic conditions and an endogenous rhythm of enhancement of immune response that is dependent on photoperiod, clocks, and melatonin.

Of particular importance is the authors' postulate that seasonal variation in certain human diseases may be caused at least in part by variation in immune responsiveness and, in turn, by variation in photoperiod. The evolutionary history of humans suggests that almost all ancestral stocks of *Homo sapiens* have lived at one time or another in situations where photoperiodic regulation of seasonal adjustments might have proven advantageous. Whereas the available evidence suggests that reproduction is not susceptible to photoperiodic regulation in humans, immune function very well may be. This possibility deserves serious consideration, and the publication of this book assures that it will receive the attention it needs.

On a still broader scale, this book will arouse interest in seasonality in general. Research on the way photoperiod regulates reproduction generated intense interest in the way reproduction is regulated seasonally by factors other than photoperiod. Likewise, interest in the way photoperiod regulates immune function seasonally will elicit the question of whether immune function varies seasonally in species that are not responsive to photoperiod and, if so, how this is done.

For those of us that have been around for awhile, it is exciting to see a renewed interest in the way animals, including humans, respond physiologically to seasonal variation and to visualize how this renewed interest might expand our knowledge in the future.

Frank Bronson
Austin, TX
July 2001

Preface

Our environment changes seasonally. From the amount of light that we are exposed to each day to the availability of food and water, we are confronted with an ever-changing environment. As a consequence of these environmental changes, all inhabitants of Earth show fluctuations in energetically expensive physiological and behavioral processes throughout the year. Thus, adaptations have evolved so that energetically demanding processes coincide with abundant resources or other environmental conditions that promote survival and, ultimately, reproductive success. The study of seasonal changes in physiology and behavior in the field is typically limited to population dynamics among nondomesticated animals. Seasonal patterns of behaviors, including those associated with reproduction, social behavior, and daily activity, as well as seasonality in the physiology that underlies these behaviors, are well documented (Bronson and Heideman, 1994). Most studies of seasonality involve some aspect of seasonal breeding, such as mating, birth, or parental care. Seasonal patterns of illness and death are equally salient among natural populations of animals in the wild, but the underlying factors driving these seasonal patterns are much less studied than the extrinsic factors driving seasonal patterns of breeding. Generally, illness and death are most common during the fall and winter, compared with spring and summer, for most nontropical species.

One goal of our book is to describe an emerging hypothesis that individuals have evolved mechanisms to bolster immune function to counteract seasonally recurrent stressors that may otherwise compromise immune function. Seasonally recurrent stressors include food shortages, low ambient temperatures, and lack of cover from hungry predators. Survival thus represents maintaining the well-known energy budget and also maintaining a balance between stress-induced immunosuppression and endogenous rhythms of enhanced immune function. In Chapter 1, we briefly review the geophysical factors leading to seasons and the development of annual and daily biological rhythms to reflect these seasons. Then, the biological and phenomenological features of biological clocks and their associated rhythms are summarized. Sufficient information will be provided to allow the reader who is naive to the field of biological timing to understand the evolutionary and physiological issues related to temporal organization that are referred to throughout the book.

The concept that seasonal patterns in the prevalence of infection exist is not new. In fact, the "flu season" has been documented for hundreds of years (Sakamoto-Momiyama, 1977). The novelty of our approach is that we consider how changes in the environment (e.g., reduced temperatures, reduced food and water availability) act as stressors to create energetically demanding conditions that can indirectly cause illness and death by compromising host immunity. Thus, we argue that mechanisms have evolved to resist seasonal stress-induced susceptibility to infection as a temporal adaptation to promote survival. To understand the immunological processes that underlie susceptibility and resistance to infection, Chapter 2 provides a detailed overview of the immune system and responses important for overcoming infection.

In Chapter 3, we review the substantial evidence for seasonal cycles in disease prevalence in both humans and nonhuman animals. Not surprisingly, most animals, including humans, are more likely to become ill and die during the winter, compared with summer. A number of factors contribute to this pattern, but we will support the notion in Chapters 3 and 4 that, without "programmed" bolstering of immune function during the winter, the incidence of disease and death would be elevated. The studies presented in Chapter 3 illustrate that seasonality in the intensity and prevalence of infection occurs in many species and may be linked to changes in the host as opposed to seasonal changes in the prevalence of the pathogen or vector. In Chapter 4, we focus on the literature that reports a seasonal change in immune function. In this chapter, we will argue that seasonal fluctuations in immune function represent changes in the host, rather than changes in the vector or presence of the pathogen. In some cases, immune function is compromised and disease rates are elevated during the winter. Although this observation may seem to rule out our working hypothesis of enhanced immune function during the winter, we will argue that seasonal stressors (reduced food availability, low temperatures, etc.) may be too pronounced to override the endogenous enhancement of winter immune function. The support for this notion is presented in Chapters 5 and 6. Many extrinsic factors co-vary across the year, but photoperiod (or day length) is an environmental cue that seems to be used by many individuals to time seasonal events. The effects of photoperiod on immune function is reviewed in Chapter 5; in virtually all reported cases, short days evoke enhancement of immune function. Melatonin, a hormone secreted from the pineal gland, mediates photoperiod, and the immunoenhancing effects of melatonin also will be reviewed in this chapter.

If animals (including humans) are capable of enhancing immune function during the winter, then why has natural selection not favored the enhancement of immune function during the summer when many more social and sexual interactions are occurring and diseases are likely to be spread? The energetic costs of maintaining immunologic function and the incompatibility between these costs and reproductive costs is discussed in Chapter 6. Also, the concept of energetic stressors and their effects on immune function is addressed in this chapter. Several hormones, including sex steroids, glucocorticoids, and prolactin, affect immune function and are affected by environmental perturbations. Thus, Chapter 7 provides evidence that steroid and peptide hormones affect the immune system and may mediate seasonal patterns in infection.

Why study seasonal patterns in immune function and illness? From a clinical perspective, understanding the role of host hormones and immunity in seasonal changes in infection may influence the development and administration of treatments for infections. From a biological perspective, the data reviewed in this book suggest that seasonal stressors impact host resistance and susceptibility to infection possibly via changes in endocrine–immune interactions. In Chapter 8, we examine seasonal changes in infection from both a clinical and basic science perspective to address why studies examining seasonal changes in immunity and infection are important. We will also examine what role these studies could play in the treatment of disease, as well as in increasing our understanding of organism–environment interactions. Perhaps more aggressive antibiotic treatment is warranted during the winter than summer for some conditions. Seasonal changes in the immune function of the host individual are rarely, if ever, considered in the clinical treatment of patients (both human and nonhuman). There are several gaps in our knowledge of seasonal patterns in population dynamics; we hope this book provides alternative hypotheses and an understanding that host immunity and susceptibility to infection play a role in mediating population fluctuations among both human and nonhuman animals.

Randy J. Nelson
Gregory E. Demas
Sabra L. Klein
Lance J. Kriegsfeld
December 2000

1

Seasonality

Live in each season as it passes; breathe the air, drink the drink, taste
the fruit, and resign yourself to the influences of each. Let them be
your only diet, drink, and botanical medicine.

Henry David Thoreau, 1906
Journals (entry for 23 August 1853)

1.1. Introduction

Life on Earth evolves in an environment with pronounced temporal fluctuations. Rivers flow to the sea; the tides ebb and rise. Light availability and
temperature vary predictably throughout each day and across the seasons.
These fluctuations in environmental factors exert dramatic effects on biotic
activities. For example, the biochemical machinery of plants and animals undergoes daily adjustments in production, performing some processes only at
night and others only during the day. Similarly, daily peaks in the metabolic
activity of homeothermic animals tend to coincide with the daily onset of
increased physical activity. Elevated activity alone does not drive metabolic
rates; rather, the general pattern of metabolic needs are anticipated by
reference to an internal biological clock. The ability to *anticipate* the onset
of the daily light and dark periods confers sufficient advantages that endogenous, self-sustained circadian clocks are virtually ubiquitous among extant
organisms (Takahashi 1996; Menaker et al. 1997).

In addition to synchronizing biochemical, physiological, and behavioral
activities to the external environment, biological clocks are important to multicellular organisms in synchronizing internal processes. If a specific biochemical process is most efficiently conducted in the dark, then individuals
that mobilize metabolic precursors, enzymes, and energy sources just prior
to the onset of dark would presumably have a selective advantage over individuals that organized their internal processes at random times. Thus, there
is a daily temporal chain, or phase relationship, to which all biochemical,
physiological, and behavioral processes are linked.

1

The scientific study of biological clocks and their associated rhythms is called chronobiology. This field borrows terms and concepts freely from engineering disciplines. For the purpose of this book, a biological clock is defined as a self-sustained oscillator (Pittendrigh 1960). A "rhythm" is a recurrent event that is characterized by its period, frequency, amplitude, and phase. "Period" is the length of time required to complete one cycle of the rhythm under study. "Frequency" is computed as the number of completed cycles per unit of time. The amount of change above and below the average value of a rhythm is called "amplitude." The "phase" represents a point on the rhythm relative to some objective temporal point associated with another cycle.

When animals, and virtually all plants, are placed in constant conditions, the period of their daily rhythms drifts slightly from 24 hours each succeeding day; these rhythms are called circadian (*circa* = about; *dies* = day [Halberg 1961]) rhythms. Individuals that are no longer synchronized (entrained) to the external factors are said to be free-running. The recurrence of free-running rhythms generated in the absence of external temporal cues is the most convincing evidence for an endogenous timekeeper. Additional compelling evidence that supports the contention that biological rhythms are generated from within the organism and not driven by unknown geophysical forces includes the observation that animals maintained in constant conditions aboard a spacecraft orbiting far above the planet (where geophysical cues are presumably absent) continue to display biological rhythms with periods similar to those observed on Earth. Other features of circadian biological rhythms are listed in Table 1.1.

The periods of some biological rhythms, including most central nervous system, cardiovascular, and respiratory rhythms, vary widely within the same

Table 1.1. *Features of Circadian Rhythms (from Pittendrigh 1960).*

1. Free-running circadian rhythms only approximate the period of the daily rotation of the Earth.
3. Circadian rhythms are endogenous.
4. Circadian rhythms are typically self-sustained oscillators.
5. Circadian rhythms are unlearned.
6. Circadian rhythms are extant at all levels of physiological organization.
7. Free-running circadian rhythms display little variance.
8. Circadian rhythms compensate for temperature.
9. Circadian rhythms are often affected by light intensity.
10. Circadian rhythms are usually unaffected by chemical perturbations.
11. Phase of circadian rhythms can be shifted by a single light (or temperature) signal.
12. Circadian rhythms can be synchronized by a limited set of environmental periodicities.

Table 1.2. *Relationship between Biological Rhythms and Environmental Cycles.*

Biological Rhythm	Environmental Cycle	Period Length	
		Entrained	Free-running
Circadian	Revolution of the Earth	24 hours	22–26 hours
Circatidal	Ebb and rise of tides	12.4 hours	11–14 hours
Circalunar	Phases of the moon	29.5 days	26–32 days
Circannual	Seasons of the year	365.25 days	330–400 days

individual. During physical exertion, for example, the period of heartbeats decreases (i.e., the frequency increases). Other biological rhythms, such as the ovarian rhythm, may remain relatively constant within the same individual, but vary significantly among different conspecific individuals. Four types of biological rhythms are typically coupled with geophysical cycles and generally do not vary under natural conditions. These relatively constant biological rhythms correspond to the periods of the geophysical cycles of night and day (circadian), the tides (circatidal), the phases of the moon (circalunar), and the seasons of the year (circannual) (Table 1.2). To appreciate the mechanisms underlying seasonal fluctuations in immune function and disease processes, the remainder of this chapter will emphasize the mechanisms of circadian and circannual rhythms as they relate to seasonality.

Not unexpectedly, circadian clocks have evolved in organisms that typically live for more than one day. Short-lived (i.e., <1 day) species, such as protists, do not exhibit true circadian rhythms. Similarly, biological clocks that measure yearly intervals are present only in animals that, on average, live longer than one year. During the course of a year, conditions can change dramatically. Seasonal fluctuations in energy availability create thermoregulatory and other physiological demands that serve as potent selective forces in the evolution of life. At high latitudes and altitudes, the seasonal fluctuations in temperature between winter and summer are substantial, and individuals that inhabit these challenging conditions have evolved mechanisms to cope with the wide alterations in ambient conditions. Animals that withstand severe seasonal fluctuations undergo such striking seasonal adaptations, that summer- and winter-captured individuals of the same species have often been mistaken for different species (Figure 1.1). Development of these seasonal adaptations permits plants and animals to exploit specific spatial niches that vary in quality over time.

Nontropical animals display many distinct behavioral adaptations to cope with the challenge of winter survival (Moffatt et al. 1993; Bronson &

Fig. 1.1. Siberian hamsters in summer and winter conditions. These animals undergo significant seasonal changes in physical appearance. In addition to the season change in coat color, which helps them blend into the background environment of summer (dark) or winter (white), many other adaptations have evolved to combat the energetic demands of winter. Some of these winter adaptations include changes in food intake and body mass, torpor, foraging, nest-building, reductions of social interactions including sexual interactions. Some animals such as collared lemmings, develop extra "digging claws" during short day exposure. These claws help these lemmings did through the winter snow to find food. (Photo by Aaron Jasnow.)

Heideman 1994). Behaviorally, individuals may hibernate, estivate, or migrate during specific energetically challenging seasons of the year. Social organization may change; highly territorial individuals that are aggressive during the breeding season may form social groups during the winter to reap the thermoregulatory benefits of huddling (McShea 1990; Ancel et al. 1997). Similarly, small mammals may join communal nests during the winter where warmth and humidity are conserved (Madison 1984). The foraging and feeding behaviors of animals may also change seasonally. For example, territorial birds may join large winter foraging flocks, or individuals may change from a nocturnal activity pattern during the summer to a diurnal activity pattern during the winter to conserve thermogenic energy while foraging

(Horton 1984). Animals may hoard food in the autumn to ensure sufficient provisions until spring (Bartness 1995). Many individuals build large, insulative nests during the winter (Dark & Zucker 1983).

Cessation of breeding during winter is perhaps the central seasonal adjustment used by animals to conserve energy. Presumably, the benefits associated with winter breeding rarely outweigh the costs. To prepare in advance for seasonal changes in ambient conditions, animals must be able to determine the season of the year. Seasonal adaptations often require significant time to develop so the ability to ascertain the time of year to *anticipate* energetically demanding conditions is critical for survival. Thus, the development of winter adaptations must begin prior to the onset of the challenging winter conditions, not in response to them. Several mechanisms have evolved that provide seasonal information. In some species, an annual clock, analogous to the circadian clock, has evolved; in other species, circadian clocks are used to determine day lengths (i.e., photoperiod), and day length information is transduced into seasonally appropriate responses (Gwinner 1986; Bartness et al. 1993). In any case, many species of animals have evolved biological clocks that time a wide variety of seasonal biological rhythms.

Post-technology humans are somewhat buffered from temporal perturbations in the environment. Despite the failure to identify adaptive benefits of seasonal breeding in humans, seasonal fluctuations in human reproduction have been reported (Roenneberg & Aschoff 1990; Bronson 1995). Although the factors driving seasonal breeding (e.g., decreased food availability, low ambient temperatures) in industrialized humans are now reduced in importance, there are no obvious forces selecting *against* seasonality. Therefore, seasonality in human reproduction persists (reviewed in Bronson 1995), and the timing mechanisms that measure yearly intervals presumably remain extant among humans (Nelson 1990; Czeisler 1995).

In addition to seasonal changes in behavior and reproductive function, several other physiological and morphological adaptations have also evolved to help individuals cope with winter energy shortages. For example, winter-evoked changes in basal metabolic rate, nonshivering thermogenesis, body mass, pelage development, gut efficiency, and endocrine function are common among small mammals (Moffatt et al. 1993). Gonadotropin and prolactin concentrations decline, sex steroid hormone production wanes, and reproductive activities stop prior to winter. Virtually all of these seasonal adaptations are initiated among small mammals in nature during autumn or in the laboratory by exposure to short (<12-hour light/day) day lengths. Taken together, these seasonal adaptations have evolved to maintain a positive energy balance during winter energy shortages, that is, when food availability is low and thermoregulatory demands are high.

Maintaining a positive energy balance is required for survival and repro- ductive success. The vast majority of studies of seasonal phenomena have focused on energetic adaptations, especially reproductive adaptations. How- ever, other threats to survival must also be met in order for individuals to survive and increase their fitness. Individuals must avoid predators, they must avoid potentially dangerous attacks by conspecific competitors, and they must avoid succumbing to pathogens or parasites. In some cases, a marginal ener- getic balance can weaken animals to the extent that they are very susceptible to disease (Berczi 1986). Immunological defense against invading organisms requires cascades of mitotic processes that presumably demand substantial energy (Demas et al. 1997; Spurlock 1997). The energetic costs associated with immunity may be a critical factor in seasonal fluctuations in immune function. Also, environmental factors that can interrupt breeding (e.g., a flood or a late blizzard), or other conditions perceived as stressful, can compromise immune function and promote opportunistic pathogens and parasites to an ex- tent that leads to premature death (Berczi 1986; Ader & Cohen 1993). Factors that interrupt breeding are often unpredictable (Wingfield & Kenagy 1991), but many potential stressful conditions, such as low ambient temperatures, reduced food availability, migration, overcrowding, lack of cover, or in- creased predator pressure, can recur on a somewhat predictable, seasonal basis potentially leading to seasonal changes in population-wide immune func- tion and death (Fänge & Silverin 1985; Lee & McDonald 1985; McDonald et al. 1988; John 1994; Lochmiller et al. 1994; Sinclair & Lochmiller 2000). Thus, in addition to the well-studied seasonal cycles of mating and birth, there are also dramatic, albeit not as well studied, seasonal cycles of illness and death among human and nonhuman animal populations (Descôteaux & Mihok 1986; Lochmiller et al. 1986; Lochmiller & Deerenberg 2000; Sinclair & Lochmiller 2000) (Figure 1.2).

Seasonal fluctuations in immune function and survivorship may not be ob- served every year or in every population. Some winters may not be perceived as stressful, either because of mild ambient conditions or because energetic coping adaptations succeed to buffer individuals from harsh conditions. Con- sequently, populations in nature may exhibit compromised, enhanced, or static immune indices during the winter, although we suspect that the literature will be biased in favor of reporting *changes* in immune function. The literature on seasonal fluctuations in immune function is summarized in later chapters.

The working hypothesis of this book is that some individuals have evolv- ed mechanisms to predict seasonal stressor-induced reductions in immune

Total Numbers

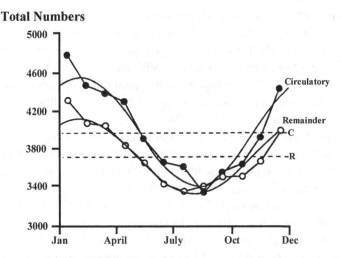

Fig. 1.2. Seasonal cycle of death. These data show the number of human deaths over the course of 1 year in Aberdeen, Scotland. Human mortality from circulatory and other causes also shows a similar seasonal fluctuation with voles, with peak number of deaths observed during the winter and relatively low mortality during the summer. C, circulatory; R, remainder. (Data redrawn from Douglas et al. 1991.)

function and make appropriate adjustments in anticipation of challenging conditions, as a temporal adaptation to promote winter survival. From an adaptive functional perspective, we propose that individuals "optimize" immune function so that they can tolerate minor infections if the energetic costs of mounting an immune response outweigh the benefits. Available energy is partitioned into competing functions where most needed for survival. Thus, when energetic requirements are high (e.g., during migration, pregnancy, territory defense, lactation, or winter), we predict that immune function should be reduced. During energetically challenging winters, energy is used for thermoregulation and maximal immune function possible, rather than growth, reproduction, or other nonessential processes. Recent evidence from bighorn sheep suggests an energetic trade-off between the costs of immune and reproductive functions (Festa-Bianchet 1989); lactating ewes exhibit higher parasitic infections, compared with nonlactating ewes. Presumably, the energetic costs of lactation reduce the amount of energy that can be allocated to immune function.

We hypothesize that exposure to short day lengths enhances immune function. Many field and clinical studies are consistent with this hypothesis (i.e., lymphatic tissue size or immune function is elevated during the winter). All laboratory studies are consistent with this hypothesis: for example, immune function is enhanced in animals housed in short days, compared with animals maintained in long day length conditions (reviewed in Nelson & Demas 1996; cf. Yellon et al. 1999). These results suggest that mechanisms have evolved to allow animals to anticipate immunologically challenging conditions by monitoring photoperiod. Presumably, this ability permits individuals to cope with these health-threatening seasonal conditions by bolstering lymphatic tissue development and immune function directly.

This book addresses the physiological mechanisms underlying the detection of and the response to seasonal environmental factors that affect immune function and disease processes. In this first chapter, the geophysical factors leading to seasonality are reviewed. We also describe the proximate mechanisms underlying seasonality. The literature on seasonal breeding is used as a departure point to understand the mechanisms of seasonality as it pertains to immune function. Although circadian rhythms in immune function have become recognized as important factors in disease and immunotherapy, this topic is beyond the scope of the present book. Interested readers are referred to several recent papers that have carefully reviewed this field (e.g., Hrushesky 1991; Petrovsky & Harrison 1998; Cardinali et al. 1999).

In Chapter 2, basic immunology is reviewed. Reports of seasonal fluctuations in disease prevalence are described in Chapter 3, followed by a review in Chapter 4 of the literature reporting seasonal fluctuations in lymphatic organ size and structure, as well as immune function. Possible interactions between lymphatic organ morphology and function and recurrent environmental stressors are also discussed in this chapter. Most research in seasonality has focused on the role of photoperiod in providing temporal information for breeding (Reiter 1993). In Chapter 5, the physiological sequelae by which photoperiod mediates the seasonal changes in immune function are assessed. The energetics of immune function will be discussed in Chapter 6. Among small mammals, short photoperiods reduce blood concentrations of sex steroid hormones (in long-day breeders) and prolactin, as well as alter the temporal pattern of pineal melatonin secretion (Goldman & Nelson 1993). The effects of these hormones on immune function and opportunistic diseases, including cancer, will be explored in Chapter 7. Melatonin will receive particular attention for its effects on immune function in Chapter 8 because this hormone has been reported to affect immune function and tumorigenesis in several model animal systems. The clinical relevance of these seasonal fluctuations in lymphatic tissue size and immune function for humans and nonhuman animals will be presented in Chapter 9. Our goal for this book is to present a comprehensive description of the field, laboratory, and clinical data on seasonal immunity and disease processes from an adaptive functional perspective.

1.2. Geophysical Factors Leading to Seasonality

To understand biological rhythms, it is helpful to understand the geophysical
fluctuations that have shaped these rhythms. In reference to the stars, the Earth
rotates on its axis every 23 hours, 56 minutes, and 4 seconds, a period of time
termed a "sidereal day" (U.S. Naval Observatory 1999). The daily rotation
(e.g., from noon to noon) of the Earth is 24 hours, 3 minutes, and 57 seconds
when measured in reference to a closer object, namely the sun; this period of
time is called the mean "solar day." The length of the solar day is not constant throughout the year because of the elliptical orbit of the Earth around
the sun. The rate of orbit tends to increase slightly when a planet is near the
sun (perihelion) and tends to decrease slightly when a planet is further from
the sun (Kepler's Second Law) (Brosche & Sündermann 1990). The actual
solar day is nearly 16 minutes shorter than the mean solar day on or about
3 November each year. The solar day length matches the mean solar day
length on only four dates: 15 April, 14 June, 1 September, and 25 December.
Although the astronomical measures of day length reveal substantial variation
in the timing of the daily rotation of the Earth, for practical purposes, 24 hours
is a reasonable approximation of the period of the Earth's daily rotation
(Saunders 1977). Although day length, as measured by the position of the
sun on each horizon, is very precise, the amount of twilight varies substantially and rarely is calculated into the amount of total day light received each
day. Despite sounding sudden, "the break of day" or "night fall" really reflects
gradual processes in changing light during dawn or dusk. The vast majority of
laboratory studies of photoperiodism do not use gradual onsets or offsets of
illumination (cf. Gorman et al. 1997); rather, they use timers that turn lights
on and off in one step. Notably, the light sensitivity of the photoreceptors that
measure day length is not characterized for most species, and the importance
of dim light in photoperiodic time measurement remains unspecified.

The Earth completes about 365.25 (365 days, 6 hours, 9 minutes, and
10 seconds) rotations during each revolution around the sun, a period of time
called the sidereal year (Brosche & Sündermann 1990). Unlike the condition
on the moon, where one side is always facing away from the sun, the daily
rotation of the Earth allows most of the surface to bask in sunlight regularly.
The shape of the elliptical revolution of the Earth around the sun explains
why winters are generally milder in the Southern Hemisphere, compared
with the Northern Hemisphere, but does not explain the basis of seasons. The
phenomenon of seasons results from the tilt of the planet's axis of rotation
(approximately 23.5°) with respect to its plane of revolution around the sun
(Figure 1.3). Thus, as the Earth revolves around the sun, the proportion of

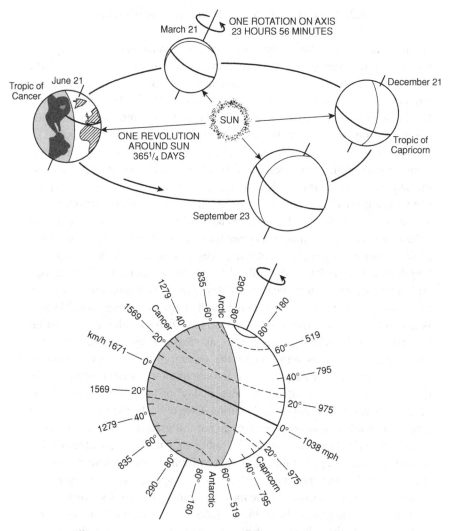

Fig. 1.3. The plane of revolution of Earth around the sun accounts for the seasons. (Reprinted from M. C. Moore-Ede, F. M. Sulzman, & C. A. Fuller, *The Clocks That Time Us*, copyright © 1982, p. 2.)

the planet exposed to the radiating energy changes (Officer & Page 1993). During the Northern Hemisphere winter, the Earth is furthest from the sun, and the planet is tilted away from the sun.

When viewed from Earth, the sun moves across the sky each day from east to west. There is also a yearly pattern of the sun's movement from day to day. The sun reaches its most northern point in the sky (i.e., declination 23.5°N) about 21–22, June; this is the date of the summer solstice in the Northern Hemisphere and winter solstice in the Southern Hemisphere (Brosche & Sündermann 1990). The sun reaches its most southern point in the sky (i.e., declination 23.5°S) about 21–22, December; this is the date of the winter solstice in the Northern Hemisphere and summer solstice in the Southern Hemisphere. The equinoxes, which occur twice a year, are the points at which the Earth's celestial equator intersects its ecliptic (Figure 1.3). In the Northern Hemisphere, the autumnal equinox is about 21–22 September, and the vernal equinox is about 21–22 March.

One very important feature of this annual progression of the sun's movement is that day length (i.e., the amount of light/day) varies. Except at the equator, day lengths during the winter are shorter than day lengths during the summer. The exact value for day length depends on latitude (Figure 1.4), with higher variation between winter and summer day lengths at high latitudes and lower variation between winter and summer day lengths closer to the equator (Brosche & Sündermann 1990). At the two equinoxes, day length is exactly twelve hours at all points on the globe. Thus, with just two pieces of information – (1) the day length and (2) if day lengths are increasing or decreasing – the exact time of year can be determined. Plants and animals have evolved mechanisms to determine the precise time of year by using day length (photoperiod) information (Sadleir 1969; Goldman & Nelson 1993; Goldman 2001).

The lengthening days of summer are associated with rising ambient temperatures, increased biotic growth, and changes in humidity (Officer & Page 1993). Similarly, the daily cycles of day and night evoke the familiar daily cycles of temperature, humidity, and photosynthesis. Life on Earth has evolved in such a manner to exploit these daily and annual cycles in energy availability. Plants and animals have evolved to fill *temporal*, as well as spatial, niches. Because so many biochemical processes are affected by light, it is relatively straightforward to imagine how circadian rhythms may have arisen early in the evolution of life. Many protists do not display circadian rhythms. This observation may reflect that many protists do not live more than a few hours, so these organisms have no reason to track daily time; alternatively, this observation may simply reflect that protists have too limited DNA stores to encode a biological clock. The ability to measure photoperiod (i.e., photoperiodism) to determine time of year, likely evolved after organisms began to survive for longer than one year. The next section describes some of the environmental cues that can be monitored by animals in conjunction with, or in addition to, photoperiod to provide information about time of year.

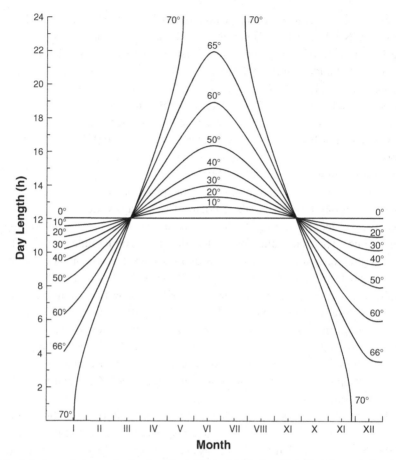

Fig. 1.4. Annual change in day length at different latitudes. Note that the variation in annual day length is highest at high latitudes, and this variation is reduced at low latitudes. Because the seasonal climatic changes are often more pronounced at high versus low latitudes, photoperiodism is more common among high versus low latitude species.

1.3. Potential Environmental Cues for Synchronization of Seasonal Cycles

Individuals inhabiting nontropical habitats experience seasonal deterioration and renewal of environmental energy resources. High thermoregulatory requirements for small mammals and birds during the winter typically coincide with low environmental food availability (Blank & Ruf 1992; Wunder 1992; Wingfield & Farner 1993). The energetic "bottleneck" created by high energy demands during times of low energy availability has led to the evolution of many adaptations that allow individuals to cope with winter. Breeding is energetically expensive, and cessation of breeding activities is central among the suite of winter-coping adaptations observed in boreal and temperate zone species (Bronson & Heideman 1994). Tactics have evolved that permit individuals to maximize the length of the breeding season without jeopardizing survival vis-à-vis energy use. Because seasonal adaptations often require time to develop, the precise timing of behavioral and physiological adaptations necessary to cope with energy shortages is a critical feature of individual reproductive success and subsequent fitness. Mechanisms have evolved to ascertain the time of year precisely to phase territorial defense, breeding, molt, migration, and other energetically expensive activities to coincide with peak energy availability and other local conditions that promote survival (Bronson 1989; Moffatt et al. 1993; Wingfield & Farner 1993; Goldman 2001).

In some cases, physiological and behavioral changes that have an obvious and immediate adaptive function may occur in direct response to environmental factors. For example, food and water may be available only during certain times of the year, and a decrease in the amount of procurable nutrients can lead to reproductive inhibition (Nelson 1987; Wingfield & Kenagy 1991; Bronson & Heideman 1994). These types of environmental factors have been termed the "ultimate factors" underlying seasonality (Baker 1938). Many animals need to forecast the onset or offset of these ultimate factors to initiate time-consuming seasonal adaptations. Therefore, seasonally breeding animals frequently detect and respond to environmental cues that accurately signal, well in advance, the arrival or departure of seasons favoring reproductive success. The cues, or "proximate" factors (Baker 1938), used to predict environmental change may or may not have direct survival value (e.g., photoperiod), and may or may not be the same as the ultimate factor (e.g., food availability).

Several environmental factors that vary across seasons could potentially serve as temporal cues for animals to discern the time of year. For example, ambient temperature, food quality and quantity, water availability, relative humidity, precipitation, and sunlight (i.e., diurnal variation in visible and ultraviolet radiation, day length, etc.) all vary on an annual cycle in most habitats. Individuals may respond to a single cue or to numerous environmental factors. For example, California voles (*Microtus californicus*) breed during the winter months along the West Coast of the United States and Mexico. Their breeding season, from November to May, coincides with the onset of the California rainy season, the subsequent availability of green food, short day lengths, and reduced ambient temperature (Lidicker 1973). California voles regress their reproductive systems in the summer when exposed to long days, high ambient temperatures, reduced green vegetation, and reduced free-standing water (Lidicker 1973). To sort out the environmental factors responsible for timing the onset and termination of breeding in this species, animals were moved into the laboratory and housed individually in either long (Light:Dark [LD] 14:10; 14 hours of light and 10 hours of dark/day) or short (LD 8:16) days, and provided with *ad libitum* food and water (Nelson et al. 1983). Voles in short days display significant regression of reproductive function, compared with their long-day counterparts; these data suggested that if these voles only responded to photoperiod, then their reproductive function would be 180° out of phase with their breeding season. Presumably, some other factor in nature masked the short-day gonadal inhibition. Because the appearance of green vegetation coincides with the onset of breeding in this species, other short-day California voles were fed fresh spinach in the laboratory. The supplemental green food blocked gonadal inhibition in short day lengths (Figure 1.5).

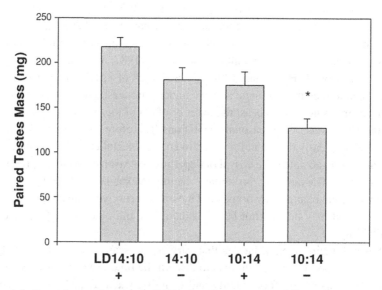

Fig. 1.5. Reproductive function in California voles exposed to different environmental factors. Mean (±SEM) paired testes size of California voles (*Microtus californicus*) in long (LD 14:10) or short (LD 10:14) photoperiods. Voles that received fresh spinach thrice weekly are noted by the "+" designation. Note that testes size (and function) was reduced in short days, but that exposure to greens prevented the short-day gonadal regression. Restricted access to water in either photoperiod suppressed testes mass. Taken together, these data suggest that the onset of the breeding season of California voles is mediated by the availability of fresh green vegetation during the California winter; cessation of breeding occurs during the annual summer drought. (Reprinted from the *Journal of Reproduction & Fertility*, Vol. 69, R. J. Nelson, J. Dark, & I. Zucker, "Influence of photoperiod, nutrition and water availability on reproduction of male California voles (*Microtus californicus*)," pp. 673–7, copyright © 1983.)

During the annual summer drought in the California and Baja Peninsula habitat, free-standing water is scarce, and California voles and other animals appear to obtain water by licking dew from plants. When water bottles were provided to laboratory-housed voles only during the 2 hours that dew was available on plants outside, the reproductive system regressed, despite exposure to long day lengths.

In sum, the breeding season of California voles is activated by the presence of green vegetation that overrides the inhibitory effects of short days on reproductive function (Nelson et al. 1983). Reproductive regression is observed during the summer in field-caught voles, and this appears to be induced by water shortages. Provision of voles with water extends breeding throughout the annual summer drought in nature (Lidicker 1973). Thus, seasonal adaptations, including reproductive inhibition, are often regulated by complex interactions among several extrinsic factors that affect survival and reproductive success. Because these extrinsic factors tend to covary in the field, laboratory studies are usually informative in teasing apart the contribution of various extrinsic factors used in the timing of seasonal adaptations.

1.3.1. Food Quality

The quality of food varies across seasons. Several components of food, including trace elements, display seasonal variation in abundance (Bronson 1989). One well-studied plant compound, 6-methoxy-2-benzoxazolinone (6-MBOA), has been implicated in the timing of rodent seasonal breeding. The precursor of 6-MBOA, 2,4-dihydroxy-7-methoxy-2-11,4-benzoxzin-3(4H)-one (DIMBOA), is abundant as the primary glucoside in young seedlings and vegetatively growing plants (Berger et al. 1981). Plant injury coincident with grazing evokes the release of enzymes that rapidly convert DIMBOA to 6-MBOA. Thus, detection of 6-MBOA in the diet is a reliable signal that the growing season has begun and that food is immediately available. 6-MBOA induces midwinter breeding in natural populations of arvicoline rodents (Berger et al. 1981). For example, two nonbreeding winter populations of montane voles (*Microtus montanus*) received supplemental feedings of rolled oats, with or without a coating of 6-MBOA. The first study was conducted in November to December and was replicated in January to February. In both cases, reproduction began after 3 weeks for 6-MBOA-supplemented voles; testicular weights doubled in the males receiving the compound. The incidence of pregnancy was 60% in the supplemented populations, compared with 0% in the unsupplemented animals (Berger et al. 1981). In laboratory studies, 6-MBOA overrides inhibitory photoperiodic information, and stimulates ovarian and uterine development in several species (Epstein et al. 1986; Korn & Taitt 1987; Schadler et al. 1988). Rodents that feed mainly on monocotyledons have been hypothesized to respond to 6-MBOA, whereas animals that feed on dicotyledons have been predicted to be unresponsive to the compound (Negus & Berger 1987). The phenomenology of a plant compound overriding photoperiodic information is fascinating, but many questions regarding the mechanisms of action for 6-MBOA remain unanswered (Nelson & Blom 1993). Although 6-MBOA has been reported to be structurally similar to melatonin (Sanders et al. 1981; Anderson et al. 1988), the interaction of 6-MBOA with melatonin receptors has not been established. The importance of any number of other trace substances in the normal seasonal timing of breeding is vastly understudied (e.g., Martinet & Meunier 1969; Bronson 1989).

1.3.2. Food Quantity

As noted previously, food availability is probably the ultimate factor driving seasonal breeding (Immelmann 1973; Bronson 1979). In many cases, food availability can also serve as a proximate factor in the timing of seasonal breeding (Bronson 1989). Severe, chronic food restriction stops reproductive activities in both male and female mammals. Most studies of the physiological effects of restricted food intake on reproduction have been conducted in the laboratory using rats. These studies generally indicate that the hypothalamic gonadotropin-releasing hormone (GnRH) pulse generator of food-deprived animals produces fewer bursts of GnRH release (i.e., reduced GnRH pulse frequency) and less GnRH per pulse (i.e., reduced GnRH pulse amplitude) (Campbell et al. 1977). It is likely that the hypothalamic GnRH system is the final common pathway through which all extrinsic factors operate on breeding (reviewed in Bronson & Heideman 1994).

The gonads are compartmentalized into two component functions: (1) gametogenesis and (2) steroidogenesis. In males, pituitary luteinizing hormone (LH), released in response to GnRH, stimulates the Leydig cells of the testes to synthesize androgens (Campbell et al. 1977). Androgens act on the hypothalamus and anterior pituitary gland to reduce secretion of GnRH and LH via a classic negative feedback loop (Figure 1.6). Prolactin is critical for testicular steroidogenesis because, among other things, it promotes LH receptor formation in the Leydig cells. Spermatogenesis is mediated by follicle-stimulating hormone (FSH) from the anterior pituitary and testosterone from the testicular Leydig cells. FSH stimulates the epithelium of the seminiferous tubules; androgens stimulate spermatogenesis. Thus, LH indirectly affects spermatogenesis because of its effects on androgen production. The regulation

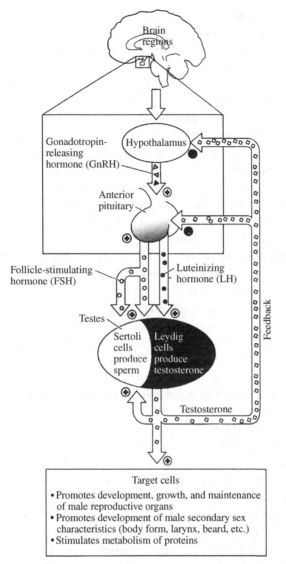

Fig. 1.6. Reproductive hormones and negative feedback. GnRH is released in a pulsatile manner from the hypothalamus. GnRH stimulates the pulsatile release of the
gonadotropins (LH and FSH) from the anterior pituitary gland. LH stimulates steroid
production from the gonads. A complex negative feedback relationship exists among
the hypothalamus, anterior pituitary, and testes. The steroid hormones from the testes
feedback to inhibit secretory activity in both the hypothalamus and pituitary gland. In
parallel, increasing LH and FSH blood concentrations feedback to slow down their secretion from the anterior pituitary, as well as GnRH from the hypothalamus. Likewise,
as GnRH is secreted, the hypothalamus responds to increasing concentrations of this
hormone by reducing its secretion.

of FSH in males remains somewhat ambiguous. In most male mammals, FSH is not released in a pulsatile manner. In contrast to LH, castration does not elicit *large* increases in FSH, although FSH is significantly elevated after castration (Wu et al. 1991).

Restricted food consumption directly reduces GnRH pulse frequency and amplitude in rats, which leads to reduced LH and testosterone concentrations; FSH appears unaffected by acute inanition in adult male rats or rhesus monkeys (reviewed in Bronson 1989). Thus, steroidogenesis is rapidly depressed, but spermatogenesis is relatively unaffected by acute restricted food intake because there is some "time lag" in the system. From an adaptive functional perspective, this probably reflects the relative energetic costs of androgen-dependent behaviors, compared with the energetic costs of sperm production. Because food shortages are often of short duration, suspension of androgen-dependent characters while hunting for food may be an adaptation to enhance survivorship for rats and rhesus monkeys. For animals that confront chronic food shortages, spermatogenesis is often suspended when food availability wanes (Blank & Desjardins 1984, 1985).

In female mammals, energetic costs often double during pregnancy and increase further during lactation (Bronson 1989). Therefore, reproductive processes of females tend to be very sensitive to reduced food intake. Although the degree of restricted calories necessary to interrupt GnRH secretion varies widely within and among species, in all mammalian species studied to date, significant caloric restriction suppresses GnRH release (Bronson 1989). Thus, restricted caloric intake reduces blood LH concentrations and suppresses ovulation (Bronson 1989).

Photoperiod can alter responses to inanition. In one study of adult male deer mice (*Peromyscus maniculatus*) housed in either long (LD 16:8) or short (LD 8:16) days, animals received food *ad libitum* or 90%, 80%, or 70% of their individual *ad libitum* food intake for 10 weeks (Nelson et al. 1992). Short-day mice displayed significant regression of reproductive function after only 10% reduction in food intake. Long-day animals, however, required a 20% reduction in caloric intake to depress reproductive function, and 30% reduction to display significant reductions in sperm numbers (Nelson et al. 1992). These data suggest that two inhibitory cues, *viz.*, short days and restricted food intake, interact to affect the mechanisms underlying reproduction.

1.3.3. Water Availability

Species that live in arid or semiarid habitats likely respond to water shortage by cessation of reproduction. Like food quality and quantity, water availability can serve as a potent cue for cessation of breeding. In many cases, water shortages can override other permissive cues, such as photoperiod, and stop reproductive activities (Nelson et al. 1995a, b).

For example, a 50–75% reduction in water intake inhibits sperm production in South Dakotan deer mice (*Peromyscus maniculatus nebrascensis*) independently from the effects of food intake, temperature, or photoperiod (Nelson & Desjardins 1987). Changes in testicular function range from no effect to a complete absence of sperm among individual males with restricted water intake. These results suggest that water availability, per se, could act as an environmental cue to regulate seasonal breeding for *Peromyscus* during otherwise optimal reproductive conditions.

In a recent study on a related subspecies, reproductive responses to a simulated drought were examined in Michigan deer mice (*Peromyscus maniculatus bairdii*) (Nelson 1993). The results indicated that, on average, water-restricted deer mice reduced the size and function of their reproductive systems relative to animals provided with water *ad libitum*. Fertility was severely limited in water-restricted male deer mice as compared to animals receiving *ad libitum* water. A wide distribution of reproductive responses to restricted water intake was evident, and mice were rank-ordered by epididymal sperm numbers. Individuals of two classes, those in the top quartile of epididymal sperm numbers and those in the bottom quartile of epididymal sperm numbers, were further compared to detect associated physiological variables in response to water restriction. In terms of mechanisms, plasma testosterone concentrations were reduced by water restriction; blood plasma LH concentrations were reduced only in water-restricted deer mice in which spermatogenesis was suspended. Carcass fat content was also reduced only in water-restricted mice that ceased sperm production. Taken together, these data suggest that water availability can serve as an environmental cue for triggering reproduction, and it appears that differential hydrolysis of body fat in response to decreased testosterone concentrations or variation in initial fat depots accounts for the variation in reproductive response to water restriction.

Although plasma concentrations of LH decrease after food or water restriction, the physiological mechanisms are not well studied and often have been attributed to "stress." That is, adverse environmental conditions, such as low ambient temperatures, reduced food quality or quantity, or limited water availability may act as stressors by activating the hypothalamus-pituitary-adrenal axis. Again, the hypothalamic-pituitary axis is the final common pathway through which food, water, and photoperiod cues are mediated (Bronson 1989). However, it remains to be tested whether increased glucocorticoid hormone secretion in response to adverse environmental cues actually

enhance steroid negative feedback. Enhanced negative feedback may mediate nonphotoperiodic (e.g., inanition mediated) reproductive inhibition, as is
the case for many instances of photoperiodic regulation (Ellis & Turek 1980;
Sisk & Turek 1982).

1.3.4. Ambient Temperature

Among mammals, ambient temperature appears to affect breeding in two
ways. Low temperatures interact with other environmental factors, such as
photoperiod or food, that signal the onset of winter and high ambient temperatures appear to suppress reproductive function directly in animals inhabiting
low latitudes (Desjardins & Lopez 1983). Low ambient temperatures do not
appear to affect reproductive function in large male mammals (Nazian &
Piacsek 1977). However, small male mammals appear to inhibit breeding
when ambient temperature is low. For example, individual deer mice
(*Peromyscus maniculatus*) that fail to inhibit reproductive function when
maintained in short day lengths display reproductive regression when exposed to low ambient temperatures (Desjardins & Lopez 1983). Similarly,
prairie voles (*Microtus ochrogaster*) housed in short days and maintained in
low ambient temperatures display significantly greater inhibition of reproductive function than short-day voles maintained at standard room temperatures
(Nelson et al. 1989). Indeed, low temperatures appear to facilitate photoperiod-
induced reproductive regression (Larkin et al. 2001). Low temperatures increase metabolic demands, but these effects can be overcome by increasing
food intake (Bronson 1989). Although strains of mice (*Mus musculus*) that
breed normally at 21°C will breed poorly at 3°C (even when provided excess food), selection for breeders will produce a strain of mice that breed
normally at 3°C within eight generations (Barnett 1973). Low temperatures
affect breeding in female *Mus* by limiting energy availability to maintain
pregnancy or lactation (Bronson 1989). Low temperatures are usually combined with other cues to inhibit breeding. High ambient temperatures usually
suppress reproductive function independent of other environmental factors.
High ambient temperatures can affect female breeding by inhibiting processes
throughout the reproductive cycle (e.g., suppression of ovulation, abortion of
embryos, and suppression of lactation). Males are also susceptible to the
effects of heat on reproductive physiology. Males suffer a direct effect of
high temperature on the gonads; sperm are very sensitive to high ambient
temperatures. When ambient temperature is >32°C, sperm production is curtailed in most species of mammals examined (e.g., Rhynes & Ewing 1973;
Hafez 1993).

Importantly, ambient temperatures that are above or below an optimal range can evoke stress responses that can interfere with seasonal breeding and affect immune function. In fact, many proximate cues of seasonality can be perceived as stressors. For example, food and water shortages, and even constant bright light exposure, can elicit a classic stress response, including hypersecretion of glucocorticoids, adrenal hypertrophy, and thymic involution. The role of stress hormones (e.g., adrenalin and glucocorticoids) in seasonal changes in immune function and disease processes will be more fully explored in Chapters 6 and 7.

1.3.5. Photoperiod

Only two types of environmental cues have been implicated in the control of seasonal cycles via an interaction with some type of internal time-keeping mechanism; these are photoperiod (or day length) and ambient temperature. Although temperature may have a very direct and obvious relationship to certain types of seasonal adaptations, such as changes in the insulating properties of the fur or the choice of a time to reproduce, photoperiod would appear to have little direct importance for most organisms. However, the annual progression in changes in both day length and temperature provide potentially valuable sources of information that can be used to determine the time of year and thereby anticipate future environmental conditions. Photoperiod is, of course, the more reliable of the two cues in this respect. Photoperiodism, or the use of day length cues to time annual cycles, is widespread among virtually all major groups of organisms, including mammals (Goldman & Nelson 1993; Goldman 2001). The use of temperature to synchronize annual cycles has been indicated more often for ectothermic vertebrates than for birds or mammals, but some evidence exists that certain hibernating mammals may use temperature cues in this way (Ueda & Ibuka 1995).

Seasonal fluctuations in reproductive activity and associated changes in the concentrations of sex steroid hormones are among the most commonly observed photoperiodic responses in mammals. Gonadal steroids, especially androgens and estrogens, have multiple actions in the regulation of a variety of secondary sexual characteristics and behaviors. Therefore, it is logical to ask how many of the various photoperiodic responses are secondary to photoperiod-related changes in gonadal steroid secretion. A blueprint for possible answers to this question is emerging from studies on the steroid feedback regulation of pituitary gonadotropin secretion in photoperiodic species. In male Syrian hamsters (*Mesocricetus auratus*), exposure to short days leads to a decrease in the circulating concentrations of both LH and FSH, leading to testicular regression and lowered concentrations of circulating testosterone (Elliott & Goldman 1981). Short-day exposed male hamsters exhibit a considerable increase in sensitivity to the negative feedback actions of testosterone, so that even very low concentrations of the androgen inhibit LH and FSH secretion (Elliott & Goldman 1981). In addition to this change in the negative feedback axis, there are also steroid-independent actions of short-day exposure. Thus, whereas castrated hamsters in both photoperiods show increased gonadotropin secretion as a result of the absence of steroid feedback, the concentrations of gonadotropin are still higher in long-day castrates, compared with short-day castrates (Elliott & Goldman 1981).

In female hamsters, exposure to short days results in the development of an anovulatory state, accompanied by extremely low concentrations of circulating LH, with the exception of a surge of this hormone that occurs every afternoon (Bridges & Goldman 1975). This pattern of LH secretion remains essentially unchanged both qualitatively and quantitatively after ovariectomy, and even following combined ovariectomy and adrenalectomy. Thus, the regulation of LH secretion in female short-day hamsters appears to be steroid independent; this is not the case for long-day females where steroid feedback effects are probably important in regulating the changes in gonadotropin secretion that occur during the various phases of the ovulatory cycle. Increases in FSH are seen following ovariectomy in both long-day and short-day female hamsters, although the degree of postovariectomy hypersecretion of FSH is greater in long-day females (Bittman & Goldman 1979). The studies in hamsters illustrate that, even within a single species, both steroid-dependent and steroid-independent mechanisms for gonadotropin regulation are components of the photoperiodic response. The relative importance of these two components may vary with sex and with the particular hormone (LH or FSH) being studied. Studies in other species confirm that both steroid-dependent and steroid-independent mechanisms are ubiquitous, but how to predict which mechanism will be most important in a given species based on either phylogenetic or ecological criteria has not been established.

Notably, steroid-dependent and steroid-independent components to a single response are frequently acting in nonreproductive photoperiodic responses as well. In male Siberian hamsters (*Phodopus sungorus*), the loss of body mass that usually occurs during exposure to short photoperiod is partly related to the decrease in circulating testosterone, because this hormone stimulates increased body mass in this species; however, an additional component of the weight loss in short days is independent of androgen action, as indicated by a further loss of body mass in castrated Siberian hamsters following exposure to short days.

Seasonal changes in androgen secretion are also related to the occurrence of daily torpor in Siberian hamsters that are exposed to short days. If relatively high concentrations of circulating testosterone are artificially maintained by implantation of constant release hormone capsules, torpor is eliminated in short photoperiod-exposed hamsters (Ruby et al. 1993). However, the decline in androgen that accompanies testicular regression in short-day males is not, in itself, sufficient to account for the thermoregulatory adaptations of these animals; daily torpor almost never occurs in long-day hamsters, even following castration. The examples cited here suggest that seasonal changes in gonadal steroid secretion frequently have some impact on nonreproductive photoperiodic responses, but that additional photoperiod-related factors are also involved in regulation of many of these seasonal adaptations. As we will discuss in Chapter 7, steroid hormones also affect immune function and may contribute to the seasonal changes in immune function and disease prevalence.

1.4. Proximate Mechanisms Underlying Seasonal Cycles

1.4.1. Circadian Rhythms

Circadian rhythms, as previously discussed, are endogenous, or self-sustaining, rhythms with period lengths of approximately twenty-four hours. The overt, or formal, properties of circadian rhythms have been studied intensely, even prior to knowledge about the physiological mechanisms underlying biological clock functions. In studies of the formal properties of circadian clocks, the clock has been treated essentially as a black box. By studying locomotor behavior, many formal properties of photoperiodic time measurement have been discerned. In the presence of a light-dark cycle, nocturnal animals limit their locomotor activities to the dark portion of the day. The onset of activity usually shows a close temporal association (or phase lock) with the onset of dark. If the light-dark cycle is phase-shifted four hours, then individuals adapt their activity rhythms to the new regimen in four or five cycles (although there are likely to be transients) during which entrainment is not perfect. Light serves as a potent environmental time cue, or *zeitgeber* (German for "time giver"), for most vertebrate species, although temperature can be an important zeitgeber for poikilothermic animals, and possibly serve as a secondary zeitgeber for homeothermic animals (Pittendrigh & Daan 1976). If nocturnal mice are placed in constant dim light, so that there is no daily zeitgeber, then the onset of their locomotor activity begins to drift at a period slightly different from 24 hours reflecting the endogenous nature of the circadian clock. Biological rhythms such as these that are not synchronized (or entrained) with environmental cues are said to be free-running. Each individual will display its own free-running period represented by the Greek letter, τ. Thus, an individual that began its activity 15 minutes later each day would have a τ of 24.25 hours.

How do circadian rhythms become entrained to precisely 24 hours? Experiments have revealed that light exposure at different times of an individual's circadian rhythm (e.g., $\tau = 24.25$ hours) has specific effects on the onset time of activity the following day. A free-running nocturnal mouse maintained in constant dark conditions can be exposed to one hour of light at different times during its subjective day (which is divided into subjective day and subjective night; the subjective night of a nocturnal mouse would begin when activity begins and ends approximately 12 hours 12.5 minutes later) (Figure 1.7). If a mouse is exposed to one hour of light any time throughout the middle of the subjective day, the next bout of locomotor activity is unaffected; i.e., for a mouse with $\tau = 24.25$ hours, locomotor activity stills begins 24.25 hours after the start of the previous bout of activity. However, if the 1-hour pulse of light is given early in the subjective night (or late in the subjective day), then the mouse does not begin its activity 24.25 hours later, but rather delays its activity onset for 1–4 hours, depending on when precisely the pulse of light occurs (Figure 1.7). If the 1-hour pulse of light is given late in the subjective night (or early in the subjective day), then the mouse does not begin its activity 24.25 hours later, but rather advances its activity onset by 1–4 hours, depending on when precisely the pulse of light occurs (Figure 1.7). Thus, the largest phase delays can be produced by light exposure early during the subjective night, when a nocturnal animal has just awakened and a diurnal animal is preparing for sleep; the largest phase advances can be produced late during the subjective night. Overall, light appears to have little effect on the circadian system during the subjective day of either nocturnal or diurnal animals. These adjustments that occur in response to light presented at appropriate circadian times serve to maintain synchrony between the environment and the endogenous clock.

Importantly, only a few seconds of light may be all that is necessary to trigger phase shifts in entrainment; so, even fossorial animals only need to sample light briefly to synchronize their circadian rhythms to 24 hours. Furthermore, although phase responses to light are best observed in free-running individuals, these effects of light also operate on individuals that are entrained to a daily light-dark cycle. Light impinges upon the phase-advance and phase-delay portions of an individual's daily responsiveness to light to reset the biological rhythm to exactly twenty-four hours (i.e., to entrain the

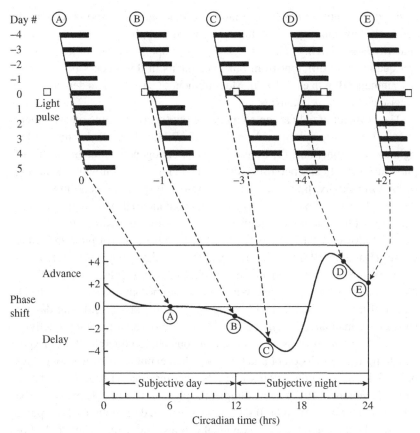

Fig. 1.7. Phase response curve for a nocturnal rodent. A free-running nocturnal rodent maintained in constant dark conditions was exposed to a 1-hour light pulse at various times during the subjective day and night. (A) When the light pulse was given in the middle of the subjective day, the subsequent activity onset time was unaffected; the middle of the subjective day is thus called the "dead zone." (B) A light pulse provided at the end of the subjective day phase-delayed activity the next day by about 1 hour. (C) With a light pulse at the start of the subjective night, a substantial phase delay (3 hours) was observed. (D) A light pulse given later during the subjective night caused a substantial phase advance (4 hours) of activity onset the next day. (E) Finally, a light pulse at the beginning of the subjective day evoked a 2-hour phase advance in activity the following day. The effect of light on an endogenous rhythm is involved in synchronizing (entraining) circadian rhythms to exactly 24 hours each day. (From Moore-Ede et al. 1982.)

rhythm). Recently, it has been established that humans respond to light of relatively low intensity (~180 lux; i.e., dim room light) to phase-shift circadian rhythms (Boivin et al. 1996). The phase response curves of humans are not qualitatively different from other mammals in the process of circadian entrainment (Boivin et al. 1996). Measurement of day length is dependent on precise circadian mechanisms.

The most commonly accepted model of photoperiodic time measurement is called the "external coincidence" model (Bünning 1973). In this model, light has two functions: (1) light entrains an endogenous circadian cycle of photoinsensitivity and photosensitivity; and (2) if coincident with the period of photosensitivity, light stimulates a long-day response. According to this model, when an animal first encounters light, an internal state of photoinsensitivity (or photo-noninducible phase) that lasts for approximately 12 hours begins, followed by a 12-hour period of photosensitivity (or photoinducible phase) (Bünning 1973). When day lengths are short (e.g., LD 11:13), the 11 hours of light are coincident with the photo-noninducible phase and short-day adaptations are maintained. With the reverse photoperiod (i.e., LD 13:11), 1 hour of light is coincident with the photoinducible phase, and long-day responses are maintained. In nocturnal rodents, the onset of locomotor activity is coincident with the onset of the endogenous photoinducible phase. Exposure to light during locomotor activity, as when exposed to typical long days or artificial day lengths, results in a long-day response (Elliott et al. 1972; Elliott & Goldman 1981). For example, the photo-noninducible phase can be set by a 1-hour pulse of light. If the animal is exposed to another 1-hour pulse of light 13 hours later, then the individual will respond to this "skeleton" photoperiod as if it was exposed to an LD 14:10 photoperiod. Physiologically, the light during the animals' subjective night suppresses melatonin (Figure 1.8). Because day length is transduced by the length of sustained melatonin secretion each night, a shortened period of melatonin secretion is interpreted by individuals as a short night or long day (Figure 1.9 and see below).

1.4.2. Circannual Rhythms

Circannual rhythms are endogenous, or self-sustaining, rhythms with period lengths of approximately 1 year. Circannual rhythmicity has been most thoroughly studied in birds and mammals, but is probably widespread among multicellular organisms. The largest body of information on mammalian circannual rhythms comes from studies of sciurid rodents, particularly from various species of ground squirrels. When golden mantled ground squirrels (*Spermophilus lateralis*) are held under constant conditions of temperature and day length in the laboratory, they exhibit recurrent cycles of body-weight

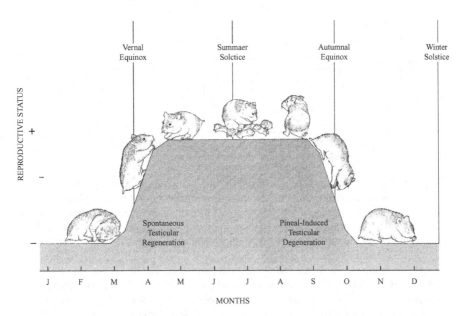

Fig. 1.8. Annual cycle of Syrian hamster seasonal adaptations. During the summer, both male and female hamsters are reproductively competent and engage in breeding. When day lengths fall below some critical value in autumn (i.e., 12 hours of light per day), the steroid negative feedback mechanism of the hypothalamus is enhanced and becomes increasingly sensitive to the effects of steroid hormones. Thus, even low concentrations of steroid hormones completely block secretion of GnRH, and consequently LH, that suppresses steroidal secretion in the testes and leads to gonadal regression. Hamsters may spend the winter in torpor in their burrow; and, in the absence of light, the gonads regrow, and begin to produce steroid hormones and gametes again in early spring. Importantly, the gonads can become functional when the day length is much less than what is required to maintain the gonads in the fall. (Reprinted from R. J. Nelson, *An Introduction to Behavioral Endocrinology*, copyright © 2000, p. 544.)

gain and loss, reproductive activity and inactivity, molting of fur, and hibernation (Gwinner 1986). All these cycles are seasonal, and they occur in a regular sequence relative to each other. In these squirrels, the period length of the circannual rhythm, whether measured for any one of these parameters or for the entire sequence of changes, is approximately 320 days. Although ground squirrels housed under constant conditions continue to show a normal sequencing of annual physiological changes, the timing of these physiological events soon loses its normal synchrony with the local annual geophysical season. Under field conditions, ground squirrels must be able to maintain an appropriate synchrony between these physiological changes and the progression of seasons to reap the benefits of the potential adaptive value of circannual rhythms.

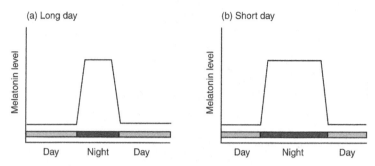

Fig. 1.9. Plasma melatonin concentrations in Syrian hamsters. (A) depicts plasma melatonin concentrations throughout the day and night of a male Syrian hamster (*Mesocricetus auratus*) housed in long days. (B) represents melatonin concentrations throughout
the day and night of a male Syrian hamster housed in short days. Note that elevated
melatonin coincides with the night length; thus, long nights (i.e., short days) result in
a relatively long duration of melatonin secretion, whereas short nights (i.e., long days)
result in a relatively short duration of melatonin secretion. The differential in the duration of the nightly pattern of melatonin secretion encodes night length and provides
the physiological signal necessary for photoperiodism. (Reprinted from R. J. Nelson,
An Introduction to Behavioral Endocrinology, copyright © 2000, pp. 91, 533.)

Circannual rhythms persist after pinealectomy (Zucker 1985; Woodfill
et al. 1994), but the ability to entrain to a simulated natural photoperiod is
lost (Hiebert et al. 2000). Similarly, in pinealectomized ewes, infusions of
melatonin that mimic durations of nocturnal melatonin secretion in natural
day lengths entrain the LH rhythm to a period of 12 months. The endogenous
pattern of melatonin secreted during the 3 months between the summer solstice and the autumnal equinox entrains the annual rhythm of LH in pinealectomized females with phase and period similar to those of intact ewes housed
outdoors (Woodfill et al. 1994; Barrell et al. 2000). Circannual oscillators
in squirrels and sheep (Woodfill et al. 1994) evidently respond differently to
melatonin at discrete phases of the annual cycle. It appears that the absolute
duration of day length and the direction of day length change, both transduced
by melatonin secretion, influence entrainment of annual sheep reproductive
rhythms (Woodfill et al. 1991, 1994).

Although hormones influence phasing, they are not essential for the generation of circannual rhythms. Ablation of several endocrine organs (e.g., testes,
ovaries, pineal gland) of ground squirrels does not interfere with rhythm
expression (Zucker 2001). Among squirrels kept in 14-hour day lengths, locomotor activity begins earlier and ends later in the day during subjective
spring, and the interval between successive recurrences of summer or winter
patterns of activity is ~10.5 months. An endogenous circannual mechanism

modulates the amount and daily timing of locomotion. Although increases in the duration and distribution of activity are correlated with the reproductive cycle and with the seasonal changes in body temperature, the circannual loco-motor rhythms persist unaltered in gonadectomized squirrels (Lee & Zucker 1995) and squirrels in which large body temperature fluctuations are elimi-nated (Freeman & Zucker 2000).

Some species that exhibit annual rhythms in the field show circannual rhythms in the laboratory only when housed under a very restricted set of con-ditions. For example, Sika deer (*Cervus nippon*) exhibit circannual rhythms of antler growth when held under constant photoperiods of LD 18:6, LD 6:18, or LL (L = hours of light; LL = constant light), but fail to show a rhythm when exposed to LD 12:12. European starlings (*Sturnus vulgaris*) show circannual rhythms of testis development when held under a 12:12 photoperiod, but not when kept in photoperiods of LD 11:13 or less, or in photoperiods of LD 13:11 or more (Gwinner 1986). Individuals of many species that are highly seasonal in the wild fail to exhibit persistent circan-nual rhythms in the laboratory under any of a very wide variety of constant conditions. Typical of this type are the many photoperiodic rodents that un-dergo winter adaptation when exposed to short photoperiods. The return to summer condition in these rodents can occur either as a result of exposure to increased day lengths or as the result of an endogenous timing process that triggers the return to summer condition following prolonged exposure to short days. The most commonly studied of these internally timed (i.e., not di-rectly cued by environmental change) responses is the so-called *spontaneous* recrudescence of reproductive organs. Spontaneous recrudescence occurs in long-day breeding rodents following several months of exposure to a constant inhibitory (i.e., short day) photoperiod. However, it seems the general rule is that most types of photoperiod-controlled winter adaptations are similarly returned to a summer state by the action of an internally timed process. When continuously held under short days, these rodents will exhibit only one cycle of change from summer to winter and then back to summer state.

Most seasonal mammals are almost continuously exposed to potentially useful environmental cues and probably make frequent use of cues during the course of each annual cycle; therefore, it is not clear whether the possession of a circannual oscillator mechanism confers any particular adaptive advantage in comparison with the mechanisms used by noncircannual seasonal animals. However, some mammals that are hibernators or estivators remain torpid for a significant portion of the annual cycle. Although these animals arouse and exhibit short bouts of euthermy (i.e., normal body temperature of about 37°C) at intervals during the torpor season, many remain in closed burrows and may not be exposed to environmental cues that would allow them to measure the passage of time. For these species, an ability to utilize an endogenous mechanism to measure time on a seasonal scale might be a significant adaptation. Although the ability to hibernate seems to have evolved on multiple occasions during mammalian evolution and occurs in only a small percentage of species, winter torpor is the general rule among terrestrial ectothermic vertebrates inhabiting temperate climes. Possibly a vertebrate mechanism for endogenous seasonal time keeping first appeared in ectotherms and was continued in the lines of evolution leading to the appearance of birds and mammals.

1.5. Physiological Bases for Seasonal Cycles: Circadian Rhythms and Photoperiodism

The formal properties of photoperiodic responsiveness, as well as the neuroendocrine, basis for photoperiodism, have been areas of intense investigation among several mammalian species. Certain common features of mammalian photoperiodism have emerged from these studies:

1. All photoperiodic mammals examined to date use an internal circadian mechanism for the process of photoperiodic time measurement to "determine" the day length.
2. Wild populations of photoperiodic mammals often display polyphenism for photoperiodic response, with some individuals being unresponsive or less responsive than the norm, to changes in day length. These interindividual differences show a high degree of heritability.
3. In all photoperiodic mammals studied thus far, including marsupials, the pineal gland is an important endocrine component of the photoperiodic mechanism. The pineal exerts its influence over photoperiodic responses via its rhythmic secretion of the hormone, melatonin.
4. All photoperiodic mammals appear not only to be capable of responding to day length cues, but they also possess an ability to carry out endogenous time keeping on a seasonal scale.

This ability of mammals to use endogenous mechanisms to track season is most obviously evidenced by the phenomenon of "spontaneous" change from winter to summer physiological conditions that occurs in virtually all photoperiodic species after a prolonged period (i.e., generally 20–25 weeks, but varying somewhat with species) of exposure to short-day photoperiod. Some, but not all, photoperiodic mammals also show a spontaneous switch from summer to winter condition following prolonged exposure to long days.

Laboratory rats (*Rattus norvegicus*) have traditionally been considered nonphotoperiodic because exposure to short day lengths, constant darkness, or blinding has minimal, if any, effects on reproductive function. However, four different experimental treatments have revealed latent photoperiodism in laboratory rats:

1. Removal of the olfactory bulbs unmasks reproductive responsiveness to short days; that is, males bulbectomized prior to puberty delay reproductive development in short days, compared with animals in long days.
2. Moderate food restriction reduces reproductive organ mass in animals housed in short, but not in long, days.
3. Male rats bearing subcutaneous testosterone implants exhibit some testicular regression when housed in short days, but not in long photoperiods.
4. Neonatal injections of testosterone result in slowed gonadal growth in short-day reared male rats as compared to their long-day counterparts.

Some of the manipulations previously listed also revealed reproductive responsiveness to photoperiod in house mice (*Mus musculus*), another rodent that has traditionally been considered to be nonphotoperiodic (Nelson 1990). The physiological means by which testosterone treatments, food restriction, or olfactory bulbectomy affect the expression of photoperiodism are unknown. Responsiveness of pelage coloration to photoperiod in rats suggests that nonreproductive traits may remain responsive to day length regardless of reproductive responsiveness (Nelson et al. 1994). The extent to which this is true for immune function in rats, mice, or even humans requires further investigation.

It is now clear that the mammalian pineal gland and its hormone, melatonin, are intimately involved in the photoperiodic process (Goldman & Nelson 1993; Hastings 1996; Hiebert et al. 2000). Pinealectomy renders a wide variety of mammals unresponsive to changes in photoperiod; indeed, pinealectomy seems to prevent most photoperiodic responses in virtually every mammalian species that has been examined. Rhythmic secretion of melatonin by the pineal gland can induce photoperiodic responses when administered in a number of ways that mimic normal circadian release. Nevertheless, particular aspects of the responses observed following constant release implants, daily injections, or daily infusions of melatonin have led to some controversy regarding the significance of various aspects of melatonin secretion. Melatonin is normally secreted in circadian fashion, with an extended peak occurring at night. The duration of this nocturnal "peak" varies inversely with day length in a variety of species, including humans. The controversy over identification of the parameter of pineal melatonin secretion that is responsible for the influence of this gland on photoperiodic responsiveness has focused mainly on the relative importance of the phase versus the duration of the nocturnal elevation of melatonin secretion. Results obtained in sheep and in Siberian hamsters strongly favor the overriding importance of the duration, as opposed to the amplitude, of the peak (Bartness et al. 1993). When pinealectomized hamsters or sheep are infused with melatonin on a daily basis, the types of responses elicited depend on the duration, but not on the phase, of each daily infusion. For example, when male hamsters receive melatonin for 6 hours or less each day, they exhibited stimulated testes (a long-day response); when melatonin is administered in daily infusions of 8 hours duration or more, animals inhibit reproductive parameters (a short-day response). These results correspond to the direct measures of blood concentrations of melatonin of hamsters housed in long or short days (Figure 1.9). Less extensive data have recently been obtained from similar experiments in Syrian hamsters (*Mesocricetus auratus*), white-footed mice (*Peromyscus leucopus*), and mink (*Mustela vison*); in each case, the results are in accord with the conclusions reached in the studies using Siberian hamsters and sheep (reviewed in Bartness et al. 1993).

The development of a method for labeling melatonin with ^{125}I has allowed the investigation of sites of uptake of melatonin by using autoradiographic techniques. Although melatonin exerts marked effects on the secretion of several anterior pituitary hormones, little uptake of radiolabeled melatonin has been observed in the pituitary, except in the case of neonatal rats and Siberian hamsters. Several species of rodents exhibit melatonin uptake in the hypothalamic suprachiasmatic nuclei (SCN); this bilateral nucleus has been established to be the primary biological clock that coordinates temporal organization among mammals and perhaps other vertebrates as well. Melatonin uptake has also been observed in several additional brain sites, including the periventricular nuclei of the thalamus and the reuniens nuclei. However, the particular sites that are labeled vary from species to species; for example, the SCN were not labeled in sheep and ferrets, although involvement of pineal melatonin in photoperiodic responsiveness has been clearly demonstrated for both these species.

The master-clock coordinating circadian rhythms appears to be located in the SCN. The evidence supporting the claim that the SCN are the biological clock timing daily rhythms include:

1. ablation of the SCN results in the abolishment of circadian rhythms (Moore & Eichler 1972; Stephan & Zucker 1972).

2. hypothalamic islands segregating the SCN from the rest of the brain generate circadian rhythms of neural activity (Inouye & Kawamura 1979).
3. a circadian rhythm of ^{14}C-labeled-deoxyglucose utilization in the SCN, but not elsewhere in the brain (Schwartz & Gainer 1977).
4. SCN slices maintain circadian organization of firing rate that can be phase-shifted (Gillette 1986).
5. transplants of fetal SCN grafts into SCN-lesioned adults result in expression of mutant circadian rhythms in the recipient, with the host adopting the endogenous rhythm of the donor animal (Ralph et al. 1990).

In the wide variety of mammals that have been subjects of melatonin uptake studies, the single site that is labeled most frequently and most strongly is the pars tuberalis. This tissue surrounds the base of the median eminence and develops during embryogenesis from the lateral wings of Rathke's pouch; the medial portion of Rathke's pouch is the anlage for the anterior pituitary. The function of the pars tuberalis is unknown, but it has been reported to exhibit seasonal histological and immunohistochemical changes. These observations, along with its avid uptake of melatonin, make it tempting to speculate that the pars tuberalis may be an important component of the photoperiodic mechanism. A number of recent studies have indicated that melatonin binding sites are also located on cells of the immune system. These findings will be described in detail in Chapter 5.

1.5.1. Reproductive Morphs

The extent to which photoperiod is a useful proximate factor depends on the year-to-year predictability of favorable environmental conditions; photoperiod is less useful as a cue for reproductive activity for animals living in irregularly changing habitats. For many rodents, with an average life-span measured in weeks or months, virtually any temperate habitat may be essentially unpredictable. Large mammals and birds, that tend to breed over several years, can rely more exclusively on photoperiod because reproductive activities will occur during the optimal environmental period, on average, over several years. A late snow or a summer drought may adversely affect breeding effort for a single year; however, the animal may recoup from an isolated reproductive failure in other years. For short-lived rodents like *Peromyscus* or *Microtus*, a single reproductive failure is likely to be disastrous in terms of evolutionary fitness. The production of offspring that vary in their responsiveness to different environmental cues may increase the fitness of an individual, compared with the production of young that rely exclusively on photoperiod to cue reproduction. For example, an animal cued by photoperiod may fail reproductively due to late snow cover, whereas its sibling may not become reproductively competent until adequate nutrition is available regardless of photoperiod; the inclusive fitness of the progenitors and the unsuccessful sibling would be enhanced by this phenotypic variation. Because each of these traits can be advantageous in certain circumstances, each phenotype can exist within a population at a stable intermediate frequency to form a balanced polymorphism (Prendergast et al. 2001). The cost-benefit equation for winter breeding must include other energetically expensive processes, including immune function. The energetics of immunity will be discussed in the context of seasonality in Chapter 6.

1.5.2. Evolution of Photoperiodic Responsiveness

There is abundant evidence that mammalian photoperiodism has a strong genetic basis and that species differences in photoperiodic responsiveness have arisen during mammalian evolution (see prior section) (Prendergast et al. 2001). It remains to be determined, however, what specific aspects of the physiological mechanisms of photoperiodism have been subject to evolutionary change. Two species that have sometimes been considered to be nonphotoperiodic – house mice and laboratory rats – possess the "typical" mammalian photoperiodic mechanism as indicated by the ability of these species to exhibit photoperiodic responses following certain experimental manipulations. Therefore, it is possible that virtually all mammals have retained the neuroendocrine circuitry that is required for measuring day length (i.e., photoperiodic time measurement) and for producing an endocrine signal that encodes the day length. This endocrine signal, in the form of the rhythmic pattern of pineal melatonin secretion, may be useful in the mediation of seasonal changes in immune function. In mammals that do not exhibit photoperiodism in the traditional sense, it may be that potential photoperiodic responses are blocked at levels downstream from these central neuroendocrine components of the photoperiodic mechanism.

Species differences in sites of melatonin binding in the brain have generated interest, but it is not apparent that these differences correlate with differences in photoperiodic responsiveness. For example, some photoperiodic species (e.g., Syrian and Siberian hamsters, white-footed mice) exhibit melatonin binding in the SCN, whereas other species that are similarly responsive to day length (e.g., sheep and ferret) do not show binding in this brain region.

Almost all mammals studied show melatonin uptake in the pars tuberalis. The species that have been studied are mostly reproductively photoperiodic. It will be interesting to determine whether any mammals fail to show evidence of photoperiodic responsiveness after thorough study. Olfactory bulbectomy, which seems to "unmask" responses in some species, should be used to determine whether these nonphotoperiodic species exhibit melatonin uptake in the pars tuberalis.

Whereas species differences in melatonin binding might form the basis for differences in photoperiodic responsiveness, there are at least two alternative possibilities: (1) it is possible that species differences in the pattern of melatonin secretion, and in particular the influence of day length on the melatonin rhythm, might underlie differences in photoperiodic responsiveness; and (2) it is also possible that there are species differences in post-receptor mechanisms that account for differences in responsiveness, including entrainment to circadian rhythms (Puchalski & Lynch 1988).

Evidence is beginning to accumulate that may eventually provide answers to these questions regarding the mechanisms of evolution of photoperiodic responsiveness in mammals. White-footed mice captured in Georgia are not photoperiod-responsive, yet they exhibit pineal melatonin rhythms that are similar to those of photoperiod-responsive mice of the same species from Connecticut (Lynch et al. 1981). Also, there are no differences in patterns of binding of radiolabeled melatonin in the brains of mice from each of these two populations (Weaver et al. 1990). Finally, in contrast to their Connecticut counterparts, Georgia mice are unresponsive to treatment with exogenous melatonin. These observations all suggest that the physiological basis for photoperiodic nonresponsiveness in the Georgia population lies at a level of the photoperiodic mechanism downstream from both the regulation of melatonin secretion and its binding to target site receptors.

Similarly, the effects of photoperiod and melatonin treatment on reproductive and immune function were assessed in males of two subspecies of *Peromyscus maniculatus* from different latitudes of origin. *Peromyscus maniculatus bairdii* (latitude $= 42°51'$N) and *Peromyscus maniculatus luteus* (latitude $= 30°37'$N) were housed in long or short days for 8 weeks. Short-day *Peromyscus maniculatus bairdii* underwent reproductive regression and displayed increased cell-mediated immune function (Demas et al. 1996). Short-day *Peromyscus maniculatus luteus* maintained reproductive function and did not display altered immune function. Melatonin treatment of male *Peromyscus maniculatus bairdii* and *Peromyscus maniculatus luteus* housed in long days caused reproductive regression in *Peromyscus maniculatus bairdii* males and increased cell-mediated immune function (Demas et al. 1996). In contrast, melatonin treatment did not affect reproductive or immune function in *Peromyscus maniculatus luteus* males. These results suggest that the effectiveness of melatonin on immune function may be linked to reproductive function.

1.6. Conclusions

In mammals, seasonal fluctuations in a variety of categories of physiological and behavioral response are regulated in whole or in part by a photoperiodic mechanism. Changes in reproductive state are among the most prominent and commonly studied of these seasonal fluctuations, and other types of changes are sometimes secondary to seasonal cycles of gonadal steroid hormones. However, most nonreproductive seasonal adaptive changes appear to be at least partly independent of fluctuations in reproductive hormones.

The annual cycle of day length change is the major environmental cue used by many temperate zone mammals to track the seasonal passage of time. In this chapter, we have emphasized that all photoperiodic mammals appear to possess an endogenous mechanism for seasonal time-keeping, in addition to their ability to measure day length. In some species, this endogenous time-keeping mechanism is expressed cyclically under constant environmental conditions, in the form of a circannual rhythm. In other species, endogenous seasonal time-keeping seems to be used only during the winter phase of the annual cycle, and photoperiod changes must be present to drive the remainder of the cycle. We suggest that a single ancestral mechanism for annual time-keeping may have served as an evolutionary precursor for both these forms of mammalian seasonal regulation (Bartness et al. 1993; Goldman & Nelson 1993).

How can the hypothesis of common evolutionary origin be tested? Despite certain similarities between circannual and photoperiodic/noncircannual species with respect to formal, or "whole animal," responses to changing photoperiod, differences also remain between the responses of these two types of species. These differences are most pronounced with respect to their behaviors when held under constant conditions. Therefore, at present it remains plausible to maintain that they represent two quite different solutions for the problem of annual time-keeping, involving distinct physiological mechanisms and with separate evolutionary origins. We expect the evolutionary relationships, or lack thereof, between circannual and photoperiodic/noncircannual patterns to become more apparent as more is learned about the anatomical and neuroendocrine bases for each respective type of mechanism. The hypothesis of evolutionary relationship elaborated in this chapter would suggest, for example, that the mechanism that times "spontaneous" return to spring/summer state in photoperiodic rodents might be the same as, or very similar to, the mechanism used for timing the various stages of circannual cycles. We would suggest that, if a component of the timing mechanism for either of these processes is discovered (e.g., through brain lesion experiments, endocrine gland ablations, etc.), this component should be examined as a potential element in the other type of system. For example, the anatomical substrate for a circannual oscillator might prove to be the same as that responsible for the timing of spontaneous recrudescence of the reproductive system in photoperiodic/noncircannual rodents (see Hiebert et al. 2000).

In sum, both circannual and photoperiodic/noncircannual species (1) use day length as a major cue in establishing synchrony between seasonal environmental time and internal physiological state, and (2) use the pineal melatonin rhythm as the major endocrine component for internal "signaling" of photoperiod change. The sites of melatonin binding in the central nervous system, especially in the pars tuberalis, should receive high priority for investigation as potential sites of interaction between endogenous seasonal time-keeping mechanisms and their synchronizing (or entraining) action by photoperiod cues. As we will discuss later, melatonin receptors in immune cells also may provide important clues about the role of this indole-amine hormone in immune function.

2

Immune Function

Self-defence is Nature's eldest law.

John Dryden, 1695
Absolom and Achitophel

2.1. Introduction

During evolution, the cells of multicellular organisms began to specialize in function to achieve a common goal of replicating shared genetic material to place into offspring. Cells within these multicellular organisms developed specific markers to identify them as part of the organism. The evolution of self-recognition molecules was likely a response to single cell organisms that began to specialize in exploiting the resources of the larger individuals for their own survival and reproductive needs (Hamilton et al. 1990). The fundamental feature of any defense against parasitism is the ability to recognize "self" from "non-self." During the course of evolution, increasingly elaborate methods have developed among parasites, bacteria, viruses, and fungi to "trick" or override the defenses of multicellular individuals to exploit their resources. In other words, these pathogens have evolved mechanisms to mimic or supersede the self signals of the host individuals. The bodies of the host individuals are the "homes" and "meals" for parasites, so parasites have evolved mechanisms to elude host defenses (Nesse & Williams 1994). In response, hosts have evolved intricate defenses against the presence of bacteria, viruses, parasites, and fungi. Immune (from the Latin, *immunis* = free from) responses have developed that can neutralize these pathogens. Many host defenses are very energetically costly, for example, regeneration of tissue to maintain the skin or maintenance of high body temperature during fever. As noted in the previous chapter, the energetic costs of these defense strategies are optimized in relation to the energetic costs of other physiological processes. Seasonal fluctuations in energy availability and requirements form the basis for seasonal fluctuations in immune function and disease processes.

To appreciate the costs associated with defense against pathogens, a basic review of immune function is presented in this chapter.

Immunology is the study of the body's defenses against infection and disease. The immune system can defend against infections arising from external sources, such as viruses, bacteria, fungi, and parasites, as well as disease arising from internal sources such as tumors and cancers. During the first decades of this century, studies of immune and endocrine function were often co-mingled. Early research classified some immune organs as endocrine glands based on their functional similarities and interactions with other endocrine glands, such as the gonads and adrenal glands (Martin 1976). For example, the atrophy observed in the thymus, bursa, and spleen, important immune organs, after puberty and the fluctuations in immune organ size in relation to seasonal breeding prompted many early hypotheses, thus suggesting that these organs were directly linked to reproductive function (Aimé 1912; Riddle 1928). Initially, the avian thymus was thought to provide the "egg envelope" (Riddle 1924) until further experiments established that thymectomy did not interfere with the production of normal eggs (Morgan & Grierson 1930; Riddle & Krizenecky 1931). Several studies initially conducted in rodents, and verified in other species, determined that the thymus was responsive to steroid hormones, such as androgens, estrogens, and glucocorticoids; for example, thymic size enlarged after castration in males (Hammar 1929; Gregoire 1945), whereas steroid replacement therapy of castrated animals caused thymic involution (Martin 1976). Additional studies in humans established that the thymus involutes with age, with marked regression observed at the time of puberty (Low & Goldstein 1984) (Figure 2.1). Importantly, the size (and presumably the function) of the thymus was shown to respond to environmental perturbations, including light, temperature, and food availability (Martin 1976). Only recently, however, has the existence of chemical communication between the neuroendocrine and immune systems been widely accepted by both immunologists and neuroendocrinologists.

Within the past 20 years, a new field of study has emerged that examines the interaction of the immune system with the brain and endocrine system. This field of study has been termed psychoneuroimmunology (Ader & Cohen 1981), and researchers in this field have established that the cells and tissues of the immune system are influenced by hormonal and neural factors of the central nervous system, as well as by psychological and social variables (see Ader & Cohen 1993, for review; Ader et al. 2001). To date, this field of study has focused primarily on using traditional experimental psychology paradigms to trace the relationship between the brain and immune system. For example, the most compelling support for neural modulation of

Proportional Size of Thymus Gland

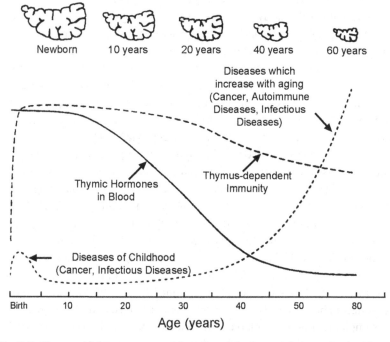

Fig. 2.1. Changes in the structure and function of the human thymus gland with age. The size of the thymus reduces, or involutes, with age in both human and nonhuman animals. Both cell-mediated immunity and production of thymic hormones decline with age. As a result, susceptibility to certain infections increase with age. (From Low & Goldstein 1984.)

immune function comes from studies demonstrating that immune responses can be classically conditioned. In this paradigm, an immunosuppressive drug (e.g., cyclophosophamide) (unconditional stimulus) is repeatedly paired with a novel stimulus, such as saccharine sweetened water (conditional stimulus). Several days later, animals are reintroduced to the saccharine-sweetened water in the absence of the immunosuppressive drug. Subsequent examination of immune function revealed that immune responses from animals reexposed to saccharine alone (i.e., conditional response) were suppressed to the level of immune responses from animals that had been repeatedly reexposed to the immunosuppressive agent alone (i.e., unconditional response). Notably, this technique has been applied with some success to the treatment of auto-immune diseases (Ader & Cohen 1982) and tumorigenesis (Blom et al. 1995) in rodents.

Another area of intense psychoneuroimmunology research is the effects of stress on the immune system. Stress-induced immunosuppression has been the subject of much research in humans, nonhuman primates, and rodents. Interest in this area stems from the rise in public concern about the increased risk of infection and development of disease in response to chronic life stressors, such as bereavement, job-related stress, divorce, and even exams (Sapolsky 1998). The first report that stress suppresses the immune system was in 1936 when Hans Selye discovered that exposure to a variety of chronic stressors caused thymic involution in rodents. Since that time, numerous studies have demonstrated that exposure to a variety of stressors in rodents (e.g., restraint, shock, isolation, and handling), nonhuman primates (e.g., isolation), and humans (e.g., bereavement, exam stress, or marital discord) inhibits immune function and increases susceptibility to infection (Glaser et al. 1987; Irwin & Hauger 1988; Laudenslager et al. 1988; Kiecolt-Glaser et al. 1991; Brenner & Moynihan 1997; Boccia et al. 1997; Kiecolt-Glaser 1999). Although exposure to chronic stressors increases susceptibility to certain infections, organisms have evolved adaptations that actually make them resistant to many stress-related illnesses (Sapolsky 1998).

Until recently, there have been few attempts to examine the adaptive function of the interactions between the immune and neuroendocrine systems as they relate to illness and infection (Ewald 1994; Nesse & Williams 1994). Most studies examining neuroendocrine-immune interactions focus on proximate explanations that address how these interactions occur (e.g., localization of hormone receptors on immune cells to identify direct links between these systems). Our approach is to provide both proximate and ultimate explanations for seasonal fluctuations in immune function and disease processes. Ultimate explanations address why the relationship between the immune and neuroendocrine systems has evolved and what adaptive function this relationship serves in terms of resistance to disease. We will use this adaptive functional approach as the framework in which to describe how and why environmental factors such as photoperiod, food resources, temperature, and social factors (including breeding and competition) influence the functioning of the immune system in both human and nonhuman animals (Zuk 1990; Nelson & Demas 1996; Sheldon & Verhulst 1996; Klein & Nelson 1999; Klein 2000a,b). To appreciate the complex interactions among the immune system and other systems in the body, an understanding of the basic concepts in immunology is useful. The goal of this chapter is to serve as a resource for readers to use as a basis for understanding the immunological assessments used in studies reviewed in later chapters.

2.2. Types of Immunity

2.2.1. Innate Immunity

Innate immunity is germline encoded and, thus, present at the time of birth. The defenses associated with innate immunity do not require prior exposure or sensitization to a particular pathogen (i.e., any foreign substance that has entered the body, also called an antigen); therefore, this type of immunity is often referred to as nonspecific. There are several types of innate immune responses (Figure 2.2). Anatomically, skin represents a physical barrier for entry of most pathogens into the body. Physiologically, the pH of skin is too acidic for most pathogens to grow and thrive. Mucosal immunity is another example of innate immunity. The mucous lining of the gut contains enzymes that can penetrate and destroy the cell walls of most pathogens. Finally, the phagocytic action of macrophages engulfing and eliminating pathogens from circulation represents another type of innate immune response. Although these innate immune responses serve to eliminate a pathogen from the body, these responses do not result in resistance to infection. Resistance is achieved by immune cells that possess the ability, in essence, to

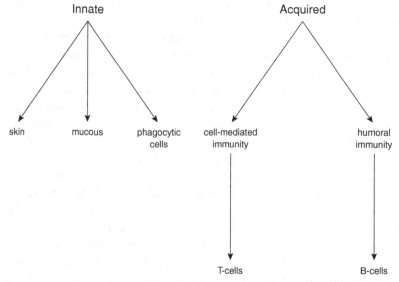

Fig. 2.2. Division of the immune system into innate and acquired immunity. Innate immunity involves responses that are nonspecific and do not require prior sensitization. Acquired immunity involves responses that are specific and acquired. Innate immunity is important for recognizing foreign pathogens, whereas adaptive immunity is responsible for overcoming infection.

"remember" prior exposure to a pathogen and improve their responsiveness to a pathogen upon subsequent exposure. This process is the cornerstone of acquired immunity.

2.2.2. Acquired Immunity

Among vertebrates, acquired immunity is both specific and learned (Table 2.1). Acquired immunity is specific because the immune cells involved possess receptors that are genetically organized to recognize and respond to a single pathogen. Acquired immunity is learned because these immune cells can only engage in an immune response based on familiarity or previous exposure to a pathogen. The importance of acquired immunity is that the recognition and elimination of a pathogen improves each time an individual is reexposed (Figure 2.3); this improvement ultimately leads to resistance against infection. For example, this memory response of the immune system is the basis for vaccinations that involves exposure of an individual to an innocuous or weakened pathogen to induce acquired immunity.

There are two types of acquired immune responses mediated by lymphocytes: cell-mediated and humoral immunity (Figure 2.2). Cell-mediated

Table 2.1. *Phylogeny of Innate and Acquired Immunity in Invertebrate and Vertebrate Animals.*[1]

	Innate Immunity		Acquired Immunity		
	Phagocytosis	NK Cells	Antibodies	T- and B- Cells	Lymph Nodes
Invertebrates					
Protozoa	+	−	−	−	−
Sponges	+	−	−	−	−
Annelids	+	+	−	−	−
Arthropods	+	−	−	−	−
Vertebrates					
Elasmobranchs (sharks, rays)	+	+	+ (IgM only)	+	−
Teleosts (common fish)	+	+	+ (IgM, other)	+	−
Amphibians	+	+	+ (2–3 classes)	+	−
Reptiles	+	+	+ (3 classes)	+	−
Birds	+	+	+ (3 classes)	+	+
Mammals	+	+	+ (7–8 classes)	+	+

[1] Defenses associated with innate immunity, including phagocytosis of foreign material by macrophages, are present in both invertebrates and vertebrates. Specific immune responses associated with acquired immunity, that enable an organism to overcome infection, are only present in vertebrates. Phylogenetically, immunity becomes increasingly specialized during evolution. +, present; −, absent. (Adapted from Abbas et al. 1994.)

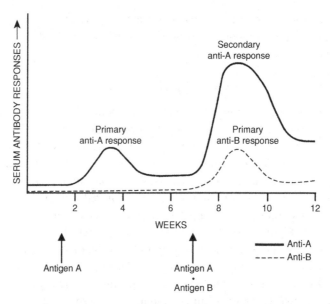

Fig. 2.3. Humoral immune response illustrates both specificity and memory. Humoral immunity is specific because antigens A and B facilitate the production of distinct antibodies. The secondary response to antigen A occurs more rapidly and for a longer period of time than the primary response to either antigens A or B. This secondary response illustrates the memory component of immunity.

immunity is modulated by activated T-cells (see next section) that can either kill a pathogen directly or release soluble factors called cytokines to attract other lymphocytes to the site of infection. Cytokines are chemical messengers that enable immune cells to communicate with one another and with neuroendocrine cells (see next section). The distinguishing characteristic of cell-mediated immunity is that it involves direct interaction between lymphocytes and the pathogen; therefore, contact with a pathogen is required for this adaptive immune response to occur. Examples of immune responses that are cell-mediated are rejection of transplanted tissue and localized inflammation in response to a foreign substance, such as tuberculin (Borysenko 1987).

Humoral immunity is another class of acquired immune responses that primarily involves activated B-cells secreting antibodies into circulation (see next section). Antibody molecules are surface, antigen-specific receptors released from B-cells in response to a particular pathogen. In contrast to T-cells, B-cells do not directly contact a pathogen, but rather the specific antibodies that are secreted come into contact with the pathogen. The antibodies bind to epitopes on the pathogen (i.e., antigens) to form inactive complexes that may either be eliminated from circulation or serve to activate the compliment cascade (a serial activation of compliment immune cells) that will destroy a pathogen by boring a hole in the cell wall. This cascade of cellular mechanisms is called compliment because the cells involved assist, or compliment, the lytic function of antibodies. Although antibodies secreted by B-cells are the key participants in humoral immune responses, cytokines secreted by T-cells and macrophages are very important for regulating humoral immunity because they serve to activate resting B-cells.

2.3. The Immune System

2.3.1. Immune Organs

The organs of the immune system are called lymphoid organs (Figure 2.4). These organs are connected to one another by an anatomical pathway called the lymphatic system. In the lymph (i.e., the fluid), lymphocytes circulate and encounter pathogens. The lymphatic system is connected to the circulatory system via the thoracic duct. Through the thoracic duct, immune cells can circulate and recirculate between lymphoid organs and blood.

Among vertebrates, there are two primary lymphoid organs. The first organ is the bursa of Fabricius in birds and the bone marrow in mammals. Among mammals, stem cells, that are the progenitor cells for all lymphoid cells, mature and become B-cells in bone marrow. The second primary lymphoid organ is the thymus, a two-lobed structure located above the heart. In the thymus, lymphoid cells that have migrated from bone marrow (in mammals) mature and develop into T-cells. These two organs are referred to as primary lymphoid organs because they are the major sites of lymphocyte development.

Cells from the primary lymphoid organs (i.e., T-cells from the thymus and B-cells from the bone marrow) migrate to secondary lymphoid organs where exposure to pathogens occurs. The secondary lymphoid organs include the lymph nodes, spleen, and gut-associated lymphoid tissues (appendix, tonsils, and Peyer's patches in the small and large intestine). In these organs, pathogens are trapped and adaptive immune responses, specific to these pathogens, are initiated by lymphocytes.

2.3.2. Immune Cells

Phagocytic cells represent an important "first line of defense" for the immune system. Phagocytosis involves the distinction between self and non-self tissues and cells, and engulfment of non-self tissue and cells. After a pathogen has been engulfed by a phagocytic cell, the cell presents a fragment of

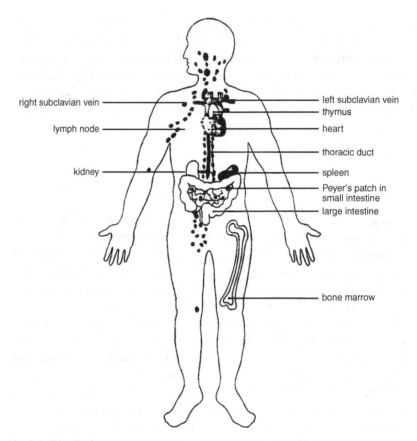

Fig. 2.4. Distribution of lymphoid organs in the body. Lymphoid organs are connected to one another by the lymph. Immune cells circulate from the lymph into the circulatory system through the thoracic duct; these cells are recirculated from the blood back into the lymph by emptying into the left subclavian vein. In mammals, the two primary lymphoid organs, where immune cells develop, are the bone marrow and thymus. Following development and maturation in these primary lymphoid organs, immune cells migrate to secondary lymphoid organs, such as the lymph nodes, spleen, and gut-associated lymphoid tissues, where these cells are activated by exposure to antigens.

the pathogen on its surface membrane through the major histocompatibility complex (MHC); thus, these cells are often called antigen-presenting cells. The MHC molecule on the surface of antigen-presenting cells serves as a membrane-bound receptor for fragments of pathogens to be recognized by surface receptors on T-cells. T-cells can only recognize the presence of non-self tissue in the body when it is presented in the cleft of an MHC molecule. The initial engulfment of a pathogen by a phagocytic cell can eliminate the pathogen from circulation. This nonspecific defense against infection is an example of innate immunity.

Phagocytic cells include granulocytes – such as neutrophils, eosinophils, and basophils – that are nonspecific in their response to a pathogen. These immune cells can phagocytize and kill pathogens effectively by releasing enzymes that degrade the ingested foreign material. Granulocytes serve an important function for the degradation of bacteria; among phagocytic cells, however, monocytes and macrophages are responsible for the "classic" innate immune responses described previously. Monocytes are essentially immature macrophages circulating in blood. Monocytes mature into macrophages, the activated form of this cell found in tissue. These cell types (i.e., monocytes and macrophages) are often referred to interchangeably. Macrophages engage in phagocytic processes throughout the body. For example, there are macrophages in the brain called microglia, macrophages in the liver called Kupffer cells, and macrophages in the lungs called arveolar macrophages. In the immune system, however, macrophages are responsible for transporting pathogens into secondary lymphoid organs to stimulate lymphocytes and generate specific, acquired immune responses.

Lymphocytes are the primary immune cells that mediate host defenses against infection. These cells possess the ability to recognize and distinguish a pathogen from self tissue, and to distinguish between different types of pathogens (this is called specificity). This function is accomplished through specific T-cell receptors located on the surface of lymphocytes that interact with MHC molecules carrying fragments of pathogen in their cleft.

Lymphocytes are of two main types: B- and T-lymphocytes. In birds, B-cells mature in the bursa of Fabricius, whereas in mammals, B-cell maturation occurs in the bone marrow (Figure 2.5). Therefore, B-cell refers to bursa- or bone marrow-derived lymphocyte. B-cells are responsible for the production and surface presentation of antibodies called immunoglobulin (Ig). When a B-cell is stimulated by exposure to a pathogen and by cytokines released from T-cells, it undergoes division; its progeny cells become either plasma cells that immediately engage in a humoral immune response or memory cells that remain quiescent in circulation until activated by reintroduction of a pathogen in the body. Stimulation of a resting B-cell results in the production of Ig against the specific pathogen. The specific Ig that recognizes a specific pathogen is called antibody. Antibody molecules on the surface of a B-cell act as specific receptors that bind to a single corresponding pathogen. B-cells are the only immune cells capable of producing and secreting antibody. B-cells play an important role in the acquired immune responses involved in humoral (or antibody-mediated) immunity.

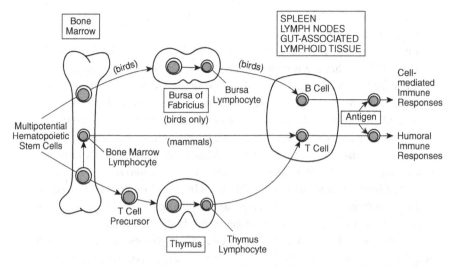

Fig. 2.5. Development and migration of T- and B-cells. All lymphocytes (T- and B-cells) originate from multipotential hematopoietic stem cells in the bone marrow. In mammals, these precursory B-cells remain in the bone marrow where they differentiate and mature into B-cells. In birds, the precursors to B-cells migrate to the bursa of Fabricius, where they differentiate into B-cells. In both birds and mammals, the precursory T-cells migrate from the bone marrow to the thymus where they differentiate and mature into T-cells. After differentiation and maturation, these lymphocytes migrate to secondary lymphoid organs where these cells encounter antigens and engage in their specific effector functions (either cell-mediated or humoral immune responses).

T-cells are the second type of lymphocytes. Because T-cells are born in the bone marrow but migrate to the thymus to mature, the term T-cell, refers to a thymus-derived lymphocyte (Figure 2.5). There are two functional classes of mature T-cells: cytotoxic T-cells and helper T-cells (i.e., Th cells). Helper T-cells are functionally and phenotypically heterogeneous, and subsets of Th cells have been classified based on their effector function. Th1 cells activate macrophages to engulf pathogens and secrete interferon-γ and interleukin (IL)-2 that induce B-cells to stop producing IgM and switch to producing IgG2a and IgG3 in rodents and IgG1 in humans. Conversely, Th2 cells activate resting B-cells to produce antibody and secrete IL-4, IL-5, IL-6, and IL-10 that facilitate B-cells to stop producing IgM and switch to producing IgG1 in rodents and IgG4 in humans. The extent to which these Ig subclasses exist in birds, reptiles, amphibia, or fish currently is not known. In mammals, subclasses of cytokines and IgG play distinct roles in response to infection (Janeway et al. 1999). Classes of T-cells are defined by membrane protein receptors expressed on the surface of each T-cell type. For example, cyto-toxic T-cells express a membrane-bound protein called CD8, whereas helper T-cells (both Th1 and Th2) express a surface protein called CD4. Conse-quently, CD stands for "clusters of differentiation" determined by the mono-clonal antibodies that recognize and bind to the same cell surface molecule (i.e., receptor) (Janeway et al. 1999). Many of the effector functions per-formed by T-cells are facilitated by the release of cytokines (detailed herein) that bind to specific receptors on other immune cells. Because all T-cell ef-fector functions involve direct cell interactions, T-cells play an important role in the acquired immune responses involved in cell-mediated immunity.

Natural killer (NK) cells are another type of immune cell that can non-specifically kill virally infected cells and some forms of tumor cells without any prior sensitization. NK cells develop from pluripotent bone marrow-derived hematopoietic stem cells. These cells are evolutionarily ancient and have been identified in both vertebrates and invertebrates (Table 2.1). NK cells are a type of effector cell distinct from cytotoxic T-cells. Whereas cytotoxic T-cells are generally involved in killing a specific foreign pathogen, NK cells are responsible for killing carcinogenic or virally infected self cells. NK cells identify infected cells by binding to antibody molecules (previously secreted by B-cells) attached to the surface of infected cells; this process of antibody coating and labeling the surface of a pathogen is called opsonization.

2.3.3. Cytokines

Cytokines are chemical messengers that are similar in function to protein hormones. Cytokines are produced by phagocytes and lymphocytes, and en-able immune cells to communicate with one another to coordinate responses to a pathogen. Cytokines work by binding to specific receptors located on the surface of target cells that facilitate autocrine, paracrine, or endocrine actions. Cytokine receptors have very high affinities for their ligand (i.e., their specific cytokines). The high affinity of the cytokine receptors mean that very low concentrations of cytokines are required to have dramatic biological effects (Abbas et al. 1994). Although cytokines can only act on their corre-sponding receptors, these receptors are located on a wide variety of cell types. Consequently, cytokine effects can be substantial.

Both innate and acquired immune responses are mediated by cytokines (see Table 2.2). Cytokines produced by macrophages are called monokines, and are generally released in response to either pathogens or stimulated T-cells. During innate immune responses, cytokines play an important role in inflammation. Cytokine secretion during acquired immune responses is

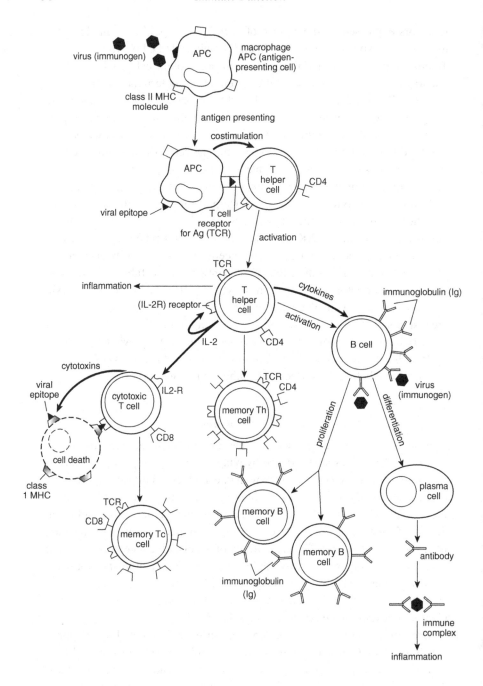

Table 2.2. *Examples of Cytokines, Their Cellular Source,*
and Primary Function.

Cytokine	Source	Function
Interferon-γ	T-cells, NK cells	Macrophage activation, increased MHC expression
IL-1 (IL-1α and -β)	Macrophages	T-cell activation, macrophage activation, fever, inflammation
IL-2	T-cells	T-cell growth and cytokine release, B-cell activation and antibody synthesis, NK cell growth
IL-4	CD4$^+$ T-cells	B-cell activation and isotype switching, T-cell activation
IL-6	T-cells, macrophages	Facilitates acute phase reactions to early infection, T- and B-cell activation
IL-10	T-cells, macrophages	B-cell activation, suppresses macrophage function
IL-12	B-cells, macrophages	NK activation, induces CD4$^+$ T-cells to differentiate into inflammatory T-cells
Tumor necrosis factor-α	Macrophages, NK cells	Inflammation, fever

mainly mediated by T-cells following specific identification of a pathogen. Such cytokines released from lymphocytes are often called lymphokines. Lymphokines generally regulate the growth and differentiation of lymphocytes during an immune response.

←——————————————————————————————

Fig. 2.6. Sequence of events involved in the "typical" immune response to a pathogen. Macrophages must first locate, engulf, and present a piece of a foreign pathogen in the major histocompatibility complex (MHC) on its surface membrane. To activate T-cells, in addition to presenting a piece of pathogen to the T-cell receptor (TCR), macrophages must also release a cytokine called interleukin-1 (IL-1). IL-1 activates helper T-cells (TH) by binding to IL-1 receptors on the surface of helper T-cells and stimulating helper T-cells to produce and release IL-2. IL-2 binds to IL-2 receptors located on the IL-2 secreting helper T-cell (autocrine action) and on cytotoxic T-cells (paracrine action). The binding of IL-2 to its corresponding receptors stimulates proliferation of cytotoxic T-cells that work to kill the foreign pathogen and proliferation of helper T-cells that either continue the production and release of IL-2 or become memory T-cells that remain dormant in circulation. IL-2 secreted by the new army of helper T-cells binds to IL-2 receptors on the surface of B-cells. The binding of IL-2 to receptors on B-cells stimulates growth and differentiation of B-cells into either plasma cells that begin to produce and secrete specific antibodies that will bind up any remaining pathogen left in circulation or memory B-cells. This entire cytokine-mediated process takes several days to accomplish.

In the course of an acquired immune response (Figure 2.6), macrophages must first locate, engulf, and present a piece of a foreign pathogen on its surface membrane. To activate T-cells, in addition to presenting a piece of pathogen on the surface, macrophages must also release a cytokine called interleukin-1 (IL-1), also called lymphocyte activating factor. IL-1 activates helper T-cells by binding to IL-1 receptors on the surface of helper T-cells and stimulating helper T-cells to produce and release IL-2, often referred to as lymphocyte growth factor. IL-2 binds to IL-2 receptors located on the IL-2 secreting helper T-cell (autocrine action) and on cytotoxic T-cells (paracrine action). The binding of IL-2 to its corresponding receptors stimulates production of other cytokines, proliferation of cytotoxic T-cells that work to kill the foreign pathogen, and proliferation of helper T-cells that continue the production and release of IL-2. IL-2 secreted by the new army of helper T-cells binds to IL-2 receptors on the surface of B-cells. The binding of IL-2 to receptors on B-cells stimulates growth and differentiation of B-cells into plasma cells that begin to produce and secrete specific antibodies that will bind up any remaining pathogen left in circulation. This entire cytokine-mediated process takes several days to accomplish.

Interferons comprise another class of cytokines. Interferons are mainly produced by T-cells and NK cells, and play an important role in inhibiting replication of virally infected cells and increasing the viral cell killing potential of NK cells. Interferons can also stimulate growth and proliferation of B-cells. Cytotoxic cytokines are another class of cytokines, and include tumor necrosis factor and its homolog, lymphotoxin. Tumor necrosis factor is released from macrophages and lymphocytes, and is important for the destruction of tumorous cells and bacteria.

2.4. Assessment of Immune Function

Throughout the book, we will refer to seasonal changes in immune function as determined by one of several standard tests of immunity. To aid the reader in the interpretation of these immune data, the methodology commonly used in these studies are reviewed here. In addition to utilizing assays of immune function, several field studies reviewed in the subsequent chapters report seasonal changes both in ectoparasite (e.g., ticks and mites) and blood parasite (e.g., protozoa and hematozoa) loads. Measures of parasite prevalence are not direct measures of immune function, because factors other than reduced immune function (e.g., changes in the vector not the host) may be contributing to the observed results. Nonetheless, these studies do provide compelling evidence for seasonal fluctuations in parasite prevalence.

Immune function can be measured in the laboratory using either *in vitro* or *in vivo* techniques. There are advantages and disadvantages to each method. *In vitro* assays are beneficial because they can provide a constant environment devoid of the fluctuations and changes in the internal *milieu* that occur continuously. Additionally, *in vitro* assessment of immune function allows us to examine the precise effects of particular neuroendocrine substances on immune responses without the confounding effects of other hormonal influences (e.g., systematically exposing lymphocytes to various hormones, individually, to assess the precise contribution of each hormone to the immune response). The assessment of isolated responses is a double-edge sword. One of the pitfalls of *in vitro* assessments of immune function is that this technique minimizes the influences of both neural and endocrine factors that are likely to have a profound impact on "natural" immune responses. Accordingly, the primary benefit of *in vivo* assessments is that the immune responses can be examined within the organism's natural internal *milieu*. Therefore, some researchers argue that *in vivo* assessments of immune function have the most information value regarding neuroendocrine–immune interactions (Maier & Laudenslager 1988). Immune assays can be developed to examine either specific immune responses against a particular antigen (or pathogen) or nonspecific immune responses, such as the replication of cells.

On a final note, in terms of immune function, "more" is not necessarily "better." For example, high immune responses can be interpreted as either competent (harmful pathogens may be eliminated from the body faster), or incompetent (a more efficient response might be one that requires activation of fewer immune cells). Therefore, as with most physiological assessments, converging evidence is required to assess fully whether the measured immune responses are beneficial or detrimental to the organism. Some of the immune parameters commonly measured in laboratory studies of human and nonhuman mammals are described here.

2.4.1. Lymphocyte Proliferation (or Blastogenesis)

Activation of lymphocytes is measured *in vitro*. This immune response is generally considered a good indicator of the mitotic ability of lymphocytes following exposure to a pathogen *in vivo* (Borysenko 1987). Following activation, lymphocytes undergo blastogenesis and begin to proliferate. The proliferation of lymphocytes is what is measured in this assay. To stimulate or activate lymphocytes, either antigens or mitogens are used. Recall, an antigen is a foreign substance that the lymphocytes have not encountered previously. A mitogen is a protein or plant lectin that stimulates lymphocytes to divide. Some mitogens, like concanavalin A (Con A) from *Canavaia ensiformis* and phytohemagglutinin from *Phaseolus vulgaris* primarily stimulate T-cells. Other mitogens, like pokeweed mitogen from *Phytolacca americana* and lipopolysaccharide from *Escherichia coli* are commonly used to stimulate B-cells. The lymphocytes used in this assay can be isolated from circulation or from specific immune organs, like the thymus or spleen. In terms of non-specific measures of lymphocytes, this measure is much more reliable and informative than merely counting numbers of different types of lymphocytes (Borysenko 1987).

2.4.2. NK Cytotoxicity

The ability of NK cells to kill tumor cells is measured *in vitro* in this assay. Measurement of the cytotoxic activity of NK cells killing tumor cells does not require prior exposure to the tumorous cells; therefore, this is considered a nonspecific measure of cell-mediated immunity. In this assay, NK cells either isolated from blood or immune organs, are incubated with tumor cells (specific for the species, e.g., mouse tumor cell lines are used to assess NK cells from mice) that have been labeled with the radioactive isotope ^{51}Cr (chromium) or a fluorescent dye. When a tumor cell is lysed by an NK cell, the tumor cell releases the chromium (or dye) into the supernatant, and it is the released chromium (or dye) that is measured and recorded as percent lysis. The advantage of this measure is that it can provide a general indication of the ability of NK cells to kill tumorous or virally infected cells *in vivo*.

2.4.3. Skin Tests for Delayed Type Hypersensitivity (DTH)

This is an *in vivo* assessment of specific cell-mediated immune responses involved in the cutaneous reaction to antigens. DTH measures the memory response of lymphocytes to a specific antigen. This response is quantified by measuring the dimensions of erythema and induration following local injection of the antigen (Borysenko 1987). This assay of immune function has direct clinical relevance. For example, DTH skin testing is the basis for diagnostic tests for such agents as tuberculin. This reaction can also be used to assess resistance to various bacterial and viral infections (Dhabhar & McEwen 1996). The manifestations of this response can also be seen in allergic reactions to such agents as poison ivy, poison oak, and various cosmetics and dyes (Dhabhar & McEwen 1996).

To assess DTH, hosts are sensitized to a novel antigen (either by injection or cutaneous application), subsequently challenged with the antigen in a different location, and then DTH is measured at the site of challenge. Immunologic familiarity (or memory) of the antigen results in swelling at the site of challenge that can be measured using calipers and quantified as an index of specific cell-mediated immunity.

2.4.4. Antibody and Cytokine Production

Antibodies, produced and released from B-cells, can be quantified using an enzyme-linked immunosorbent assay (ELISA). The ELISA tests for the presence of antibodies primarily in plasma or serum, following exposure to a novel antigen (i.e., a foreign substance that the organism has not been exposed to previously). The basic premise of the ELISA involves the binding of antibodies in circulation to a known amount of antigen. This antibody–antigen complex is then labeled with a secondary antibody, followed by a nonradioactive tag. This nonradioactive enzyme changes optical density (OD; color) when it binds to the secondary antibody that is bound to the antigen–antibody complex. It is the colorimetric change that provides quantitative information. This assay provides a direct assessment of an immune response mounted against a specific pathogen; therefore, this measurement is considered a good indicator of specific humoral immune function.

Other soluble factors, such as cytokines, can be quantified in a similar manner as antibodies, using an ELISA. For these assays, a known amount of antibody generated against the cytokine of interest (e.g., anti-IL-2 antibody) is coated onto ELISA plates and samples (i.e., from circulation or immune organs), and a series of standards (i.e., recombinant cytokines) are then added so that the cytokines can bind to the primary antibody on the microtiter plate. Then, using a secondary, enzyme-linked antibody against the cytokine of interest (e.g., biotinylated anti-IL-2 antibody) and an enzymatic substrate buffer, the amount of bound cytokine (from both test samples and standards) can be quantified based on the enzymatic change in OD. The concentration of cytokines from the standards are used to construct a standard curve that can be used to infer the amount of cytokine in test samples based on the OD reading for each sample. This ELISA procedure can be used to quantify both ligand and receptor concentrations.

Although the ELISA measures antibody and cytokine concentrations, it does not provide information about the cells producing these proteins. In other words, the ELISA does not distinguish between a few cells that are producing a lot of protein and many cells that are each producing a little protein. To examine the immune cells that are secreting antigen-specific antibodies or cytokines, the enzyme-linked immunospot (ELISPOT) assay can be used.

The ELISPOT is a colorimetric assay that uses isolated immune cells (from circulation or immune organs), specific secondary antibodies, and an enzymatic reaction the causes antibody or cytokine secreting cells to be colored, resulting in the presence of "spots" that can be counted manually or with an automated reader. Quantitative reverse transcription-polymerase chain reaction (RT-PCR) also can be used to examine mRNA levels. These assays provide information about how much cytokine an immune cell can produce. Although these techniques are widely used in clinical research using humans and mice, these procedures have not been adequately developed for use in nonmammalian species.

2.4.5. Serological Measures

Using classic methods of hematology, numbers and types of immune cells – such as granulocytes, macrophages, and lymphocytes – can be assessed in blood smears. Additionally, parameters such as hematocrit, hemoglobin, and red blood cell counts can be assessed using this procedure. One problem with this immune measure is that it is unclear how numbers of cells in circulation translate into an assessment of immune function. More recently, immune cell subpopulations in blood and tissue have been assessed using fluorescence markers for specific immune cells and a cell sorter, called a flow cytometer. Using this technique, both relative and absolute numbers of macrophages, T-cells, B-cells, and NK cells can be quantified. Additionally, subpopulations of cells can be quantified. For example, of T-cells in circulation, the ratio of $CD4^+$ (i.e., helper T-cells) to $CD8^+$ (i.e., cytotoxic T-cells) cells can be determined. In many immunodeficiency diseases, such as human immunodeficiency virus (HIV), the ratio of helper T-cells to cytotoxic T-cells is altered. Accordingly, this immune measure has important clinical and basic science implications.

2.5. Summary

The goal of this chapter is to provide a basic reference to help interpret results from studies presented in later chapters. Immune responses occur over time and involve complex interactions among many cells that in turn requires substantial energy (Demas et al. 1997). The cornerstone of the immune system is the ability of immune cells to: (1) exhibit specificity and (2) remember pathogens to overcome infection. With these two simple features, the immune system can respond to an infinite number of pathogens. The cost of this feat is that mechanisms involved in specificity, such as the development of receptors, recruitment of additional cells, and mitotic activities, take time and energy to complete. Therefore, the entire acquired immune response can often take several weeks, upon initial exposure to a pathogen. Because memory processes are also involved in acquired immunity, upon subsequent reexposure to a pathogen, the acquired immune response will result in faster elimination of a pathogen from circulation and subsequently involve the utilization of less energy.

These responses are remarkable in their own right, but what has become apparent in recent years is that other systems in the body, such as the neuroendocrine system, interact with the immune system to alter the course of immune responses and the relationships among immune cells (see Chapter 7). In the following chapters, we will present data that suggest that host defenses against infection represent the dynamic interaction between the neuroendocrine and immune systems. Thus, the neuroendocrine system may function to balance energetic utilization among thermoregulation, reproduction, growth, and immune function. Additionally, we will provide evidence that the interactions of these systems can be further influenced by exposure of an organism to various extrinsic factors. These findings will be presented from an adaptive-functional perspective emphasizing the trade-offs among energy-dependent processes of the immune system and other physiological needs. Seasonal changes in immune function and disease processes reflect these energetic shifts across the year when energetic availability and requirements fluctuate. As noted in Chapter 1, the use of the annual cycle of changing day lengths to anticipate the time of year requires the precise coordination of the neuroendocrine system.

3

Seasonal Fluctuations in Disease Prevalence

All diseases occur at all seasons of the year, but certain of them are
more apt to occur and be exacerbated at certain seasons.

Hippocrates, 400 B.C. *Aphorisms*

3.1. Introduction

Seasonal fluctuations of illness and death among humans and nonhuman animals have been recognized for centuries. In most cases, people assumed that factors associated with the changing seasons brought about illness and death. Seasonal patterns were observed in the timing of disease onset and in the severity of symptoms. It was noted that there is "a time to be born," as well as "a time to die," and "a time to heal" (Ecclesiastes 3:2–3). In early Greek medicine, seasonal changes in climate or other environmental factors were thought to be central in causing many afflictions, including consumption (pulmonary tuberculosis), boils, asthma, ulcers, and lesions. Hippocrates observed that, "in autumn, diseases are most acute, and most mortal, on the whole. The spring is most healthy, and least mortal." In his *Aphorisms*, Hippocrates noted many relationships between the weather, seasons, and temperament of the patient, and the onset or severity of various diseases. For example, he wrote, "if the winter be of a dry and northerly character, and the spring rainy and southerly, in summer there will necessarily be acute fevers, opthalamies, and dysenteries, especially in women, and in men of a humid temperament." Although the associations observed by early physicians were often based on isolated cases, these correlations between disease and environmental conditions continued to influence medicine for the next 2000 years. Many superfluous correlations were established, and these correlations often evolved into erroneous causative arguments. With the widespread adoption of the germ model as the etiological heuristic in the nineteenth century,[1] the influences of seasons and patient personality in disease processes were virtually abolished in Western medicine.

The relationships that Hippocrates and other early Greek physicians noted between environmental conditions and disease were mainly correlations based on astute observations, not cause–effect relationships. Malaria was originally thought to be caused by stagnant swamp air. In fact, the word, malaria comes from the Italian word for "bad air." We now know that malaria is caused by the parasite *Plasmodium falciparum* that is carried to humans by the mosquito vector. Many other correlations among diseases, death, and changing seasons have been noted. Recently, it was reported that the month of birth influences life expectancy after the age of 50. Europeans who were born in autumn (October–December) live longer than those born in spring (April–June) (Doblhammer & Vaupel 2001). Australians show the same pattern, but shifted 6 months so that individuals born in April–June outlive those born in October–December. Importantly, individuals who migrated from Great Britain to Australia retained the European pattern of longer lives if born in autumn (Doblhammer & Vaupel 2001). The incidence of several neurological diseases is affected by month of birth. Epilepsy appears to have the most consistent pattern, with an excess of births in winter (Torrey et al. 2000). Multiple sclerosis (MS)(Salemi et al. 2000), Parkinson disease, and amyotrophic lateral sclerosis appear to have an excess of spring births (Torrey et al. 2000). Studies of cerebral palsy are not conclusive, although it is possible that individuals with summer births are more likely to be afflicted than winter births. Alzheimer disease, congenital malformations of the central nervous system, and mental retardation do not seem to be influenced by birth month (Torrey et al. 2000).

Earlier relationships between season and disease were also noted. For example, the cold winds of winter were associated with the onset of influenza in China. The Cantonese word for the constellation of flu-like symptoms roughly translates as "hurt by the cold wind." The equivalent Mandarin word translates as "cold breeze." Of course, the English word "cold" means both low ambient temperature and the symptoms associated with the flu. The synonym linking low temperature to the symptoms of influenza appears in many languages, including Latin, French, Italian, Spanish, German, Dutch, Irish, and Vietnamese. The correlation between cold weather and the onset of infection

---

[1] In its extreme interpretation, the germ model presumes that all diseases are caused by a pathogen. Thus, it is the role of medical research to determine the causative pathogens to all disease. This model was based on the stunning successes of Pasteur, Lister, and Koch in determining microbial etiologies for several common afflictions. Western medicine relies heavily on this model despite recent failures (e.g., the inability to find viral causes for most human cancers).

with new strains of influenza likely reflects the annual custom of allowing swine, ducks, geese, and other farm animals into human dwellings in Asia as the weather deteriorates in the fall (Nesse & Williams 1994). Thus, a mutated influenza virus is transferred from the farm animals to humans, who then pass the virus onto other humans throughout the winter. A new strain of influenza virus usually arrives in Europe in late winter. The peak flu season in North America is late winter and early spring.

The first formal study of the seasonal variation in human mortality entitled, *Natural and Political Observations Made upon the Bills of Mortality*, was published in 1662 by John Graunt. Graunt compiled mortality data from residents of London and made many observations regarding the rate of infant mortality, the rate of mortality from various causes, and the variation between mortality in city residents, compared with rural residents. Graunt also categorized the mortality data according to season. Generally, seventeenth century mortality in London showed both summer and winter peaks. In recent years, most seasonal diseases among humans display only a winter peak in occurrence (Sakamoto-Momiyama 1977). As socioeconomic conditions improve, many human diseases undergo what has been called deseasonality. Of all the factors examined, the most highly correlated factor with deseasonality of human disease appears to be improved housing conditions or more specifically, societal acquisition of central heating (Sakamoto-Momiyama 1977). This finding is consistent with our hypothesis that immune function responds to energy costs.

As noted in Chapter 2, the symptoms of a given disease usually reflect both the effects of our immune system fighting off an infection (e.g., fever) and the specific adaptations of the pathogen to spread itself to a new host (e.g., sneezing or open pustules). Natural selection is operating on pathogens as well as their hosts, and most pathogens become exquisitely adapted to their specific niches (i.e., hosts). The generation time of pathogens is usually less than the generation time of their hosts, so evolution through natural selection is much faster among pathogens than hosts. Thus, many pathogens are able to evolve distinct adaptations to undermine the immune defenses of their hosts and exploit individuals of specific species; because of the selection pressure to keep up with the genetic differences of the descendants of the hosts, pathogens rarely develop the ability to infect other species. Sexual reproduction, with its generational shuffling of genes, may have evolved as a mechanism to prevent pathogens from specializing on the genetic machinery of individuals and their descendants (Hamilton et al. 1990). Occasionally, mutations occur in pathogens that permit exploitation of new host species. But, the mere development of such a mutation is insufficient to exploit a new

Table 3.1. *Human Diseases and Their Likely Nonhuman*
Ancestral Disease Origin.

Human Disease	Likely Ancestral Disease in Nonhumans
Influenza	Influenza (swine/ducks)
Malaria	Avian malaria (chickens/ducks)
Measles	Rinderpest (cattle)
Pertussis	Pertussis (dogs/swine)
Smallpox	Cowpox (cattle)
Tuberculosis	Tuberculosis (cattle)

host species successfully; the mutated pathogens must also come into contact with potential hosts. Many of the most virulent diseases among humans probably originated in domesticated animals (see Table 3.1). Pathogens that had specialized in cattle, pigs, chickens, ducks, and dogs were recently (i.e., in the evolutionary time scale) introduced to humans.

Humans began domesticating animals 10,000–15,000 years ago. One criterion important for early humans in domesticating animals was the ability for the animals to tolerate one another in relatively small spaces. Most of the species that were successfully domesticated were socially tolerant animals that lived in herds or flocks (e.g., dogs, cattle, horses, ducks, turkeys, and chickens) (Nesse & Williams 1994). The natural history of diseases of animals that live in high-density groups is different from the diseases of solitary animals or animals that live in low-density groups. As all other organisms subject to natural selection, pathogens have been selected to create as many successful offspring as possible in the shortest amount of time; consequently, infections of hosts living in high densities are spread quickly. These infections may cause what are considered "herd" or "flock" diseases. Infected individuals either develop sufficient immunity and overcome infection or die. In response to host defenses, the pathogen may either become extinct or remain dormant and await introduction into a new group of unexposed hosts. This repopulation may take several years for long-lived species like humans or may require only 1 year for short-lived species like rodents. Thus, annual or multiannual epidemics are common among high-density, group-living species.

Population densities of humans increased with the domestication of plants and animals, causing humans to become targets of opportunity for herd infections (Nesse & Williams 1994). The change in human social organization from small bands of hunter-gatherers to densely populated agricultural centers has been associated with increased prevalence of epidemic diseases, many of which have "crossed-over" from other species to infect humans. Humans tend to increase their exposure to domesticated animals, as well as to each other, during the winter when cross-species infections are easily contracted. Thus, infection rates among humans of many herd diseases are predicted to be associated, in a complex manner, with the annual fluctuations in the infection rates among the animal pathogen pools, and the annual commensal housing of humans with their domesticated animals. The remainder of the book focuses on how seasonal factors influence immune function and disease processes.

Modern descriptive studies of seasonal fluctuations in illness and death have been useful, in some cases, to determine the extent to which the seasonal pattern reflects seasonality in the pathogen, host, or some combination of factors involving both pathogen virulence and host defenses. These studies have revealed that (1) the immune system is responsive to changing seasons, (2) the onset of many diseases fluctuate seasonally, and (3) both human and nonhuman animals show similar seasonal variation in susceptibility to disease, suggesting that common underlying mechanisms are involved. In this chapter, we report that seasonal fluctuations in disease prevalence and death are widely observed among humans and nonhuman animals. The human clinical studies will be followed by a discussion of the descriptive and laboratory studies of seasonal changes in disease prevalence and death among nonhuman animals.

3.2. Disease Prevalence in Humans

Seasonal changes in infection rates have been reported for many human diseases. To make predictions about illness and to establish the causes of seasonal fluctuations, it is important to determine the extent to which seasonal fluctuations in disease reflect seasonal changes in vector activities or host susceptibility to disease. For some infections, the evidence for seasonality is not particularly robust (e.g., human immunodeficiency virus). Other illnesses, such as allergies, occur with seasonal precision; however, the seasonal variation in allergies is due mainly to annual fluctuations in the prevalence of the allergen, not necessarily in host responsiveness to the allergen. In other cases, there is substantial evidence for seasonality that results from seasonal changes in the host's immunity (e.g., respiratory tract infections) or a combination of changes in the prevalence of the vector and host susceptibility to the disease (e.g., malaria). These seasonally fluctuating diseases fall into several categories: infectious diseases, autoimmune diseases, cardiovascular diseases, and cancers. Seasonal relationships that defy explanation are also prevalent. For example, the influence of birth month on neurological disease risk (Torrey et al. 2000). Data from the studies reviewed in this chapter suggest that humans undergo seasonal changes in susceptibility to many diseases. A sample of these seasonal diseases is described in Table 3.2.

Table 3.2. *Seasonal Variation in Peak Prevalence of Human Illness and Disease.*

Disease	Peak Prevalence	References
Malaria	Winter–early spring	Chougnet et al. 1990
Leishmaniasis	Winter–early spring	Villaseca et al. 1993
Influenza	Winter–early spring	Glezen et al. 1982
Human reovirus	Winter	Kapikian et al. 1976
Respiratory syncytial virus	Winter–early spring	Hall 1991
		Miller 1992
	Summer	Sakamoto et al. 1995
Coronaviruses	Winter–early spring	Cavallaro & Monto 1970
		Hambre & Beem 1972
		Hendley et al. 1972
Enteroviral infection	Summer	Glimåker et al. 1992
		Martino et al. 1987
Tuberculosis	Winter	Pietinalho et al. 1995
Legionnaire's disease	Summer	Arderiu et al. 1979
		Edson et al. 1979
Brucellosis	Spring–early summer	Dajani et al. 1989
Pneumonia	Winter–spring	Dagan et al. 1992
		Eskola et al. 1992
Mycosis	Winter–spring	Chariyalertsak et al. 1996
Coronary heart disease	Winter	Douglas et al. 1990
Stroke		
Cerebral infarction	Spring–summer	Biller et al. 1988
	Winter–spring	Sobel et al. 1987
Ischemic attacks	Winter–spring	Dunnigan et al. 1970
		Azevedo et al. 1995
Intracerebral hemorrhage	Winter–early spring	Biller et al. 1988
		Azevedo et al. 1995
Transient ischemic attacks	Summer	Sobel et al. 1987
MS	Spring–summer	Bamford et al. 1983
		Sibley & Paty 1980
IDDM	Autumn-winter	Blom & Dahlquist 1985
Rheumatoid arthritis	Fall–winter	Rosenberg 1988
Breast cancer		
No. of cases diagnosed	Winter	Ownby et al. 1986
Initial detection	Spring–summer	Chleboun & Gray 1987
		Cohen et al. 1983
		Kirkham et al. 1985
		Jacobsen & Janerich 1977
		Mason et al. 1990
	None	Galea & Blamey 1991
Risk of death	Summer	Sankila et al. 1993
Season of removal	Winter	Ownby et al. 1986
Season of birth	Summer	Yuen et al. 1994
Lung cancer	Summer–fall	Tang et al. 1995
Melanoma	Spring–summer	McWhirter & Dobson 1995
Urinary bladder carcinoma	Fall–winter	Høstmark et al. 1984

3.2.1. Infectious Diseases

Infectious diseases are one of the leading causes of death among humans worldwide. A disease is considered infectious if it is caused by a pathogen and is potentially detrimental to the survival of the host (Kaufmann 1993; Ewald 1994). Infectious diseases can arise from parasitic, viral, bacterial, or fungal origins, and can activate an array of host immune responses. Typically, seasonal prevalence of infectious diseases has been considered to represent annual fluctuations in either the activity of the pathogen or vector, or to represent seasonal variation in exposure and contact of the host with the pathogen or vector. For example, river blindness is caused by infection with the nematode, *Onchocerca vovlus* that is transmitted to humans through contact with an intermediate host, the blackfly (*Simulium damnosum*). The peak prevalence in river blindness corresponds with peak breeding activity of the blackfly during the wet season (Tomkins 1993). Infection with other parasites – such as *Dracunculus medinensis, Schistosoma mansoni, Schistosoma haematobium,* and *Schistosoma japonicum* – is also elevated during the rainy season, compared with the dry season; yet, the cause of infection is hypothesized to be increased contact of humans with contaminated water supplies during the wet season (Tomkins 1993). This epidemiological approach has been fundamental to the understanding and control of environmental factors that influence the course of infection (Tomkins 1993). However, this approach to seasonality of infection and disease has neglected to address the extent to which seasonal changes in resistance and susceptibility of the host contribute to annual fluctuations in disease prevalence.

Parasite Infections

Many parasite infections show seasonal patterns. These annual fluctuations are caused by seasonal changes in the prevalence of the parasite, as well as by seasonal changes in the resistance or susceptibility of the host to these parasite infections. For example, malaria is a common seasonal disease among humans residing in the tropics (Theander et al. 1990), with peak prevalence generally occurring during the wet season (i.e., winter to early spring) (Chougnet et al. 1990). Accompanying the increased prevalence of the parasite *Plasmodium falciparum* are changes in both cell-mediated and humoral immunity of the host. The extent to which the changes in immunity reflect seasonal changes in the parasite, seasonal changes in the host, or an interaction between the two has yet to be resolved.

Cell-mediated immunity against malaria appears to fluctuate with season. For example, lymphocyte proliferation in response to malaria antigens in healthy individuals is compromised during the wet season (i.e., malaria transmission season) compared with the dry season (Chougnet et al. 1990; Theander et al. 1990; Abu-Zeid et al. 1992; Riley et al. 1993b). Lymphocyte proliferative responses to non-malaria-related mitogens, such as phytohemagglutanin, do not show similar seasonal changes (Theander et al. 1990; Abu-Zeid et al. 1992). The reduced cellular responses coincident with the malaria season may be caused by decreased $CD8^+$ and natural killer cell numbers (Chougnet et al. 1990). Another cellular change associated with the malaria season is higher cytokine [i.e., interferon-γ (IFN-γ)] production during the wet season than the dry season (Riley et al. 1993a).

Reports of humoral responses against malaria-related antigens are inconsistent. Some reports suggest that antibody levels against malaria-related proteins remain stable across seasons (e.g., Chougnet et al. 1990; Riley et al. 1993a), whereas others suggest that antibody responses against some *P. falciparum* proteins are greater during the wet season than the dry season (e.g., Scarselli et al. 1993). In terms of resistance to malaria, healthy individuals with sickle cell hemoglobulin (i.e., carriers of the sickle cell trait) have lymphocyte proliferative responses that are elevated during the wet season, possibly because the sickle cell trait is thought to inhibit growth of *P. falciparum* (Abu-Zeid et al. 1992). The increased prevalence of the sickle cell trait in regions where malaria is endemic (i.e., naturally occurring) may represent a host-mediated adaptation against *P. falciparum*.

Another protozoan parasite that shows seasonal fluctuations in terms of host infection is *Leishmania*. Sandflies (*Phlebotomus martini*) are the primary vector of this disease, and they infect the host with *Leishmania* while extracting blood. In the tropics, incidence of cutaneous leishmaniasis is highest during the wet season (i.e., winter to early spring) (Villaseca et al. 1993), possibly reflecting seasonal variations in natural resistance to *Leishmania*, prevalence of the parasite, or biting activity of sandflies (Gomez & Hashiguchi 1991).

Viral Infections

Viral infections also display seasonal fluctuations. For example, influenza and other respiratory tract infections show seasonal incidence that may reflect either seasonal changes in the mechanisms of transmission (e.g., social contact) or seasonal changes in host resistance. In some developing countries, respiratory infections peak during the dry season, which may be caused by either the impact of the dry climate on the epithelium of the respiratory system (Tomkins 1993) or the effect of annual variation in host resistance to respiratory infection. In the United States, the epidemic peak for influenza virus activity ranges from winter to early spring; deaths attributed to pneumonia (i.e., incidence of disease) follow this peak in influenza virus activity by approximately 2 weeks (Glezen et al. 1982). However, this seasonal pattern shows variation worldwide; in Asia, the occurrence of influenza currently peaks in the spring and summer (i.e., April–August), in Europe influenza virus activity currently decreases during the spring (i.e., April–May), in Central and South America outbreaks of influenza are currently highest in the summer (i.e., June–August), and in South Africa influenza activity peaks in the spring and summer (i.e., May–August) (Centers for Disease Control and Prevention 1991). Additionally, these reported seasonal patterns in influenza activity can differ, depending on the strain of the virus (Centers for Disease Control and Prevention 1991).

In addition to seasonal changes in the course of the influenza virus, there also are seasonal fluctuations in the resistance of humans to the influenza virus. For example, febrile reaction to the influenza vaccination is higher in healthy males administered the vaccination in the winter, compared with the summer months (Shadrin et al. 1977). Similarly, after vaccination, there is increased antibody production against various influenza strains in the winter, compared with the summer months (Shadrin et al. 1977).

The incidence of other viruses responsible for respiratory infections also varies on a seasonal basis. For example, respiratory syncytial virus, responsible for bronchiolitis and pneumonia in children, shows peak prevalence in the winter and early spring (Hall 1991; Miller 1992). Respiratory infections caused by adenovirus infections generally peak in the summer; although prevalence of some adenovirus strains peaks in the winter (Sakamoto et al. 1995). In Norwegian adults diagnosed with respiratory viruses, antibody production against adenovirus antigens is higher in the winter than summer months (Omenaas et al. 1995). Coronavirus infections cause upper respiratory illnesses, and occur primarily in the winter and early spring (Cavallaro & Monto 1970; Hambre & Beem 1972; Hendley et al. 1972). Another disease that shows a seasonal pattern is gastroenteritis, caused by a retrovirus-like agent. This virus is a major cause of gastroenteritis in children and shows a peak in infection rate during the winter months (Kapikian et al. 1976).

Enteroviruses show a seasonal pattern of infection, with the prevalence of infection being highest during the summer months. The prevalence of enterovirus infections in the summer months have led people in countries, such as Japan, to assume that diseases arising during the "hot" season are caused by enterovirus infections. Subacute thyroiditis, an inflammatory disease of the thyroid, is often caused by enterovirus infection and has a peak prevalence of infection during the summer (Martino et al. 1987). Enterovirus infection can also cause aseptic meningitis. In patients diagnosed with aseptic meningitis, the production of IgM antibody against enterovirus antigens is elevated in the summer, compared with other seasons of the year. In patients diagnosed with aseptic meningitis arising from other viral infections (i.e., nonenteroviral origins), there is no seasonal pattern in the production of IgM (Glimåker et al. 1992) (Figure 3.1). Similarly, in healthy blood donors, there is increased production of IgA antibody against enterovirus antigens in blood samples obtained in the summer, compared with the winter months (Cello & Svennerholm 1994). Taken together, these data suggest that seasonal changes in infection can influence the evolution of seasonal adaptations in host immune responses to viral infections; such that, even in healthy individuals, immune responses are elevated during the time of increased prevalence of enteroviral agents.

Bacterial Infections

Many bacterial infections fluctuate seasonally. Bacterial infections can be of intracellular (i.e., enter the host cells) or extracellular (i.e., remain in

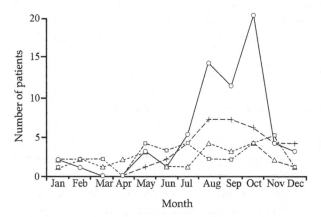

Fig. 3.1. Number of patients diagnosed with aseptic meningitis originating from enteroviral infection (o) and nonenteroviral infections (Δ), and number of patients with IgM antibody produced against enteroviral antigens (+) and nonenteroviral antigens (□). Seasonal variation is only seen in aseptic meningitis cases arising from enteroviral infection, with the peak number of infected patients in the summer and fall. Antibody production against enterovirus antigens parallels the distribution of patients with enteroviral infection, in which highest antibody production is observed in blood samples obtained in the summer and fall. [Reprinted from the *Journal of Medical Virology*, vol. 36, M. Glimåker, A. Samuelson, L. Magnius. A. Ehrnst, P. Olcén, & M. Forsgren, "Early diagnosis of enteroviral meningitis by detection of specific IgM antibodies with a solid-phase reverse immunosorbent test (SPRIST) and μ-capture EIA," pp. 193–201, copyright © 1992, with permission from Wiley-Liss, Inc., a division of John Wiley & Sons, Inc.]

circulation) origin. Intracellular bacteria underlie a variety of infectious diseases, including tuberculosis, Legionnaire's disease, and brucellosis, each of which show pronounced seasonal fluctuations. Extracellular bacterial infections, including cholera and pneumococcus, also show seasonal patterns in occurrence.

Mycobacterium tuberculosis usually enters the host through the lungs, the primary site of tuberculosis infection. Seasonally, this infection is most pronounced during low temperatures. For example, the frequency of tuberculosis increases during the winter and often during unseasonably cool summers in Japan (Pietinalho et al. 1995). The fatality rate of tuberculosis may be related to the season of birth. In Tokyo, deaths from pulmonary tuberculosis are highest in women born between January and June (i.e., winter and spring); however, there is no seasonal pattern in fatalities among males based on season of birth (Miura et al. 1992). The season of birth may be an important determinant of susceptibility to tuberculosis because it may influence in utero nutrients, fetal or neonatal exposure to infection, or possibly fetal or neonatal immunologic resistance to infection (Miura et al. 1992). Data from nonhuman animals suggest that the physiological signals of season (e.g., photoperiod) that an individual receives in utero can affect immune function later in life (Blom et al. 1994).

Seasonal fluctuations in immune responses to the bacterium, *Legionella pneumophila*, the agent that causes Legionnaire's disease, have been reported. For example, among healthy residents of Michigan, the prevalence of individuals with serum antibodies to Legionnaire's disease is higher in the summer (i.e., August–September) than in the winter (i.e., February–March) (Edson et al. 1979). Similarly, the frequency of Legionnaire's disease is highest in the summer in Spain (Arderiu et al. 1979).

Brucella bacteria are spread to humans through contact with animals and can cause a variety of systemic diseases. Infection with *Brucella melitensis*, transmitted through contact with goats and sheep, shows a seasonal pattern of incidence (Dajani et al. 1989). In Jordan, over a 2-year study period, both the incidence and frequency of infection with *B. melitensis* increase during the spring and early summer (i.e., March–July) and decreases during the fall (Dajani et al. 1989). Although contamination of unpasteurized milk products may be involved in this seasonal trend (Dajani & Halabi 1986), host immune responses to *B. melitensis*, involving activation of T-cells (i.e., cell-mediated immunity), may fluctuate seasonally as well.

There also is seasonal variability in infection with *Vibrio cholerae*, the bacterium that causes cholera. Infection with this bacterium generally occurs following ingestion of contaminated water, and the infection usually settles in the gut where the organism undergoes replication. The seasonal pattern in cholera outbreaks in South Africa parallels the pattern of temperature and rainfall in the area, with peak incidence of cholera occurring during the months with highest temperatures and heaviest rainfall (Küstner & Du Plessis 1991). Although resistance to cholera involves activation of the host's immune system, the extent to which this pattern is due to seasonal changes in the immunity of the host or changes in exposure to the vector remains unspecified.

Pneumococcal infections also show pronounced seasonal patterns. *Streptococcus pneumoniae* causes numerous infections, including pneumonia and meningitis. Exposure to this bacterium can also cause lower respiratory infections. Among Finnish children, there is a bimodal seasonal pattern in the number of pneumococcal infections, with peaks during the spring (i.e., May) and fall (i.e., November) (Figure 3.2a; Eskola et al. 1992). A similar distribution of pneumococcal infection has been observed among Israeli children, with peaks during the spring (i.e., March–May) and winter (i.e., November–February) (Figure 3.2b; Dagan et al. 1992). Although the cause of this seasonal variation remains unclear, seasonal changes in respiratory infections may precede and, thus, affect susceptibility to *S. pneumoniae* (Dagan et al. 1992).

Helicobacter pylori infection rates peak in October, but this infection is not correlated with the seasonal fluctuation in the diagnosis of epigastric ulcer disease during a 3-year prospective study of 1,076 patients in Germany (Raschka et al. 1999). In contrast, neither sex, age, nor gastric acidity correlated with *H. pylori* infection.

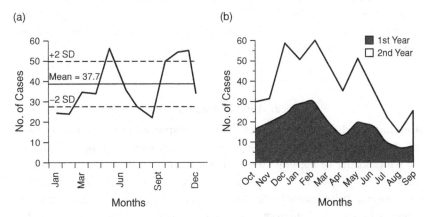

Fig. 3.2. Seasonal fluctuations in the number of Finnish children (a) and Israeli children (b) with pneumococcal infections. There is a bimodal seasonal pattern in the number of pneumococcal infections within each country, with peaks during the spring and fall. (From Eskola et al. 1992 and Dagan et al. 1992.)

Fungal Infections

Infectious diseases, such as mycosis, that are caused by fungi also show seasonal variation. Immunity against fungal diseases generally involves activation of cell-mediated immune responses. Consequently, fungal diseases are most detrimental to individuals with compromised cell-mediated immunity, such as acquired immunodeficiency syndrome (AIDS) patients. Among AIDS patients in northern Thailand, infection with the fungal pathogen, *Penicillium marneffei*, is more pronounced during the rainy season (i.e., May–October) than during the dry season (i.e., November–April) (Chariyalertsak et al. 1996). Bamboo rats (*Rhizomys sumatrensis* and *Cannomys badius*) are the natural reservoirs for *P. marneffei*. Consequently, seasonal patterns in *P. marneffei* infection may represent variation in the prevalence of the pathogen, changes in human contact with rodents, changes in host immunity against infection, or a combination of these factors.

In sum, infectious diseases display seasonal fluctuations in occurrence. Overall, infectious diseases show the highest prevalence during the winter months. The extent to which this pattern reflects responsiveness to annual changes in photoperiod, temperature, or other environmental factors (e.g., rainfall) remains unclear. These data are consistent with the hypothesis that human immune function is compromised during energetically challenging conditions, subsequently leading to seasonally recurrent diseases. We hypothesize that mechanisms have evolved to shunt energy to immune function during times of the year when exposure to infectious diseases is elevated. Presumably, increased numbers of people would be stricken by these seasonally recurrent diseases in the absence of enhanced immunity. Additional studies are necessary to ascertain if immune function in humans is elevated in anticipation of seasonal disease onset (see Chapter 4).

3.2.2. Autoimmune Diseases

Autoimmune disease arises when immune responses are mounted against self tissues and cells. For example, in multiple sclerosis, immune cells attack the white matter of the brain and spinal cord, whereas connective tissue in the joints is attacked by immune cells in rheumatoid arthritis. Autoimmune responses can be triggered by genetic (e.g., major histocompatibility complex genotype), as well as environmental (e.g., stressors and viruses) factors. In essence, like allergic responses, these diseases arise from overactivation of immune responses. These responses are often persistent and can lead to long-term destruction of tissues and cells, because the "antigens" mediating these immune responses are the host's own tissues and cells.

MS

MS involves demyelination of neurons in both the brain and spinal cord, by immune cells (both T- and B-cells) that can penetrate the blood-brain barrier (Steinman 1996). The etiology of this autoimmune disease is still unknown, yet numerous studies suggest that environmental factors can affect progression of MS. The association between latitude and MS is consistent with the notion of photoperiodic influence on human immune function (Davenport 1922; Limburg 1950; Kurtzke 1975, 1980). The prevalence of MS increases at higher latitudes, both north and south (reviewed in Rosen et al. 1991). Examination of the worldwide distribution of MS cases reveals that MS occurs at the highest frequency in southern Canada and the northern United States, northern and central Europe, areas of the former Soviet Union, and in New Zealand and southern Australia (Kurtzke 1980). A consistent correlate with MS is the amount of December solar radiation; that is, a high number of sunny hours in December is associated with low numbers of MS cases in the region (Acheson et al. 1960). In addition to geographical influences, there are seasonal variations associated with MS. The onset of MS occurs most frequently in the spring and summer (Bamford et al. 1983), whereas the incidence of relapse of MS is greatest in the spring and lowest in the autumn (Wuthrich & Rieder 1970). Individuals with MS exhibit episodes of increased IFN-γ secretion from peripheral blood mononuclear cells in response to mitogens, and IFN-γ worsens the symptoms of MS (Beck et al. 1988). Production of IFN-γ displays a seasonal cycle with high levels produced in the fall and winter and low levels produced in spring and summer in MS patients, but not in individuals without MS (Balashov et al. 1998). The authors speculate that the seasonal fluctuations are the result of local environmental factors, such as temperature, ultraviolet radiation, or viral infections. These results are also consistent with the effects of changing photoperiod. Furthermore, some have speculated that the seasonal exacerbation of MS is associated with seasonal changes in melatonin levels, because serum melatonin levels in humans are low in the spring and high in the winter (Arendt et al. 1977; Beck-Friis et al. 1984). Therefore, the decrease in melatonin levels in the spring may underlie the onset of MS symptoms during this season (Sandyk & Awerbuch 1993).

Insulin-Dependent Diabetes Mellitus (IDDM)

IDDM involves the destruction of insulin-producing pancreatic β-cells, by the immune system. In the Northern Hemisphere (e.g., Denmark, Finland, Sweden, United States), the highest incidence of IDDM occurs in the autumn and winter (Sterky et al. 1978; Christau et al. 1979; Fleegler et al. 1979; Blom & Dahlquist 1985). Similarly, in Southern Hemisphere countries (i.e.,

Chile, Australia), onset of IDDM, as well as the number of IDDM cases diagnosed, is highest in the winter and lowest in the summer (Durruty et al. 1979; Fleegler et al. 1979). Coincidentally, the seasonal component of IDDM onset is associated with seasonal fluctuations in certain viral infections that can often initiate the destruction of pancreatic β-cells (i.e., cells that secrete insulin) (Blom & Dahlquist 1985). The common viral etiology of IDDM may explain the similar seasonal distributions of IDDM in different regions of the world (Fleegler et al. 1979.)

Rheumatoid Arthritis

Rheumatoid arthritis is a systemic autoimmune-inflammatory disease primarily of the joints that also has a seasonal component. Although sensitivity to the pain associated with rheumatoid arthritis does not change seasonally (Hawley & Wolfe 1994), the onset of rheumatoid arthritis varies seasonally with peak incidence occurring from November to January (Rosenberg 1988).

No consistent seasonal pattern is obvious among autoimmune diseases. The season of peak prevalence for autoimmune disease depends on the disease in question; the prevalence of MS is highest during the spring and summer, whereas the prevalence of IDDM and rheumatoid arthritis peaks in the fall and winter. Seasonal fluctuations in autoimmune diseases, that involve over-activation of host immune responses, illustrate that human immune function is responsive to seasonal information. Future studies will need to elucidate the precise seasonal information, such as temperature or photoperiod, that triggers the seasonal fluctuations in the onset and incidence of autoimmune diseases.

3.2.3. Heart Disease and Stroke

Human mortality shows a seasonal pattern in the Northern Hemisphere (e.g., United States and Scotland), with the highest number of deaths occurring in January (i.e., winter) and the lowest number of deaths in July and August (i.e., summer) (Figure 1.2). Of these seasonally occurring deaths, 52% are caused by circulatory diseases, including ischemic heart disease and cerebrovascular disease (Figure 3.3) (Bundeson & Falk 1926; Douglas et al. 1991). A similar seasonal pattern is seen in the Southern Hemisphere (e.g., New Zealand), but 180° out of phase with the Northern Hemisphere in which mortality from coronary heart disease and cerebrovascular disease peaks around July (i.e., winter) and declines to its lowest rate in February (i.e, summer) in both men and women (Figure 3.4) (Douglas et al. 1990). Whereas deaths from ischemic heart disease show a single peak in the winter, incidence of ischemic heart disease show two peaks; that is, one in the winter and another peak in the

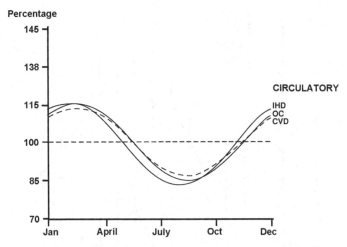

Fig. 3.3. Seasonal distribution of deaths, in the Northern Hemisphere, due to ischemic heart disease (IHD), cerebrovascular disease (CVD), and other circulatory diseases (OC). The highest number of deaths from heart diseases occurs in the winter and the lowest number of deaths occurs in the summer. (From Douglas et al. 1991).

spring (Dunnigan et al. 1970). The seasonal pattern of deaths from circulatory disease, occurring both in the Northern and Southern hemispheres, may be caused by changes in temperature. Deaths from coronary disease increase by 40% at low temperatures, compared with mild temperatures; however, the seasonal rise in deaths from coronary diseases during the winter is reportedly independent of temperature, especially among men and women younger than 60 years of age (Enquselassie et al. 1993).

In addition to heart disease, there are seasonal patterns in incidence of and death from stroke. A 7-year study in Iowa reveals that the incidence of cerebrovascular disease is highest in the summer and fall, and lowest in the spring (Biller et al. 1988). Peaks in the incidence of stroke often depend on the specific type of stroke being examined. Occurrence of transient ischemic attacks is highest in the summer months, whereas cerebral infarctions peak in the winter and spring (Sobel et al. 1987). This seasonal variation in stroke incidence may be related to changes in temperature. Specifically, in Portugal, hospital admission for cerebrovascular disease, intracerebral hemorrhage, ischemic stroke, and transient ischemic attack is increased when temperatures are reduced (Figure 3.5; Azevedo et al. 1995). Taken together, the incidence of stroke is highest in the winter and may be related to the lower temperatures during the winter months. Alternatively, incidence of stroke may not be related to direction of temperature change, but rather to the degree of

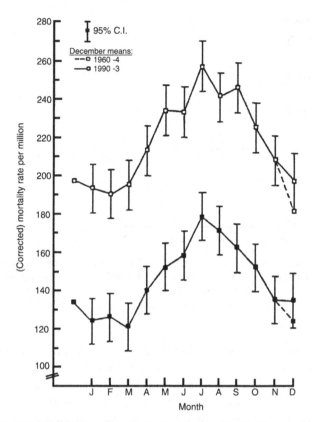

Fig. 3.4. Seasonal distribution of deaths, in the Southern Hemisphere, due to coronary heart disease. The highest number of deaths from coronary heart disease is seen in the winter, and the lowest number of deaths is seen in the summer. C.I., confidence interval. (From Douglas et al. 1990).

temperature change, because a higher incidence of stroke has been reported during extremely warm weather (Rogot & Padgett 1976; Berginer et al. 1989). Recently, the question of bacterial infection, viz. *Helicobacter pylori* as a cofactor in cardiovascular disease has been raised (Gasbarrini et al. 1999). Until the contribution of bacterial infection in cardiovascular disease is determined, the seasonal infection rates will remain unknown.

In sum, peak prevalence of most cardiovascular diseases occurs during the winter months. The extent to which this seasonal pattern reflects changes in host responsiveness to fluctuating environmental conditions remains to be empirically tested. Additionally, the extent to which seasonal fluctuations in heart disease represent alterations in immune function must also be addressed. For example, the role of inflammatory responses, such as the infiltration of

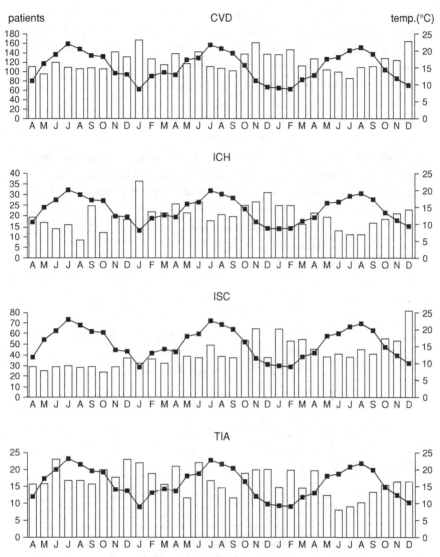

Fig. 3.5. Monthly distribution of the average temperature and the number of patients in Portugal diagnosed with cerebrovascular disease (CVD), intracerebral hemorrhage (ICH), ischemic stroke (ISC), and transient ischemic attack (TIA). For each disease, more patients are diagnosed with heart disease or stroke when temperatures are low. (From Azevedo et al. 1995.)

immune cells to the site of damage and subsequent release of cytokines, has been hypothesized to be involved in the accumulation of arterial plaque (Serrano et al. 1997). Seasonal variation in immune-induced arterial plaque build-up should be considered as a possible mechanism involved in seasonal patterns of heart disease.

3.2.4. Cancers

Seasonal variation in the initial detection and occurrence of malignant breast cancer is widely reported. Tumor detection increases in the spring and summer (Jacobsen & Janerich 1977; Cohen et al. 1983; Kirkham et al. 1985; Chleboun & Gray 1987; Mason et al. 1990). This seasonal variation has been observed across many countries, including Australia (Chleboun & Gray 1987), England (Kirkham et al. 1985), Israel (Figure 3.6; Cohen et al. 1983),

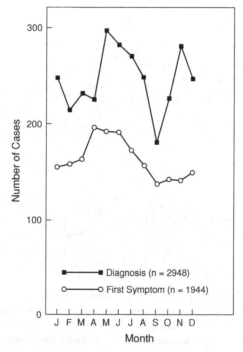

Fig. 3.6. Distribution of diagnosed breast cancer patients and month of first detectable symptoms in Israeli women. Both detection and occurrence of breast cancer are highest during the spring and summer. (Reprinted from *Cancer Research*, vol. 43, P. Cohen, Y. Wax, & B. Modan, "Seasonality in the occurrence of breast cancer," pp. 892–6, copyright © 1983, with permission from the American Association for Cancer Research.)

and the United States (Mason et al. 1985). This seasonal trend is only true of
primary tumors; no seasonal variation is reported for the time of recurrence
(Mason et al. 1987). Additionally, this seasonal trend in detection of malignant breast cancer has been observed mainly among premenopausal women
(Cohen et al. 1983; Kirkham et al. 1985; Mason et al. 1985), suggesting that
the changes in steroid hormones that occur during menopause may influence
tumor detection. For example, in women younger than 50 years of age, tumor
detection in the spring or summer is associated with increased survival if the
tumor is estrogen and progesterone receptor positive. Conversely, in women
over age 50, tumor detection during the spring or summer is only associated
with increased survival if the tumor is estrogen and progesterone receptor-
negative (Mason et al. 1990). Seasonal variation in relative risk of death data
has been examined and suggests that tumor detection in the summer (i.e., July
or August) is associated with greater risk of death than for tumors detected
in the spring (i.e., March) or fall (i.e., November). Studies of seasonal variation in tumor size suggests that larger tumors are detected in the spring and
summer; although seasonal variation is present in the incidence of tumors in
both breasts, it is more pronounced in the right, compared with the left, breast
(Kirkham et al. 1985). Relative risk of death can also vary as a function of
season in the initial detection, with the greatest risk of death associated with
detection in August and lowest risk of death in March (Figure 3.7; Sankila et al.

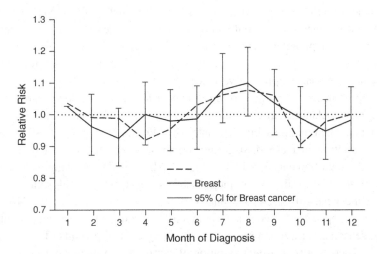

Fig. 3.7. Relative risk of death from breast cancer based on the month in which patients
were initially diagnosed with breast cancer. The relative risk of death from breast cancer
is highest when the cancer is detected in the summer. C.I., confidence interval. (From
Sankila et al. 1993.)

1993). Regardless of the substantial evidence suggesting the importance of season in the initial detection of tumors, some studies suggest there is no effect of season (Galea & Blamey 1991), whereas others acknowledge confounding variables that could be associated with seasonal trends observed for initial detection (e.g., social and psychological factors) (Kirkham et al. 1985).

In addition to the season of detection being an important prognosis factor in breast cancer, the season of tumor removal and even the season of birth can influence the course of breast cancer. Patients with tumors removed in the fall and spring are more likely to have larger tumors than those removed in the winter or summer (Ownby et al. 1986). Additionally, inflammatory responses are more likely to be elevated for tumors removed in the winter, compared with those removed during other seasons (Hartveit 1984; Ownby et al. 1986). General immune responses can be influenced by the season of tumor removal as well. Patients with tumors removed in the spring have lower IgM levels, compared with patients with tumors removed during the fall or winter (Ownby et al. 1986). Similarly, inhibition of leukocyte migration is greatest in patients with tumors removed in the winter and lowest in patients with tumors removed in the summer (Ownby et al. 1986). As previously suggested, the season of birth can also be an important factor in breast cancer, but not necessarily other forms of cancer. According to a retrospective study of birth dates from women who developed breast cancer in Sweden, there is an increased risk of breast cancer in women born in June and a decreased risk of breast cancer in women born in December (Yuen et al. 1994). Seasonal changes in melatonin or other hormones in utero could influence the risk of breast cancer in women, suggesting that prenatal environment can affect adult disease processes (Yuen et al. 1994). In fact, there is substantial evidence that prenatal photoperiod influences postnatal reproductive processes (Weaver et al. 1987).

Other cancers in addition to breast cancer also show seasonal variation. Malignant melanoma in children under 15 years of age shows the lowest incidence when diagnosed in the winter (McWhirter & Dobson, 1995). Seasonal variation in DNA damage and repair was reported in patients with nonmelanoma skin cancer, which appeared to be linked to day length (Moller et al. 1998). The binding of carcinogens to DNA in peripheral leukocytes from lung cancer patients shows seasonal variation, in which binding to DNA is higher in biopsies obtained in the summer and fall among lung cancer patients only; seasonal variation is not observed among control patients (Tang et al. 1995). There is also seasonal variation of carcinomas of the urinary bladder in terms of occurrence and severity of symptoms; the majority of patients discover the first symptoms during the fall to early winter (i.e., September–December), and the prognosis of tumors appearing during these seasons is of a higher histological grade (i.e., worse) than tumors removed during other seasons (Høstmark et al. 1984). Some forms of cancer, such as testicular and endometrial, do not generally show seasonal influences (Prener & Carstensen 1990; Dhom 1991).

There is also seasonal variation in the effectiveness of chemotherapy for certain forms of cancer. Pharmacological agents, such as doxorubincin and *cis*-diamminedichloroplatinum, that are used for treatment of metastatic ovarian carcinoma or urinary bladder cell carcinoma, show seasonal variation in their effectiveness. The toxic effects of these two agents on white blood cells is greater when administered in the winter, compared with the spring, summer, or fall (Hrushesky 1983). Data such as these have spurred much interest from medical and scientific communities, and have facilitated the acceptance of chronotherapy as an important factor in disease treatments.

In sum, there is substantial evidence that season of the year interacts with some forms of tumorous cancers. This appears to be particularly true of breast, lung, and bladder cancers. The physiological mechanisms linking season of the year, endocrine function, and cancers are described in later chapters.

3.3. Disease Prevalence in Nonhuman Animals

In common with seasonal changes in disease prevalence in humans, animals inhabiting temperate latitudes exhibit annual changes in the occurrence of a variety of diseases. In fact, the seasonal occurrence of influenza in avian populations is so pronounced that an "influenza season" has often been noted in birds (Sinnecker et al. 1982). There are numerous examples of seasonal changes in parasite prevalence in animals. Likewise, several investigators note a seasonal relapse of a number of diseases (e.g., avian malaria). Again, whether these changes in disease reflect variation in the prevalence of the vector or the ability of the host to defend against invading pathogens remains an open topic of debate. Evidence will be presented herein to argue that seasonal occurrence of disease and infection, in part, represents seasonal changes in the immunocompetence of the host.

3.3.1. Viral and Bacterial Infections

Numerous studies on the seasonal occurrence of infectious disease have been conducted in avian species. For example, one study determined that there is an increased risk of sudden death syndrome in male broiler chickens in January, with the lowest risk occurring in July (Gardiner et al. 1988). Likewise, a 12-year study of several subtypes of aquatic birds in eastern Germany revealed the existence of an "influenza season," beginning in August and continuing throughout the winter months (Süss et al. 1994). Strong seasonal patterns of influenza have also been reported in ducks, domestic turkeys, and waterfowl in the United States (Halvorson et al. 1985; Alfonso et al. 1995). In addition to uncovering a seasonal occurrence of influenza, these studies also provide convincing evidence that aquatic birds (e.g., ducks) may represent the primary source of influenza viruses to other species, including humans (Halvorson et al. 1985). The influenza A virus is capable of replicating in the intestines of ducks without producing any detectable disease. The virus is readily excreted in the feces, making feral ducks an ideal intermediate host for the transmission of the disease. It remains possible that seasonal prevalence of influenza in humans and other species is a direct result of transmission of this virus by aquatic birds. Whether the seasonal occurrence of the virus in these birds represents seasonal variation in immune function of the host or a fixed interval in the progression of the infection requires further research.

Studies in other avian species have revealed seasonal changes in the rate of viral reinfection. For example, 9 of 10 carrier birds infected with laryngotra-cheitis, a respiratory disease of chickens caused by an α-herpesvirus, shed the virus during egg-laying (Hughes et al. 1989). During the previous 3.5 weeks, only two birds had shed the virus. Similarly, homing pigeons (*Columbia livia*) maintained under seminatural conditions, latently infected with herpes virus, increase the rate of viral shedding during breeding (Vindevogal et al. 1985).

A study of fish farms and rivers of northeast Spain revealed seasonal patterns of bacterial and viral diseases (Ortega et al. 1995). Bacterial diseases were more prevalent during periods of high temperature, whereas viral infections were highest during seasons with low temperatures. Importantly, the periods of greatest risk for any diseases were times of substantial temperature change (i.e., spring and autumn). The stress of temperature change leads to reductions in immune function among numerous species of fishes (Van Muiswinkel et al. 1991). Thus, complex interactions between ambient temperature and altered immunocompetence throughout the seasons likely leads to seasonal changes in disease and death among fishes.

The severity of lesions of both atrophic rhinitis and pneumonia vary with season among slaughter swine (Cowart et al. 1992). Lesions of atrophic rhinitis are more severe among hogs slaughtered in the summer, whereas lesions of pneumonia are more severe among animals killed in the winter. Seasonal differences in the incidence of these infections are likely due to seasonal changes in the immune responses that underlie susceptibility to these diseases. Alternatively, because housing conditions also differ for slaughter swine during winter and summer, this factor may contribute to the yearly pattern of infection.

The prevalence of *Campylobacter* infections in rhesus monkeys (*Macaca mulatta*) living in outdoor enclosures at the Yerkes Regional Primate Center in Atlanta, Georgia, show pronounced seasonal fluctuations. *Campylobacter* causes diarrhea; *Campylobacter* infection rates vary annually, with highest levels observed during the spring (Mann et al. 2000a, b).

3.3.2. Parasite Infections

Seasonal changes in parasite prevalence have been well-documented for numerous species. In birds, parasitemia tends to peak during the spring and summer (i.e., coincident with migration and breeding). For example, the number of birds infected with avian lice peaks during the summer (Chandra et al. 1990). Likewise, the incidence of numerous blood parasites (i.e., *Plasmodium*, *Leucocytozoon*, *Haemoproteus*, *Listeria*, and *Trypanosoma*) increases during the spring and summer (Box 1966; Beaudoin et al. 1971; Fenlon 1985; Alexander & Stimson 1988; John 1994). In domestic ducks, recurrence of avian malaria (*Leucocytozoon simondi*) during February and March does not appear to be the result of reinfection because artificial increasing of day lengths during the winter could hasten the onset of the disease, and infection is coincident with the onset of egg-laying (Chernin 1952; Figure 3.8). These data suggest that reinfection is the result of a change in immunocompetence rather than a change in the dispersion of the parasite.

Taken together, these data suggest that the energetic stress of migration, egg-laying, and possibly sex steroid hormone alterations during breeding may lead to an increased risk of infection due to a suppression of immune function. Several studies provide evidence that this may indeed be the case. Avian glucocorticoid production has an annual rhythm, with peak activity in the spring of each year (John 1966). Because the peak in the yearly glucocorticoid rhythm coincides with the peak incidence of infection in birds, seasonal changes in adrenal corticoids may underlie season variations in the rates of many diseases. It is also possible that increased rates of infection during the spring and summer are due to concomitant increasing concentrations of sex steroids. Spleen weight is lowest during the period of gonadal recrudescence, just prior to the breeding season (Oakeson 1953; Silverin 1981). Likewise, blood smears collected from birds during winter in central Europe did not contain *Haemoproteus* gametocytes, but gametocytes appeared in both migratory and nonmigratory birds just prior to nesting. A more extensive discussion of

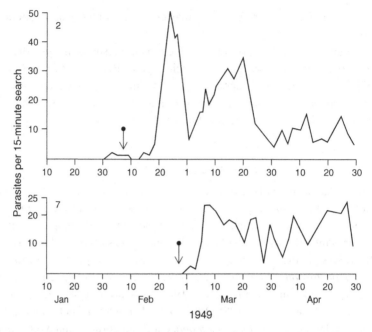

Fig. 3.8. Time of onset and the course of relapse of *Leucocytozoon simondi* in domestic ducks either artificially exposed to 12-hour light/day (no. 2) or a natural light regime (no. 7). Data are presented as the number of parasites recovered during each 15-min search. (From Chernin 1952.)

the effects of hormones on seasonal changes in immune function will be presented in Chapter 7.

Numerous studies concerned with the seasonal occurrence of disease and parasite prevalence in mammals have been conducted on livestock used for commercial sales. The obvious goal of these studies is to determine the factor(s) influencing the seasonal variation in disease in order to predict and control these factors. Several studies in cattle have monitored the course of parasitic infection throughout several years (Hoover et al. 1984; Baker & Fisk 1986; DeBont et al. 1995). Most of these studies are consistent with the hypothesis that seasonal changes in the abundance of the vector account for seasonal variations in parasitemia and disease. For example, cattle grazing in the California Sierra foothills exhibit an increased incidence of nematode infections during the summer months (Baker & Fisk 1986). This increased rate of nematode infections appears to involve seasonal fluctuations in the pattern of infective nematode larvae infesting herbage ingested by grazing cattle. Similarly, the incidence of *Haemonchus contorus* infection in sheep and goats exhibits a peak during the dry season (i.e., October–January) relative to the wet season (i.e., March–May) in Sierra Leone (Asanji 1988). This finding suggests that increased rates of infection during the dry season likely follow the population explosion of nematodes resulting from larvae that develop on vegetation during the rainy season. Likewise, bovine anaplasma infections are highest during the spring and summer in sentinel cattle occupying southeastern Oregon, northwestern Utah, and central Idaho (Zaugg 1990). The author suggests that this spring and summer peak in infection rate is likely to be due to the increased arthropod activity during these seasons. Although seasonal variation in the vector likely accounts for annual variation in infection rates, it remains plausible that seasonal variation in parasite prevalence and disease in cattle may be due to yearly alterations in immune function. Further studies are necessary to investigate this possibility.

Similar findings have been reported for horses and other livestock. Specifically, postmortem analyses of 150 horses from Victoria, Australia, revealed a seasonal change in gastrointestinal parasite prevalence with mean intensities peaking in summer and autumn (Bucknell et al. 1995). This annual variation in parasitemia parallels the prevalence of the intermediate hosts (e.g., *Musca domestica* and *Stomoxys calcitrans*) of some gastrointestinal parasites in Australia. The intermediate hosts carry more infective larvae during the summer than any other season (Dunsmore & Jue Sue 1985). These findings suggest that, as in cattle, seasonal occurrence of parasitic infections in horses are likely the result of changing viability of the vector. More recent studies on thoroughbred stud farms in New South Wales, Australia, suggest that seasonal changes in disease rates also occur in horses (Gilkerson et al. 1994). In this study, the seasonal occurrence of the equid herpes virus-4 was investigated. Twenty-six foals were identified with the virus by examining monthly nasal swabs for the presence of equid herpes virus-4 antibodies. All of the seropositive animals were identified in the months of January, February, or March (Gilkerson et al. 1994). These data indicate that seasonal changes in infection rates of horses may result from seasonal alterations in immune function.

Female white-tailed deer (*Odocoileus virginianus*) exhibit a seasonal pattern of gastrointestinal nematode abundance, with the highest rates of infections of both *Haemonchus contortus* and *Oesophagostomum venulosum* occurring during summer (Waid et al. 1985). This peak in nematode infestation coincides with a time when female deer are in "suboptimal condition," due to the energetic stress resulting from the demands of lactation. Summer is also a time of maximum abundance for the most common gastrointestinal helminths in Texas. Taken together, these findings suggest that a combination of decreased immune function in summer among female white-tailed deer

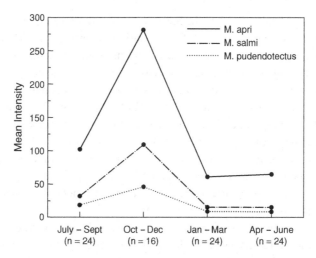

Fig. 3.9. Seasonal prevalence of infections with *Metastrongylus apri*, *Metastrongylus salmi*, and *Metastrongylus pudendotectus* in feral swine from southern Florida between 1978 and 1979. (Reprinted from the *Journal of the American Veterinary Medical Association*, vol. 181, D. J. Forrester, J. H. Portor, R. C. Belden, & W. B. Frankenberger, "Lungworms of feral swine in Florida," pp. 1278–80, copyright © 1982.)

along with increased abundance of an opportunistic parasite likely lead to a peak incidence of infection during the summer months.

Lungworm infections of feral swine in Florida vary across seasons, with peak rates of infection occurring from October through December (Forrester et al. 1982; Figure 3.9). It is unclear whether this seasonal occurrence of lung nematode infections is the result of changes in immune function across seasons or changes in the optimal conditions for the vector. Because the peak incidence of infection coincides with the optimal season for earthworms (i.e., the intermediate host for lungworms), the authors suggest that annual change in lungworm infections in swine is likely due to a change in the vector.

Populations of small rodents experience pronounced seasonal and yearly fluctuations in population density. These seasonal changes in local abundance are likely due to individuals responding to intrinsic or extrinsic factors that regulate population density annually via altered survival (possibly through altered immunocompetence) and reproduction (Lochmiller et al. 1994). Because cotton rats (*Sigmodon hispidus*) in central Oklahoma exhibit dramatic fluctuations in density both within and between years (Odum 1955), the possibility that annual changes in population density are due to seasonal alterations in immune function (and concomitant disease) was investigated (Lochmiller et al. 1994). During the first year of the study, several immune parameters varied on a seasonal basis, with the highest immune responses occurring in the winter (Lochmiller et al. 1994). It remains plausible that seasonal alterations in parasite-induced immunosuppression lead to a decreased ability to battle invading organisms, thereby leading to increased morality during a time of high parasitemia (Seed et al. 1978). Declines in a population of meadow voles (*Microtus pennsylvanicus*) are associated with evidence of infection by Theiler's encephalomyelitis virus and reovirus-type-3 (Descôteaux & Mihok 1986). These studies lend further support to the hypothesis that population fluctuations of small mammals may be the result of opportunistic pathogens striking while immune function is impaired.

Voles and shrews investigated for the presence of helminth parasites in Wales show a seasonal pattern of infestation (Lewis 1966). One study at Bogesund, near Stockholm in south-central Sweden, reports seasonal variation in the ability of bank voles (*Clethrionomys glareolus*) to infect larval ticks (*Ixodes ricinus*) with Lyme disease (Tälleklint et al. 1993). Although larval infections of bank voles are at a peak in June and July, nearly 70% of all Lyme disease infections occur during August and September; virtually no infections occur during the winter. Further research is necessary to determine whether these data reflect a seasonal change in the immune function of the host, or reflect the latency to infection from year to year.

In urban areas of India, house rats (*Rattus rattus*) exhibit seasonal patterns of parasitic infection, with a greater incidence of *Hydatigera* infections occurring during the spring and winter than any other time of year (Malhotra 1986). The season of greatest infection coincides with a time when ambient temperature is not extreme. Although environmental factors, such as temperature, can affect immune parameters of the host, it remains possible that annual alterations in parasitemia reflect the effects of temperature on the proliferation of the helminths in India. Thus, the increasing adverse conditions of summer in India (i.e., high temperatures and food shortage) tend to cause high mortality rates among the parasites (Malhotra 1986).

Seasonal changes in parasitic infection also occur in fishes. A study of perch (*Perca fluviatilis*) captured in Lough Neagh, the largest lake in the British isles, revealed seasonal variation in the number of fish infected with the metacercarial cysts of *Cotylurus variegatus* (Faulkner et al. 1989). Other studies of fishes in Russia, Lithuania, and the United States have confirmed that parasitic infections peak in spring and summer (Rautskis 1970; Amin 1978; Malakhova 1961). Highest mean worm burdens are recorded between May and June, coinciding with the breeding season in this species. Female perch have a greater infection of the visceral cavity than males during this time (Faulkner et al. 1989; Figure 3.10). This sex bias in worm burden is likely due

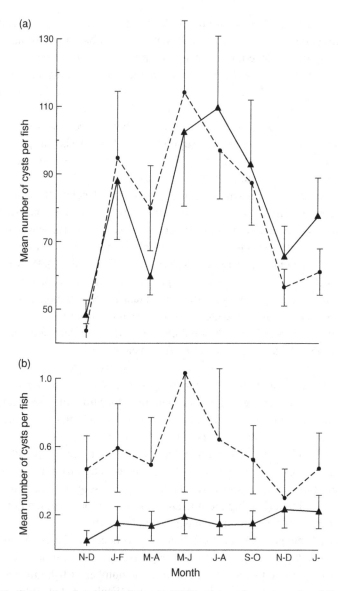

Fig. 3.10. Seasonal variation in the mean (±SEM) number of metacercariae of *Cotylurus variegatus* recovered from (a) the pericardial cavity and swim bladder and (b) the visceral cavity of male (▲) and female (●) perch. (Reprinted from the *International Journal of Parasitology*, vol. 19, M. Faulkner, D. W. Halton, & W. I. Montgomery, "Sexual, seasonal and tissue variation in the encystment of *Cotylurus variegatus* metacercariae in perch, *Perca fluviatilis*," pp. 285–90, copyright © 1989, with permission from Elsevier Science.)

to decreased immune function in females during the breeding season (Leigh 1960). Taken together, it appears that the breeding season for perch represents a time when reduced immune function leads to increased susceptibility to parasites, and this decrease in immunocompetence is greater in females. Alternatively, the warmer water temperatures in the spring and summer may promote parasite reproduction and survival during this time.

Seasonal cycles in parasite prevalence in fish may be dependent on water temperature. Seasonal fluctuations in parasite prevalence in largemouth bass (*Micropterus salmoides*) sampled in South Carolina is independent of ambient temperature (Eure 1976). Fish were sampled from two different stations, some stations were heated to 10°C above normal for the area, whereas the other stations were left unheated. Regardless of water temperature, a similar seasonal pattern of *Neoechinorhynchus cylindratus* infections is observed for both areas, with numbers being low during the summer and early fall in both heated and unheated areas. However, parasite density is higher in fish maintained in the heated areas. A combination of seasonal alterations in temperature (favoring parasite survival or affecting host immunocompetence), along with altered host immune function, likely accounts for annual changes in parasitemia among fish inhabiting nontropical latitudes.

3.4. Conclusions

Seasonal patterns of illness and death exist among both human and nonhuman animals. In some cases, yearly fluctuations in infection appear to represent seasonal changes in the prevalence of the vector; many parasites fluctuate in number annually due to variation in local environmental conditions. In other cases, there is increasing evidence that seasonal patterns of disease likely reflect yearly alterations in the resistance and susceptibility of the host to disease. A number of studies have revealed seasonal alterations in resistance to several diseases during particular times of the year. In humans, seasonal patterns are observed in the incidence of a variety of infectious diseases, in addition to cancer and heart disease. For example, in humans, malaria prevalence is highest during the wet season coincident with an increase in the prevalence of the *Plasmodium* parasite. Along with this increase in malaria infection rates, decreased cellular immune responses to malaria antigens occur during the wet season, likely leading to increased prevalence of the disease. Likewise, increasing evidence from laboratory, field, and agricultural studies suggests that susceptibility to a variety diseases varies seasonally.

In nonhuman animals, disease onset is often associated with events that require substantial amounts of energy. For example, in birds, disease onset often occurs during reproduction, egg-laying, or migration, all of which are processes that are energetically challenging. Likewise, in fishes, periods of increased viral infections are associated with times of water temperature changes, suggesting that temperature-associated stress (and concomitant alterations in energy balance) may lead to reduced immunocompetence and increased disease. It is unclear whether seasonal alterations in human immune function and disease are due to yearly alterations in energy availability. Future studies of preindustrial human societies with greater exposure to annual changes in environmental conditions and where reproduction (Bronson 1995) and food availability occur on a more seasonal basis are necessary to uncover a relationship between season, energetics, immune function, and disease. The energetics of seasonal changes in immune function is discussed at length in Chapter 6.

This chapter presents evidence of yearly cycles in disease prevalence. We also argue that these fluctuations in infection rates are not merely due to changes in local environmental conditions that can lead to decreased immunocompetence and subsequently increased vulnerability to disease. It is likely that a complex interaction between seasonal changes in immune capacity, along with changes in the abundance of opportunistic pathogens, drives annual cycles of illness and death. The degree to which these annual alterations in disease susceptibility and death rates reflect seasonal fluctuations in immune function is discussed in Chapter 4.

4

Seasonal Changes in Immune Function

Another year! – another deadly blow!

Wordsworth, November, 1806

4.1. Introduction

In contrast to other physiological processes, immune function has been assumed to remain relatively constant across the seasons (see Sheldon & Verhulst 1996). Thus, seasonal fluctuations in death and disease were, historically, attributed to seasonal cycles in pathogen virulence and not to changes in host immunity (see Chapter 3 for details). Recent evidence, however, suggests that immune function varies substantially on a seasonal basis (Lochmiller et al. 1994; Nelson & Demas 1996). Maintaining maximal immune function is energetically expensive; the cascades of dividing immune cells, the onset and maintenance of inflammation and fever, and the production of humoral immune factors all require significant energy (Chapter 6; Maier et al. 1994; Demas et al. 1997). Energy utilization involves elevation of basal metabolic rate, blood glucose, and free fatty acid levels. Therefore, mounting an energetically costly immune response has the potential to compromise the ability to preserve protein stores in muscle (Beutler & Cerami 1988). Additionally, mounting an immune response requires resources that could otherwise be allocated to other biological functions (Sheldon & Verhulst 1996). Thus, it is reasonable to consider immune function in terms of energetic trade-offs. Individuals may partition resources among the immune system and other biological processes, such as reproduction, growth, or thermogenesis. Consequently, animals may maintain the highest level of immune function that is energetically possible, given the constraints of processes essential for survival, growth, reproduction, thermogenesis, foraging, and other activities (Festa-Bianchet 1989; Richner et al. 1995; Deerenberg et al. 1997). The observations that immune function fluctuates seasonally and is compromised during times of breeding, migration, and molt are consistent with this idea (Zuk 1990; John 1994; Zuk & Johnsen 1998; Merino et al. 2000).

"Stress" is a notoriously ethereal concept that has been used to describe any factor that increases glucocorticoid secretion, including injury, pain, infection, overcrowding, harsh ambient temperature, food deprivation, noise, restraint, and social interactions. The ecological literature illustrates that environmental factors perceived as stressors, such as low food availability, low ambient temperatures, overcrowding, lack of shelter or increased predator pressure, recur seasonally and correlate with seasonal fluctuations in immune function among individuals and seasonal changes in populationwide disease and death rates (Lochmiller et al. 1994; Lochmiller & Deerenberg 2000). Laboratory studies have established that various stressful stimuli inhibit immune function (Keller et al. 1983; Laudenslager et al. 1983; Glaser et al. 1987; Irwin & Hauger 1988; Brenner & Moynihan 1997; Boccia et al. 1997; Madden 2001) and that the effect is mediated by stress-induced increases in circulating glucocorticoid concentrations (Claman 1972; Monjan 1981; Mac-Murray et al. 1983; Hauger et al. 1988; Besedovsky & del Rey 1991; Ader & Cohen 1993; Black 1994; Biondi 2001). Thus, winter survival, at least among nonhuman animals, is hypothesized to require a positive balance between short-day enhanced immune function and glucocorticoid-induced immunosuppression (Nelson & Demas 1996; Sinclair & Lochmiller 2000). The balance between short-day enhanced immune function (i.e., to the point where autoimmune disease becomes a danger) and stressor-induced immunosuppression (i.e., to the point where opportunistic pathogens and parasites overwhelm the host) must be met for animals to survive and become reproductively successful (Lochmiller & Deerenberg 2000; Merino et al. 2000). The role of glucocorticoids in modulating this delicate energetic balance between immune function and other seasonally occurring activities will be further discussed in Chapter 7.

This link between stress and disease was first identified in the modern era by Hans Selye who demonstrated that exposure to stressors evoked elevated glucocorticoid secretion, adrenal hypertrophy, and thymic involution (Selye 1956). Selye termed this nonspecific response the general adaptation syndrome. According to Selye, general adaptation syndrome consisted of three stages: (1) the initial *alarm* stage in which the stressor is recognized; (2) the *resistance* stage that consists of successfully overcoming any short-term physical insult; and (3) the stage of *exhaustion*, in which disease ensues because the stressor becomes chronic and exhausts all physiological systems. Although Selye's ideas were the foundation of early stress research, there were flaws in his reasoning. Disease does not develop, as Selye believed, because bodily defenses become depleted; stress-related disease arises because with chronic stress the physiological responses themselves become damaging (Sapolsky 1992).

This widely publicized link between immune function and "stress" can be reinterpreted in the context of an energetic perspective. Stressors disturb energy homeostasis. Because restoring homeostasis requires more energy than maintaining homeostasis (Sapolsky 1998), exposure to a stressor increases energy demands on individuals. Sapolsky (1998) has suggested that individual differences in coping with similar stressors reflect differences in perception and processing of information about stressful situations. However, given the link between immune function and energy balance, it is also possible that individual differences in coping with stress may reflect differences in energy partitioning or energetic efficiency.

During long-term perturbations in energy homeostasis, such as during winter when temperatures are low and food is scarce, glucocorticoids are released and energy is mobilized to restore homeostatic equilibrium. There may be competition for energy among several physiological processes, including growth, cellular maintenance, immune function, thermogenesis, and reproductive processes (Merino et al. 2000; Sinclair & Lochmiller 2000) (Figure 4.1).

Behavioral ecologists generally measure the costs of immune function in the currency of natural selection, viz. the quantity and quality of offspring. Thus, high infection rates are usually correlated with reduced reproductive success, measured in litter or clutch size (see Tella et al. 2000). This relationship is often established experimentally by manipulating reproductive effort (adding or subtracting offspring from a nest) (e.g., Nordling et al. 1998) and observing changes in immune status, or manipulating immune function (activating specific immune responses) (e.g., Horak et al. 2000) and assessing reproductive success. Conspecific females apparently attend to signals of immune function to choose mates (Hamilton & Zuk 1982). These signals more accurately reflect immune status than the innocuous parasite burdens (e.g., fleas and ticks) often reported by behavioral ecologists (Moller et al. 1998). Although only about a dozen species have been studied thus far, a positive relationship has been found between high immune status and morphological displays, such as color or size of male secondary sexual displays (reviewed in Moller et al. 1999). Males with high immune status also tend to exhibit "attractive" behavioral sexual displays (Saino & Moller 1997; Soler et al. 1999). For example, male black wheatears (*Oenanthe leucura*) engage in a remarkable sexual display of "weight lifting" (Soler et al. 1999) by carrying stones (mean = 3.1 kg/season for this 35 g song bird) into nooks and crannies located in caves. Apparently, females choose their mates based on who has lifted the greatest number of heavy stones. Males that carried heavy

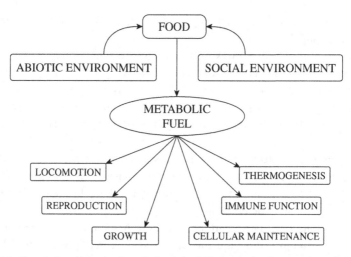

Fig. 4.1. Competing behavioral and physiological demands that require substantial metabolic energy. When demands are higher than the energy immediately available, individuals engage in trade-offs among various energy demands to reduce total energy needs. During energetic shortages, locomotion, reproduction, and growth are curtailed with little effect on survival. As energy becomes increasingly scarce, survival may be compromised because of reduced energy allocation to thermogenesis and immune function. Extrinsic factors, such as social environment, are hypothesized to significantly influence the partitioning of metabolic reserves to these various behavioral and physiological processes. (Adapted from Wade and Schneider 1992. Reprinted from *Reviews of Reproduction*, vol. 4, S. L. Klein & R. J. Nelson, "Influence of social factors on immune function and reproduction," pp.168–78, copyright © 1999.)

stones showed more robust T-cell responses to phytohemagglutinin (PHA) injections than birds that carried relatively light stones (Soler et al. 1999). Thus, this weight-lifting display reflects male cellular immune function. Several studies have documented that male sexually selected traits display seasonal fluctuations that mirror immune function (Zuk & Johnsen 1998; Merino et al. 2000). It is important to recall that behavioral ecologists are working at a different level of analysis, in terms of immunological "costs," than physiologists (see Lochmiller & Deerenberg 2000). Throughout the book, we will be discussing the energetic (kcal) costs of immune function, rather than the fitness costs.

We, as well as others, propose that individuals of some species use photoperiod information to initiate or terminate specific seasonal responses that maintain energy homeostasis, and these same mechanisms also affect both reproduction and immune function (Chapter 1; Bartness & Goldman 1989; Heldmaier et al. 1989; Saafela & Reiter 1994). Accordingly, animals may have evolved mechanisms to combat seasonal stress-induced reductions in immunity as a temporal adaptation to promote survival. As proposed in Chapter 1, we hypothesize that exposure to short (winter-like) day lengths enhances immune function. Clinical, field, and laboratory data will be reviewed in this chapter that are consistent with the hypothesis that immune function is bolstered prior to and during the winter.

The primary proximate cue used by nonhuman animals to ascertain time of year (i.e., season) is photoperiod (reviewed in Chapter 1). In this chapter, we will review data from nonhuman animals that clearly illustrate photoperiod, alone or in combination with other supplementary cues (e.g., temperature, food availability), mediates seasonal changes in immune function. The question of whether photoperiod affects human immune function remains open. The current dogma is that humans are not photoperiodic, at least in terms of reproductive behavior (Bronson 1995; Reiter 1998). Despite their failure to respond to changes in day length, seasonal rhythms in human conception, mortality, body mass, and suicide have been widely reported (reviewed in Ascoff 1981 and Chapter 3). Because humans are exposed to both natural and artificial environmental fluctuations, it is currently very difficult, if not impossible, to delineate the underlying causes of seasonal cycles among humans. Although the mechanisms that underlie seasonality in humans remains unspecified, we do know that seasonal cycles in behavior and physiology occur (Bronson 1995; Reiter 1998). As reviewed in Chapter 3, seasonal changes in disease prevalence and host resistance (i.e., immune function) among humans are well known. The purpose of this chapter is to explore the hypothesis that seasonal changes in death and disease represent fluctuations in host immunity among both human and nonhuman animals. This hypothesis is supported if immune function changes seasonally and if these seasonal patterns in immunity exist independent of infection; in other words, we address the question: do healthy individuals exhibit annual changes in immune cell numbers and effector functions? There is overwhelming evidence to suggest that, among both humans and nonhuman animals, immune function varies seasonally and this variation occurs independent of fluctuations in pathogen prevalence. Thus, data presented in this chapter suggest that seasonal changes in immune function may have either evolved independent of seasonally occurring pathogens or co-evolved as an adaptation to overcome seasonally occurring pathogens.

4.2. Seasonal Changes in Humans

As presented in Chapter 3, many infections occur seasonally in both human and nonhuman animals. Seasonal changes in infection and disease prevalence may represent fluctuations in either the host or the pathogen. Our working

Table 4.1. *Seasonal Changes in Immunity of Healthy Individuals.*

Response	Season of Elevated Activity	Reference
Serum immunoglobulin concentrations		
IgA	None	Lyngbye & Krøll 1971
		MacMurray et al. 1983
		Nordby & Cassidy 1983
		Stoop et al. 1969
IgM	None	Lyngbye & Krøll 1971
		MacMurray et al. 1983
		Nordby & Cassidy 1983
		Stoop et al. 1969
IgG	Winter	MacMurray et al. 1983
		Wagnerová et al. 1986
	Summer	Lyngbye & Krøll 1971
	None	Nordby & Cassidy 1983
		Stoop et al. 1969
Total no. of circulating lymphocytes	Winter	Reinberg et al. 1977
		Maes et al. 1994
% of peripheral	Winter	Bratescu & Teodorescu 1981

hypothesis is that seasonal variation in disease prevalence is often due to changes primarily in the host, not the pathogen. In Chapter 3, we presented substantial data that suggest changes in host resistance mediate seasonal variation in infection and illness; however, a definitive test would be to document the existence of seasonal changes in host immune function that are independent of infection status. Accordingly, seasonal changes in immune function have been established in healthy individuals and are reviewed in this chapter. Indeed, measurements of both cellular and humoral immunity display seasonal variation (see Table 4.1).

4.2.1. Cell-Mediated Immunity

In Chapter 2, cell-mediated immunity was defined as adaptive immune responses mediated by lymphocytes that either directly kill a pathogen or release cytokines to attract other lymphocytes to the site of infection. The distinguishing characteristic of cell-mediated immunity is that it involves direct interaction between lymphocytes and pathogens. Aside from anecdotal observations by Greek philosophers and physicians (see Chapter 3), seasonal variation in immune function in humans has only been systematically studied for the past 20 years and has paralleled growing interest in the relationship between organisms and their environment. In one of the first studies of seasonal variation in immunity among healthy individuals, circulating lymphocyte numbers were counted and were reportedly highest in the winter months among healthy subjects in France (Reinberg et al. 1977). Although this study did not assess phenotypic variation of lymphocytes (i.e., variation in the different subsets of lymphocytes), it was one of the first to examine seasonal variation in the immune system of healthy human subjects.

Subsequent studies, characterizing changes in subsets of lymphocytes, have supported the initial finding that the number of immune cells in circulation changes seasonally. The seasonal rhythm of circulating lymphocytes, however, varies depending on the subset of cells examined. For example, in six study populations of healthy subjects in the Western United States, although the percentage of viable B-cells was significantly elevated in the winter, the percentage of T-cells was reduced in the winter, compared with the summer (MacMurray et al. 1983). T-cells can be subdivided into cytotoxic T-cells and helper T-cells, and labeled using monoclonal antibodies that recognize the cell-surface receptors, CD8 and CD4, respectively. In a study of healthy Ph.D. students in the Netherlands, the number of circulating CD8[+] T-cells was reportedly lowest during the winter months (Figure 4.2; Van Rood et al. 1991). In healthy Belgian subjects, the total number of lymphocytes was higher in blood samples collected in the winter than summer months (Maes

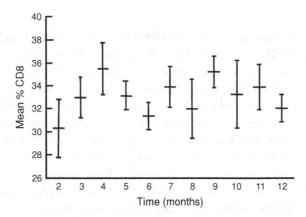

Fig. 4.2. Mean (±SEM) percentage of circulating CD8[+] T-cells per month from healthy Ph.D. students in the Netherlands. The percentage of CD8[+] cells is lowest in the winter and highest in the summer months. (Reprinted from *Clinical and Experimental Immunology*, vol. 86, Y. Van Rood, E. Goulmy, E. Blokland, J. Pool, J. Van Rood, & H. Van Houwelingen, "Month-related variability in immunological test results: Implications for immunological follow-up studies," pp. 349–54, copyright ©1991, with permission from Blackwell Science Ltd.)

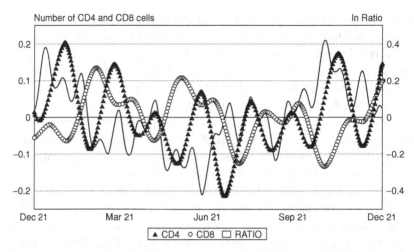

Fig. 4.3. Seasonal variation in the number and ratio of CD4[+] and CD8[+] T-cells from healthy Belgian subjects. Both the number and ratio of cells are higher in the winter than in the summer months. (Reprinted from *Experientia*, vol. 50, M. Maes, W. Stevens, S. Scharpé, E. Bosmans, F. De Meyer, P. D'Hondt, D. Peeters, P. Thompson, P. Cosyns, L. De Clerck, C. Bridts, H. Neels, A. Wauters, & W. Cooreman, "Seasonal variation in peripheral blood leukocyte subsets and in serum interleukin-6, and soluble interleukin-2 and -6 receptor concentrations in normal volunteers," pp. 821–9, copyright ©1994.)

et al. 1994). In this same study, the number of circulating CD4[+] and CD25[+] T-cells, CD20[+] B-cells, and the CD4[+]/CD8[+] ratio also peaked in the winter and was lowest in the summer (Figure 4.3; Maes et al. 1994). Serum concentrations of interleukin (IL)-6 (a cytokine that stimulates T- and B-cell growth and differentiation) and soluble IL-6 receptor concentrations were also higher in the winter than summer months; soluble IL-2 receptor concentrations did not change seasonally (Maes et al. 1994). In a study of healthy West African children, the percentage of total lymphocytes was lower during the rainy (i.e., June–October) than dry (i.e., November–May) season; the percentage of CD8[+] T-cells, however, was higher during the rainy than dry season (Lisse et al. 1997). Taken together, these data indicate that subsets of lymphocytes shift seasonally. Specifically, cells involved in cell-mediated immunity, or more specifically cytotoxic T-cells, are reduced in the winter and lymphocytes that are important for antibody-mediated immunity, or more precisely B-cells and helper T-cells, are elevated in the winter.

The environmental factors that signal seasonal changes in host immune cell numbers in humans are not known; however, studies have examined the potential role of temperature and photoperiod as mediators of seasonal fluctuations in immunity. In one study, four subjects were followed for 7 years, three subjects worked daytime jobs, and one subject was a night-shift worker. Each subject donated blood monthly over the 7-year study. On average, the percentage of circulating B-cells was highest during the winter and lowest during the summer for the three subjects working day jobs (Paglieroni & Holland 1994). In addition, the percentage of CD8$^+$ T-cells was lower, and the percentage of CD4$^+$ T-cells was higher in the winter than summer months in the three day workers (Paglieroni & Holland 1994). In contrast, the shift worker had no observable seasonal pattern in any of the lymphocyte subsets examined over the 7-year study (Paglieroni & Holland 1994), suggesting that exposure to light during the dark phase of the light:dark cycle may regulate seasonal rhythms in immunity (Plytycz & Seljelid 1997).

In another study of six blood donors, absolute numbers and percentages of B- and T-cells in peripheral blood were examined over the course of 1 year (Bratescu & Teodorescu 1981). Although the total number of lymphocytes and leukocytes did not vary throughout the year, the proportion of B-cells was nearly doubled during the winter months, compared with the summer months (Bratescu & Teodorescu 1981). To determine whether this seasonal variation in the percentage of circulating B-cells was temperature-dependent, some participants were exposed to low temperatures ($-25°C$) for 35 minutes before blood collection, and the percentage of B-cells in peripheral blood was examined. Low temperatures did not affect B-cells; that is, proportions were higher in the winter than summer regardless of temperature (Bratescu & Teodorescu 1981). In this same study, subpopulations of helper T-cells were also assessed. Specifically, Th1 and Th2 cells, that are mediators of inflammation and B-cell activation, respectively, were examined. The ratio of Th1:Th2 cells was increased in the spring/summer; once again, this seasonal pattern was observed regardless of temperature (Bratescu & Teodorescu 1981).

In addition to seasonal variation in the number and percentage of immune cells, the effector function of immune cells varies seasonally. For example, there are seasonal differences in mitotic activity of normal human peripheral blood lymphocytes (Boctor et al. 1989). In healthy males and females in the United States, proliferative responses of peripheral blood lymphocytes to the T-cell mitogens, Con A and PHA, were higher in the summer months, compared with the winter months (Boctor et al. 1989). Conversely, there was no seasonal variation in proliferative responses to the B-cell mitogen, pokeweed mitogen (PWM), suggesting that the mitotic activity of peripheral B-cells is unaffected by season (Boctor et al. 1989). Contradictory findings were reported for healthy Ph.D. students in the Netherlands, in which proliferative responses of T-cells to the mitogens, Con A and PHA, were higher in the

winter than in the spring and summer months (Van Rood et al. 1991). However, in common with the United States study (Boctor et al. 1989), there was
no seasonal variation in proliferative responses to the B-cell mitogen, PWM
(Van Rood et al. 1991). In addition to seasonal fluctuations in the mitotic
ability of circulating lymphocytes, a study of Japanese children revealed that
lymphocytes isolated from the tonsils and stimulated with Con A proliferate
and differentiate more in the summer than in the winter months (Komada
et al. 1989). Additionally, when the cells were labeled with monoclonal antibodies and subsets of T-cells were examined, the proportion of $CD4^+:CD8^+$
T-cells is higher in the summer and lower in the winter months (Komada
et al. 1989). The discrepancy in T-cell function among these populations may
reflect differences in environmental factors (e.g., latitude) or physiological
factors (e.g., energy balance).

As described in Chapter 2, B-cells produce antibody, and helper T-cells
serve to activate resting B-cells to begin antibody production; hence, these
two cell types are intimately involved in antibody-mediated (i.e., humoral)
immunity. Conversely, cytotoxic T-cells that actively eliminate infected cells
are key players in cell-mediated immunity. Data presented in this section suggest that because B-cells and helper T-cells are elevated in the winter and because cytotoxic T-cells are higher in the summer, seasonal trade-offs may exist
between antibody-mediated and cell-mediated immunity. Antibody-mediated
immunity is essential for overcoming bacterial infections. Based on the data
reviewed in the previous chapter, bacterial infections are most prevalent in
the winter. Thus, seasonal variation in antibody-mediated immunity may have
evolved as a coadaptation to seasonal changes in bacterial pathogens. Conversely, cell-mediated immunity is higher in the summer and is critical for
killing viral pathogens; there are no obvious seasonal trends, however, in
the prevalence of viral infections (Chapter 3; Centers for Disease Control
and Prevention 1991). These findings suggest that seasonal variation in host
cell-mediated immune responses are independent of seasonal changes in the
pathogen. Future studies will need to examine whether variation in host immune function co-evolved as an adaptation to annual changes in the prevalence
of certain pathogens.

4.2.2. Humoral Immunity

Humoral immunity is defined as primarily involving B-cell production of antibody against a specific pathogen (see Chapter 2). Several studies of healthy humans suggest that there is seasonal variation in antibody production; however,

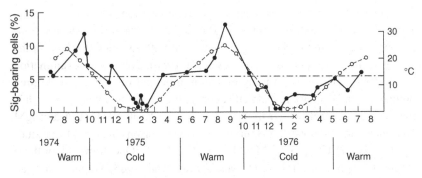

Fig. 4.4. Seasonal variation in the presence of surface immunoglobulin (S-Ig) molecules on B-cells from healthy individuals in Japan. The percentage of B-cells with S-Ig is higher in the summer, compared with the winter (•), and this seasonal pattern in the presence of S-Ig parallels seasonal changes in monthly temperature (○). (Reprinted from *Clinical and Experimental Immunology*, vol. 33, T. Abo & K. Kumagai, "Studies of surface immunoglobulins on human B lymphocytes. III. Physiological variations of sIg+ cells in peripheral blood," pp. 441–52, copyright © 1978, with permission from Blackwell Science Ltd.)

these studies have yielded conflicting results. Antibody production involves the activation of resting B-cells to begin secreting surface immunoglobulins (Igs) against a specific pathogen. Once an Ig molecule is secreted from a B-cell, that molecule is called antibody. Each Ig molecule has two distinct polypeptide chains (i.e., heavy and light chains) that are connected by a disulfide bond. A study conducted using blood samples from healthy adults in Japan revealed that the presence of Ig molecules on the surface of resting B-cells varies seasonally, in which the percentage of B-cells with surface Ig is higher in the summer, as compared with the winter, and this seasonal pattern occurs for both heavy and light chains (Figure 4.4; Abo & Kumagai 1978). These data suggest that seasonal variation in humoral immunity may be the result of initial differences in the number of resting B-cells with Igs, as opposed to seasonal changes in the ability, or effector function, of B-cells to produce antibody. Thus, the number of resting B-cells capable of antibody production upon activation may be reduced in the winter.

Several studies suggest that antibody production also varies seasonally. In one study, blood samples of patients from five different Veterans Administration hospitals were tested and revealed significantly higher IgG concentrations in samples collected in the winter, compared with those collected in the summer for four of the five hospitals (MacMurray et al. 1983). In terms of environmental factors modulating this seasonal variation, the average monthly temperature was inversely related to IgG concentrations (MacMurray et al. 1983). Similarly, healthy children in Prague had higher IgG concentrations in samples collected in the autumn, compared with samples obtained in the spring over a 2-year period (Wagnerová et al. 1986). Conversely, examination of seasonal variation in a variety of serum proteins from adult and children outpatients in Denmark revealed that concentrations of IgG were higher during the summer than winter months (Lyngbye & Krøll 1971). IgA and IgM concentrations did not differ significantly across seasons in any of these studies (Lyngbye & Krøll 1971; MacMurray et al. 1983). One drawback of these studies (Lyngbye & Krøll 1971; MacMurray et al. 1983; Wagnerová et al. 1986) is that they examined total, nonspecific antibody present in circulation (i.e., all antibody produced against all pathogens in circulation). In one study, however, IgG production against specific mycobacterial and leishmanial antigens [*Mycobacterium leprae* soluble antigen (MLSA) and *Leishmania mexicana* (LV4), respectively] was examined in healthy subjects vaccinated with inactive/killed bacteria (Sharples et al. 1994). Both anti-MLSA and anti-LV4 IgG production were higher in the summer, compared with the winter months (Figure 4.5; Sharples et al. 1994). These data suggest that, in addition to seasonal variation in circulating concentrations of total antibody, host production of specific antibody varies seasonally as well (Sharples et al. 1994).

Finally, there are several studies that report no seasonal variation in humoral immunity. For example, in a study of healthy adults and children in the Netherlands, no significant seasonal changes were observed in serum IgG, IgM, or IgA concentrations, regardless of the age or sex of the subjects (Stoop et al. 1969). Similarly, serum sampled over a 24-month period from healthy adults and children in Michigan revealed no seasonal changes in Ig concentrations, although serum IgM showed the greatest variability in the fall–winter period (Nordby & Cassidy 1983).

In summary, these studies illustrate that seasonal variation in humoral immunity is present in humans and in some cases may underlie seasonal variation in host defense against pathogens (Sharples et al. 1994). The extent to which seasonal variation in humoral immunity serves to "protect" the host from infection remains unknown, but the previously presented data support the hypothesis that seasonal changes in infection and disease reflect variation in host immunity. Future studies must determine whether elevation of nonspecific antibody production precedes seasonal exposure to and occurrence of pathogens. In other words, do nonspecific changes in host immunity serve to bolster immune function in anticipation of exposure to infectious pathogens?

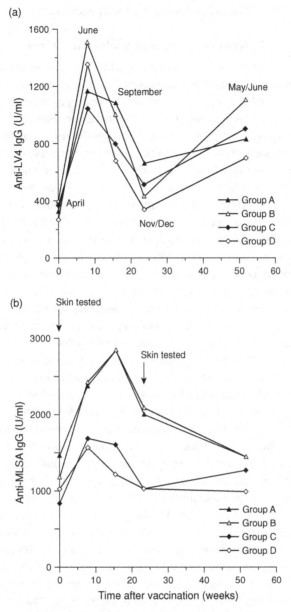

Fig. 4.5. Serum IgG production against specific (a) leishmanial and (b) mycobacterial antigens [*Leishmania mexicana* (LV4) and *Mycobacterium leprae* soluble antigen (MLSA), respectively] in healthy subjects vaccinated with inactive/killed bacteria in England. Both anti-MLSA and anti-LV4 IgG production are higher in the summer, compared with the winter months. (Reprinted from *Vaccine*, vol. 12, C. E. Sharples, M. Shaw, M. Castes, J. Convit, & J. M. Blackwell, "Immune response in healthy volunteers vaccinated with BCG plus killed leishmanial promastigotes: antibody responses to mycobacterial and leishmanial antigens," pp. 1402–12, copyright © 1994, with permission from Elsevier Science.)

4.3. Seasonal Changes in Nonhuman Animals

As discussed previously, many clinical studies report seasonal changes in immune function in specific human populations. Some of these changes may be due to seasonal changes in environmental conditions, whereas others are more likely caused by psychological and/or social variables that covary with seasonal changes in the environment. Although the results of these studies are intriguing, such studies prove to be problematic because it is difficult to identify the precise variables that contribute to changes in immune function observed across the season. In general, humans are not seasonal animals, with most individuals buffered from extreme seasonal fluctuation in the environment, including fluctuations in ambient temperature and even day length, due to our primarily "indoor" existence. Any effects of season on humans are likely to be subtle, involving complex interactions with multiple variables.

In addition to the changes in immune function seen in human populations discussed previously, considerable field research has also revealed seasonal changes in immunity in nonhuman animals. Because most nontropical species of animals, unlike humans, are not buffered from potentially dramatic seasonal changes in their environment, these studies allow for a more precise determination of the seasonal variables affecting immune function. Despite this benefit, virtually all previous studies of seasonal changes in nonhuman mammalian species have placed considerable emphasis on seasonal fluctuations of reproduction and energy balance (Bronson & Heideman 1994). For example, many studies have demonstrated striking seasonal patterns of mating and birth (Bronson & Heideman 1994; Bronson 1995). Salient cycles of disease and death also exist; these cycles, however, are much less commonly studied. For example, disease and death rates typically increase during the winter, compared with the summer for individuals of many vertebrate species (John 1994; Lochmiller et al. 1994). Many of these animals presumably become sick and die from a direct failure to balance their energetic demands due to exposure to extreme weather or starvation. Many animals, however, die indirectly from opportunistic diseases that seem to overwhelm immunological defenses, presumably at times when these defenses are low (Lochmiller et al. 1994; Nelson & Demas 1996; Lochmiller & Deerenberg, 2000). Several epidemiological studies have provided evidence for reduced immune function and increased death rates from infectious disease during the winter (reviewed in Boctor et al. 1989; Afoke et al. 1993; John 1994; Lochmiller et al. 1994; Nelson et al. 1995; Nelson & Demas 1996; Sinclair & Lochmiller 2000). A direct link between seasonal fluctuations in the environment and specific immunological changes has been established in several studies.

Table 4.2. *Seasonal Changes in Immunity of Nonhuman Animals.*

Response	Season of Elevated Activity	Reference
Lymphoid tissue mass		
Spleen	Summer	Newson 1962
Gut-associated lymphatic	Summer	Shivatcheva & Hadjioloff 1987a
tissue		Shivatcheva & Hadjioloff 1987b
Thymus	Summer	Aimé 1912
		El Ridi et al. 1981
		El Ridi et al. 1981
		Hussein et al. 1979
	Winter	Nakanishi 1986
	None	Olsen et al. 1993
Immune cell no.		
White blood cells	Summer	Lochmiller et al. 1994
	Fall	Gould et al. 1998
Immunoglobulins	Summer	El Ridi et al. 1981
		Hussein et al. 1979
		Lochmiller et al. 1994
		Nakanishi 1986
		Sealander & Bickerstaff 1967
		Sidky et al. 1972
		Stone 1956
	Winter	Dobrowolska et al. 1974
		Dobrowolska & Adamczewska-Andrzejewska 1991
T-cell proliferation	Summer	Shifrine et al. 1980a, b
	None	Olsen et al. 1993

4.3.1. Gross Lymphatic Tissue

Many of the sophisticated assays for assessing immune function in nonhuman animals are relatively recent in their development and are difficult to implement in the field. Because of these limitations, immune status has traditionally been assessed indirectly, via changes in gross lymphatic tissue mass (e.g., spleen, bursa of Fabricius, thymus), that presumably reflect immunologic activity of the tissue. Despite more recent evidence suggesting a less than adequate correlation between tissue mass and immune function, most of the early field work focused on tissue mass; thus, a review of these studies is presented in this chapter. Most field studies examining immune parameters across seasons have reported reduced lymphatic tissue size during winter (reviewed in Nelson & Demas 1996). Indeed, seasonal patterns in the development, regression, and regeneration of the thymus, spleen, and bursa of Fabricius have been described in many vertebrate species.

Ectotherms

Much of the early work on seasonal changes in lymphoid tissue has been conducted on ectothermic animals. All ectotherms possess cells involved in recognition and processing of antigens, although their immune systems appear to be more primitive than mammals (Zapata et al. 1992). A relatively large number of studies have reported seasonal changes in lymphoid tissue, with a trend toward regression of lymphoid tissue during the mating period and winter, compared with other times of the year. Regression of thymic tissue has been noted in fish species that die after their first spawning (Robertson & Wexler 1959; Honma & Tamura 1984). In fish species that live after spawning, however, no such regression of the thymus is seen. These species do undergo seasonal changes in thymic mass independent of breeding, with the largest masses seen in June and July (Honma & Tamura 1984). In brown trout (*Salmo trutta*), production of thymic tissue is highest in spring and summer, and degeneration of lymphoid cells can be found staring in autumn and lasting throughout the winter (Alvarez et al. 1988). More recent research has demonstrated seasonal fluctuations in lymphocyte number in the thymus, spleen, and pronephros, with maximal numbers in spring and summer and the lowest numbers during winter and summer (Alvarez et al. 1998). Seasonal changes in lymphoid tissue have been reported for anurans as well. For example, in frogs (*Rana temporaria*), maximal development of the thymus occurs in the summer, with marked involution of this organ in the winter. The summer thymus has many small thymocytes in the cortex and a central medulla. Winter atrophy leads to a lack of distinction between the cortex and medulla of the thymus, as well as a general reduction in thymic size (Plytycz et al. 1995). In older animals, seasonal differences still occur between the cortex and medulla, but there are no changes in thymic size. This effect appears to been independent of ambient temperature, as experimental manipulation of this variable does not alter the seasonal change in the thymus. Additionally, frogs maintained on a summer-like activity pattern during the winter still maintain a winter-type thymus, whereas animals kept in a winter-like hibernating state during the summer maintain a summer-like thymic pattern (Bigaj & Plytycz 1984). These results suggest that the environment does not trigger seasonal changes is thymic mass in this species; rather, they are regulated by some form of endogenous circannual rhythm (Zapata et al. 1992).

Seasonal changes in lymphatic organ mass are observed among individuals of reptilian species (reviewed in Zapata et al. 1992). Reptiles generally possess a well-defined thymus during autumn, which gradually undergoes involution throughout the winter (Zapata et al. 1992). In summer, however, there is large species variability in thymic response with a highly developed thymus found in lizards, but an involuted thymus found in some snakes and turtles. Several initial reports of this kind were conducted in turtles (*Clemnys leprosa, Testudo mauritonica*) (Aimé 1912). In turtles, thymic mass is reduced during winter estivation and regenerates in early spring. Thymic mass continues to increase throughout the summer, reaching maximal size immediately prior to estivation (Aimé 1912). In the turtle *Mauremys caspica*, the thymic cortex becomes involuted in the spring and increases in size during the early summer and autumn with a small involution occurring again in the winter (Leceta et al. 1989). Some of the seasonal changes in the thymus of turtles can be attributed to changes in steroid hormones. For example, there appears to be a relationship between the development of lymphoid tissue during the summer and spring, and circulating testosterone and corticosterone concentrations in *Mauremys caspica* (Leceta & Zapata 1985). In addition, in June when circulating concentrations of testosterone begin to increase from previously low values, turtles undergo lymphocyte mobilization from the thymus and spleen with a decrease in the size of the thymic cortex, decreased number of cortical lymphocytes, but an increase in the number of medullary lymphocytes (Varas et al. 1992). Testosterone treatment triggers the decrease in cortical lymphocytes earlier and also decreases medullary lymphocytes (Varas et al. 1992). These results suggest that some of the seasonal changes in the thymus are due to changes in circulating steroid hormones.

Recent studies have also demonstrated seasonal changes in the thymus in lizards. Thymuses of male and female lizards (*Scincus scincus*) become extremely involuted, and thymic cells become undifferentiated during the winter, compared with the summer (Hussein et al. 1979) (Figure 4.6). Thymuses begin to regenerate in the spring, leading to redifferentiation of thymic tissue into a distinguishable medulla and cortex. Seasonal fluctuations in lymphoid cell subpopluations have also been reported in lizards. For example, the number of thymocytes and splenocytes begin to decrease at the beginning of autumn and disappear completely by winter. In spring, these cells gradually reappear, reaching a seasonal maximum by the end of spring (Saad et al. 1987; El Ridi et al. 1988). Additionally, the number of thymic-like cells are maximal in late spring and summer, decrease during early autumn, and are completely undetectable by winter (Saad & El Ridi 1988). A similar finding has been reported for Colubrid snakes (*Psammophis schokari*) inhabiting the Egyptian deserts (El Ridi et al. 1981). As discussed previously, seasonal changes in

Seasons	No. birds	Age (mo.)	Weights (grams) of			
			Body	Gonads	Liver	Spleen
Males						
Winter	128	15.7	160	1.022	2.870	.037
Spring	142	14.6	158	1.225	2.956	.039
Summer	98	17.6	155	1.142	3.169	.042
Autumn	131	17.9	160	0.912	3.047	.037
Females						
Winter	129	15.6	158	.345	3.146	.046
Spring	112	14.8	155	.363	3.402	.053
Summer	83	15.5	153	.328	3.342	.048
Autumn	120	17.7	157	.307	3.291	.044

Fig. 4.6. Seasonal changes in splenic mass in male and female longhorn doves. The spleens of both males and females are smaller during the autumn and winter, compared with the spring and summer. (From Riddle 1928.)

lymphoid tissue morphology do not necessarily reflect seasonal changes in immune *function*.

Birds

Splenic and thymic sizes have been reported to be minimal in several avian species when the gonads begin the process of undergoing vernal recrudescence (e.g., Krause 1922; Riddle 1928; Höhn 1947, 1956; Oakeson 1953, 1956; Fange & Silverin 1985; John 1994) (Figure 4.7). Mallard ducks (*Anas platyrhynchos*) undergo thymic involution at puberty (Höhn 1947). There is also a pronounced regeneration of thymic tissue at the end of each breeding season in mid-summer in both male and female mallards; thymic regression occurs prior to autumnal migration in this species, presumably as a result of stress (Höhn 1947). Similar observations have been made in house sparrows (*Passer domesticus*) and robins (*Turdus migratorious*) (Höhn 1956).

White-crowned sparrows (*Zonotrichia leucophrys gambelli* and *Zonotrichia leucophrys nuttalli*) undergo a reduction in relative splenic size at the beginning of the breeding season in western North America (Oakeson 1953, 1956); splenic regression in this particular case cannot be attributed to the stress of migration because both migratory and nonmigratory populations display identical patterns of splenic development (Oakeson 1953, 1956). Migratory pied flycatchers also display a seasonal pattern of splenic development, with regression occurring prior to the onset of the breeding season in Sweden and splenic regeneration occurring during subsequent incubation and feeding of the hatchlings (Fänge & Silverin 1985).

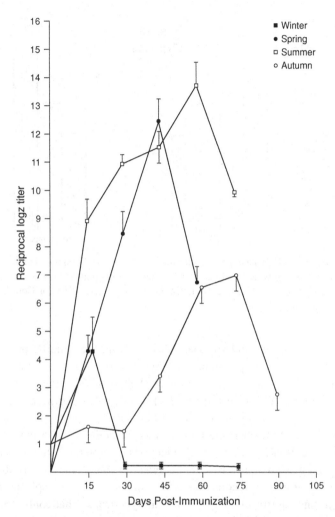

Fig. 4.7. Seasonal variation in the antibody response to rat red blood cells in lizards (*Scincus scincus*). Serum agglutinins were virtually nonexistent in the winter, whereas a vigorous antibody response is seen during the summer. (Reprinted from the *Journal of Experimental Zoology*, vol. 209, M. F. Hussein, N. Badir, R. El Ridi, & S. El Deeb, "Effect of seasonal variation on immune system of the lizard," pp. 91–2, copyright ©1979, with permission of Wiley-Liss, Inc., a subsidiary of John Wiley & Sons, Inc.)

Mammals

Seasonal changes in lymphatic organ mass have also been reported in individuals of mammalian species. For example, in short-tailed voles (*Microtus agrestis*), splenic mass and splenic reticular cell counts are reduced in winter,

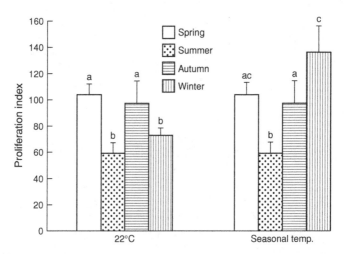

Fig. 4.8. Mean (±S.E.M.) maximal proliferation index of peripheral blood mononuclear cells collected during winter and summer for male rhesus monkeys (*Macaca mulatta*) housed under natural photoperiod after stimulation with IL-2, PHA, or Con A. After Mann et al. (2000).

compared with spring and summer (Newson 1962). Similarly, pine voles (*Microtus pinetorum*) display increased thymic masses in early autumn when reproductive organ masses are declining (Valentine & Kirkpatrick 1970). Red-backed mice (*Clethrionomys rutilus*) wild-trapped at different times of the year display peak thymic masses in February and peak splenic masses in September and October; the lowest mass for both organs occurs in July (Sealander & Bickerstaff 1967). Reductions in splenic lymphoid tissue in European ground squirrels (*Citellus citellus*) (Shivatcheva & Hadjioloff 1987a, b) occur in winter, compared with summer values. The spleens of immature ground squirrels undergo a gradual hypertrophy that continues until mid-summer, followed by an involution by the end of summer and the beginning of autumn. The largest degree of involution seems to occur at the start of the hibernation period in November and December (Shivatcheva & Alexandrov 1986; Shivatcheva & Hadjioloff 1987a, b). Adult cotton rats (*Sigmodon hispidus*) display seasonal cycles of thymic development and involution with reduced thymic masses during the summer and maximal sizes occurring in the winter (Lochmiller et al. 1994). No such pattern, however, is seen in juvenile cotton rats. These results suggest that energetic investment in immunity is reduced during times when reproductive effort is high.

Seasonal changes in lymphatic organ mass occur in hibernating mammals. For instance, European ground squirrels (*Citellus citellus L.*) display a circannual rhythm in splenic morphology, as well as in the lamina propria of the mucosa and Peyer's patches (Shivatcheva & Hadjioloff 1987a, b). These lymphatic tissues are regressed in the autumn in both hibernating and non-hibernating squirrels; but, regression is more complete in hibernating animals. Hypertrophy of splenic and gut-associated lymphoid tissue is seen prior to arousal from hibernation in the spring (Shivatcheva & Hadjioloff 1987a, b).

4.3.2. Cell-Mediated and Humoral Immunity

In common with the seasonal changes in lymphatic organ mass among birds, mammals, and reptiles, there also exist seasonal changes in specific immune parameters, as well as immune function (reviewed in Nelson & Demas 1996).

Ectotherms

Seasonal changes in immune function appear common among reptiles (reviewed in Zapata et al. 1992). For example, male and female Colubrid snakes (*Psammophis schokari*) display enhanced humoral antibody response to injections of rat red blood cells and human serum albumin in summer, compared with winter (El Ridi et al. 1981). In addition, snakes tested in summer display a stronger lymphoproliferative response to mixed leukocyte cultures, compared with winter-caught animals (Farag & El Ridi 1985). Male and female lizards (*Scincus scincus*) captured in winter fail to form humoral antibodies to rat red blood cells; antibody responses are low in spring, but becomes elevated in summer (Hussein et al. 1979).

Adult male tortoises (*Mauremys caspica*) maintained under natural light and temperature conditions display seasonal changes in antibody production (Leceta & Zapata 1986). After primary immunization with sheep red blood cells (SRBCs), tortoises tested in the autumnal condition had the appearance of 2-mercaptoethanol-sensitive antibodies, whereas animals maintained in the simulated summer condition had no such antibodies. Furthermore, splenic plaque-forming cells (PFCs) were present in fall-, but not summer-simulated animals (Leceta & Zapata 1986). Antibody levels and PFCs were present in both fall- and summer-simulated tortoises after secondary immunization with SRBCs.

Mammals

Seasonal changes in both cell-mediated and humoral immunity have also been demonstrated in mammalian species. Cattle (*Bos taurus*) produce fewer antibodies in response to an antigen (J substance) in the winter, compared with summer (Stone 1956). Lymphocyte proliferation to the T-cell mitogens,

Con A and *Phytolacca americana*, an extract of PWM, are elevated in February relative to other months of the year (Lochmiller et al. 1994). In European voles (*Microtus arvalis*), circulating Ig concentrations were elevated in autumn and winter, compared with summer-trapped voles (Dobrowolska et al. 1974; Dobrowolska & Adamczewska-Andrzejewska, 1991). Both peripheral white blood cells and neutrophil counts mimic this pattern in some years, whereas the pattern was reversed in other years. Ground squirrels also display lower levels of hemagglutinins raised against SRBCs in winter (Sidky et al. 1972). Red-backed mice (*C. rutilus*), wild-trapped near College, Alaska, at different times throughout the year, also display seasonal changes in blood reticulocyte counts (Sealander & Bickerstaff 1967). Specifically, reticulocyte counts display a primary peak in early winter, with lesser peaks occurring in late winter and mid-summer. In contrast, short-tailed voles (*M. agrestis*) display lowest blood reticulocyte counts in winter, compared with summer (Newson 1962).

In a field study of bank voles (*Clethrionomys glareolus*) in northwestern Italy, males showed impaired humoral immune responses during the breeding season (Saino et al. 2000). In contrast to predictions, when the population densities were experimentally reduced, males did not enhance immune function, and females displayed *compromised* immune function (Saino et al. 2000). These field data are consistent with the results of a laboratory study in which prairie voles (*M. ochrogaster*) were maintained individually for 10 weeks either in long (LD 16:8) or short (LD 8:16) photoperiods in rooms with either high (10.96 animals/m^3) or low (0.18 animals/m^3) population densities (Nelson et al. 1996). In both photoperiodic conditions, basal blood corticosterone concentrations were higher in voles from low-density, compared with high-density, rooms. Splenic masses were unaffected by day length, but were elevated among high-density animals. Similarly, serum IgG levels were elevated among high-density animals. Taken together, these studies support the hypothesis that population fluctuations of small mammals may be the result of opportunistic pathogens striking while immune function is impaired.

In a field study that examined the seasonal variation in photoperiod, temperature, and population density on immune response, three prairie vole (*M. ochrogaster*) populations were examined over a year (Sinclair & Lochmiller 2000). Multiple regression analyses indicated that the model containing relative spleen mass, cytokine-stimulated T-cell proliferation, and *in vivo* hypersensitivity explained a significant amount of variance in population density, whereas cytokine-stimulated T-cell proliferation and relative thymus mass explained a significant amount of variance in survival rate (Sinclair & Lochmiller 2000). Taken together, these results indicate that seasonal environmental changes enhance components of host immune function under field conditions, and these field conditions may counteract the immunoenhancing effects of short days in laboratory studies.

Cotton rats (*S. hispidus*) display a reduced number of peripheral white blood cells, but an increased number of PFCs in response to SRBCs injection in winter, compared with summer values (Lochmiller et al. 1994). Additionally, the total peripheral blood leukocytes, packed cell volume, and total serum proteins were examined in this species. Significant fluctuations were seen in leukocyte numbers and packed cell volume, but not serum proteins – although these fluctuations were independent of specific seasons (Lochmiller et al. 1994). It has been hypothesized that seasonal changes in humoral and cell-mediated immune function in cotton rats are due to changes in specific splenocyte subpopulations (Davis & Lochmiller 1995). To test this hypothesis, unique splenocyte subpopulations were identified using fluorescein-conjugated cell surface markers (e.g., Con A, agglutinins, PWM) throughout the year. All splenocyte subpopulations were more prevalent in the fall and winter, compared with spring and summer, supporting the notion that it is these specific subpopulations underlying seasonal changes in immune function (Davis & Lochmiller 1995). Another study examined the seasonal change in PFCs in response to SRBC injections in two mouse strains (Ratajczak et al. 1993). Interestingly, there is a strain difference in immune response, with antibody formation highest in CD1 mice during the spring and highest during the summer for C57/B6C3F1 mice (Ratajczak et al. 1993). More recent research on female C57/B6C3F1 mice has demonstrated significant seasonal and circannual variation in immune responsiveness (Dozier et al. 1997). For example, the peak PFC levels occurred during the summer, whereas the heaviest spleen and thymus masses were found in the spring (Dozier et al. 1997). As in humans, these results suggest that different immune parameters may vary independently on a seasonal basis. Cattle (*Bos taurus*) raise fewer antibodies in response to an antigen (J substance) in the winter, compared with summer (Stone 1956). Thus, it is possible that disease resistance to specific pathogens (e.g. viruses) may be high during times of the year when resistance is low against other types of pathogens (e.g., ectoparasites). Although this hypothesis is intriguing, more research is necessary to test it.

Seasonal changes in immune parameters have also been observed in nonhuman primates (Mann et al. 2000a). There is a seasonal shift in the frequency of cells expressing Th1 cytokines (IL-2 and interferon-γ), and those expressing Th2 cytokines (IL-4) in peripheral blood mononuclear cells that were collected from rhesus monkeys (*Macaca mulatta*) during the winter and summer. Lymphoid cells collected during the summer tend to synthesize the Th1 cytokines, IL-2, and interferon-γ, than those cells collected during the winter. Cells from animals during the summer also display higher natural killer (NK)-mediated lysis than those cells from animals in the winter. Conversely, the *in vitro* proliferation response of peripheral blood mononuclear cells is higher in the winter than in the summer. A related study showed that circulating numbers of white blood cells and neutrophils, and lymphocyte proliferation in response to mitogens, are higher in the winter than in the summer (Mann et al. 2000b).

Collectively, these results provide strong evidence for seasonal changes in immune function. However, they do not allow the determination of the specific environmental variables that may underlie these seasonal changes in immunity. To provide a more controlled setting in which to study seasonal changes immunity, outbred beagle dogs (*Canis familiaris*) were maintained in open colonies. These dogs display reduced lymphocyte responses to the B-cell mitogen PHA-P and the T-cell mitogen Con A in winter relative to summer (Shifrine et al. 1980a, b). In another nonhuman animal study, seasonal fluctuations in IgM and IgE antibody-forming cells, E-rosette-forming cells, as well as blood and lymphoid NK and K-cell activity was examined in rats, rabbits, and dogs (Melnikov et al. 1987). Significant seasonal changes were found in specific immune responses (e.g., IgM antibodies and K-cells), with enhancement reported during autumn and winter, compared with spring and summer; less dramatic changes were reported for IgE or E-rosette-forming cells (Melnikov et al. 1987). Many of the seasonal changes in immune function may be regulated by the pineal hormone melatonin (see Chapter 8); but, for animals maintained in constant conditions, the source(s) of seasonal information is not obvious. For example, injections of exogenous melatonin can enhance antibody-dependent cellular cytotoxicity in laboratory-housed mice in the summer, but not the winter (Giordano et al. 1993). These results suggest that there is a seasonal fluctuation in the sensitivity of immune tissue to the immune-enhancing properties of melatonin. In another study, lymphocyte proliferation in response to both T- and B-cell mitogens varied seasonally in house mice, despite the fact that the animals were maintained in a constant environment (Planelles et al. 1994). Specifically, proliferation was highest in autumn and summer and lowest in spring and winter. The proportion of B-lymphocytes was lowest in spring, whereas T-lymphocytes did not show any significant seasonal variation. A similar circannual rhythm in immune function has been reported with respect to NK cell activity, as well as mitogen-stimulated lymphocyte proliferation in mice (Pati et al. 1987). Seasonal variations in NK cell activity, lymphocyte responsiveness to

Con A, PHA, and lipopolysaccharide were examined in this study. NK cell activity was highest during January–February and lowest during July–August; in contrast, proliferation to all mitogens was highest in April–June and lowest in January–February (Pati et al. 1987). Collectively, these results suggest that seasonal changes in immune cell number and function may be regulated by the internal biological rhythms of the animal (Planelles et al. 1994).

Taken together, the results of animal studies of seasonal changes in immune function suggest that immunity is generally compromised during the breeding, compared with the nonbreeding season. However, it should be noted that significant variations in both the direction and extent of seasonal changes in immune function exist, depending on the immune parameter assessed and the species tested. One important confound with field studies of seasonal changes in immunity is that a wide variety of environmental factors beyond experimental control can influence immune function. Although field studies are necessary to examine seasonal changes in immunity in an ecological context, discovery of the causative agents, as well as the interactive nature of these agents on immunity, requires controlled laboratory studies.

4.4. Conclusions

As the research discussed in this chapter has suggested, it is apparent that immune function can vary across the year in both humans and nonhuman animals. These seasonal changes in immunity range from changes in the size and gross morphology of immunologic tissue (e.g., spleen, thymus) to more subtle changes in circulating antibody concentrations and specific immune cell types. Changes in immune function may be due, at least in part, to environmental factors that fluctuate seasonally, such as day length, ambient temperature, humidity, rainfall, and food availability.

The majority of human and nonhuman animals likely experience potentially dramatic seasonal fluctuations in both energy expenditure and energy availability. For example, most nontropical animals experience high thermoregulatory requirements in winter when energy (food) availability is typically at a seasonal low. The result is an "energetic bottleneck" (i.e., increased energy demands during times of reduced energy availability). All animals, including humans, require a positive energy balance to survive and reproduce. Like all other physiological processes, immune function is energetically expensive (Demas et al. 1997). Mounting immunological defenses against an invading pathogen involves a cascade of cellular processes that demand substantial energy. Although the research discussed in this chapter demonstrated seasonal changes in immunity (and not pathogen prevalence per se) underlying seasonal variation in disease prevalence, presumably animals are more susceptible to opportunistic pathogens that cause disease and premature death during times of reduced immune function. This idea is discussed in more detail in Chapter 6.

Most animals use environmental factors to determine the time of year. Many environmental factors can be perceived as stressful, such as low ambient temperatures, reduced food availability, overcrowding, or lack of cover – leading to glucocorticoid secretion and immune suppression. Seasonally recurring changes in these potential stressors can lead to seasonal changes in immune function and presumably disease susceptibility. Seasonal fluctuations in immune function may not be observed every year or in every population studied. Some winters may not be perceived as stressful, either because of mild ambient conditions (e.g., an "Indian summer") or because winter coping strategies successfully buffer the individual from harsh conditions. Thus, animal populations studied in their natural environment may exhibit compromised, enhanced, or static immune function across the seasons. Controlled laboratory studies are necessary to determine the precise contributions among various environmental factors and immune function. The present chapter summarized the existing literature on seasonal fluctuation in immune function. The succeeding chapters will examine the precise environmental factors mediating these seasonal changes in immunity, as well as the potential underlying physiological mechanisms for these changes.

5

Photoperiod, Melatonin, and Immunity

In winter I get up at night, And dress by yellow candle-light,
In summer, quite the other way, I have to go to bed by day.

Robert Louis Stevenson, 1885
A Child's Garden of Verses

5.1. Introduction

In the previous two chapters, we documented seasonal patterns of immune function, disease, and death in a wide variety of species. Seasonal phenomena may be imposed on animals by their environments. Alternatively, seasonal phenomena may reflect an interaction between individuals intrinsic seasonal clocks and the environment. If the second alternative is true, then individuals must coordinate internal seasonal rhythms with the environment. Most formal studies of seasonality have focused on day length, i.e., photoperiod, as the environmental cue used by animals to coordinate intrinsic seasonal rhythms with extrinsic seasonal environmental changes. In this chapter, we will review the formal properties of photoperiodism, describe how the pineal gland and its primary hormone, melatonin, mediate the physiological effects of day length, and finally document how photoperiod and melatonin influence immune function.

Photoperiodism, first proposed by the botanists Garner and Allard in 1922 to describe the response of plants to the length of day and night, currently is defined as the ability to determine day length in both plants and animals. Photoperiodism has evolved in virtually all taxa of plants and animals that experience seasonal changes in their habitats. Among vertebrate animals, photoperiodism is linked to a number of seasonal adaptations, including reproductive, metabolic, morphological, and, most germane to the topic of this book, immunological adaptations to cope with seasonal changes in ambient conditions. The first demonstration of photoperiodic time measurement regulating vertebrate reproduction was established in birds (Rowan 1925). During the winter, juncos (*Junco hyemalis*) were maintained in outdoor aviaries in Edmonton, Alberta, and exposed to several minutes of artificial light after the onset of dark each day (lights were illuminated at sunset). Under these artificial lighting conditions, these birds came into reproductive condition despite the harsh Canadian winter temperatures. In comparison, juncos living in the relatively mild "Riviera-like" climate of Berkeley, California, but exposed to normal winter day lengths, remained in nonreproductive condition (Rowan 1925). Thus, Rowan concluded that the number of hours of light per day, not ambient temperature or food availability, regulated the annual breeding cycle of juncos. The first demonstration of photoperiod regulating reproduction in mammals was reported for European field voles, *Microtus agrestis* (Baker & Ranson 1932). The role of photoperiod in mediating seasonal adaptions has since been documented for hundreds of vertebrate species (Goldman & Nelson 1993).

Syrian, or golden, hamsters (*Mesocricetus auratus*), represent the most common mammalian model used in laboratory studies of photoperiodism. Hamsters, in common with most small mammals, are "long-day breeders." Gestation is relatively brief in these animals; mating, pregnancy, and lactation occur during the long days of late spring and early summer. Adult male hamsters undergo gonadal regression when day lengths fall below 12.5 hours of light (Elliott 1976). The minimum day length that supports reproduction is called the "critical day length" (Figure 5.1). Critical day length is not a fixed variable, but differs among populations of animals living at different altitudes and latitudes. For example, Siberian (*Phodopus sungorus*) hamsters occupy very different habitats than Syrian hamsters; Siberian hamsters live in high latitude habitats, and in the laboratory they display a critical day length for reproductive function of 13 hours of light/day (Goldman 2001).

When Syrian or Siberian hamsters are maintained in day lengths <12.5 or 16 hours of light, respectively, blood concentrations of gonadotropins and sex steroid hormones decrease, accessory organ mass diminishes to about 10% of the original size, and reproductive behaviors stop. Male hamsters remain reproductively quiescent for approximately 16–20 weeks, a period of time that roughly corresponds to the duration of short days experienced in the wild during autumn and winter. Hamsters are photorefractory during the recrudescent phase; that is, gonadal condition becomes unlinked from photoperiodic inhibition. Photorefractoriness permits attainment of fully functional gonads in the spring before environmental photoperiods attain the day length necessary for gonadal maintenance in the autumn. This is an important adaptation that allows burrowing animals (that live in constant dark conditions) to anticipate spring conditions with the development of fully functional reproductive

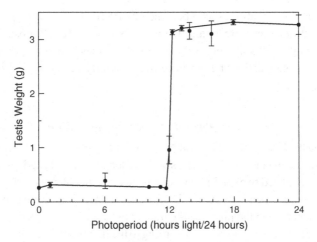

Fig. 5.1. *Critical day length for maintenance of gonadal function.* In Syrian hamsters, testicular development in response to photoperiods ranged from 0 to 24 hours of light/day. Each circle represents the mean (±SEM) paired testes mass (g) of groups of hamsters housed for approximately 3 months in these photoperiods. The minimum day length that supports reproduction is called the "critical day length" and is between 12 and 12.5 hours of light/day for these hamsters. (Reproduced with permission from Elliott 1976.)

systems without long-day exposure. Chronic maintenance in the long days of summer is necessary to "reset" or reestablish the photoperiodic mechanism so that short days can again attain inhibitory status. Immune function is enhanced by exposure to short days. One important question that has not yet been addressed experimentally is whether hamsters that have undergone gonadal recrudescence after long-term exposure to short days continue to show enhanced immune function, or does the immune system of hamsters undergo a refractoriness to chronic short days?

Sheep (*Ovis aries*) are another important model species in studies of mammalian photoperiodism. In common with other ungulates and most large mammals, sheep are so-called "short-day breeders." Mating typically occurs in autumn; females are pregnant throughout the winter, and the lambs are born and nursed in the spring when food and other conditions are most conducive for survival. As day lengths decrease in late summer, the rate of hypothalamic pulses of gonadotropin-releasing hormone (GnRH) secretion increases, which eventually stimulates increased gonadotropin secretion that initiates reproductive function (Xiong et al. 1997). In Suffolk ewe, seasonal reproductive transitions primarily appear to reflect changes in the responsiveness of the GnRH neurosecretory system to the negative feedback influence of estradiol. After mating, sheep become photorefractory; that is, short days lose their stimulatory effects and mating behavior wanes. Exposure to the long day lengths of summer is not necessary to reestablish responsiveness to short days, suggesting the presence of an underlying circannual cycle of photosensitivity.

5.1.1. Hourglass Model of Photoperiodism

As previously described, the critical problem for any photoperiodic individual is to discriminate between long and short day lengths. Two broad hypotheses have been advanced to explain how animals respond to changes in day length: (1) the response depends on the total number of hours of light per day or (2) the response depends on the phase of the light exposure relative to some internal rhythm of photoresponsiveness. According to the first model, animals monitor the accumulation (or depletion) of some physiological agent during one part of the light-dark cycle; this process is reversed during another portion of the cycle. This time measurement hypothesis is referred to as the hourglass model. The hourglass model appears to be used by several invertebrate species (Vaz Nunes & Saunders 1999), but few, if any, vertebrate species (Ruis et al. 1990).

Alternative models exist to explain how organisms respond to day length. These so-called "coincidence models" rely on phase relations between internal circadian oscillators or phase relations between external factors and endogenous circadian cycles of receptive states. There are two types of coincidence models of photoperiodic time measurement: (1) the external coincidence model and (2) the internal coincidence model. Some of the signal transduction features of melatonin secretion appear to be consistent with the external coincidence model.

5.1.2. Coincidence Models of Photoperiodism

External Coincidence Model

The second hypothesis of photoperiodic time measurement was originally formulated to explain flowering in plants. According to Bünning, "... the physiological basis of photoperiodism ... lies with the endogenous daily rhythms ... " (Bünning 1973). The crux of Bünning's hypothesis is the assumption of an endogenous circadian rhythm of subjective day (photoinsensitivity/noninducibility, with a duration of about 12 hours) and subjective night (photosensitivity/inducibility, also with a duration of about 12 hours). Light was postulated to serve two functions: (1) light entrains (i.e., synchronizes) the endogenous circadian rhythm of subjective day and subjective night, and

(2) light stimulates photoperiodic responses if it is coincident with the subjective night (i.e., the photoinducible phase). Thus, when first exposed to light, the 12-hour subjective day is set; light intruding during the next 12 hours will not maintain the reproductive system. For example, a photoperiod of 6 hours of light (L) and 18 hours of darkness (D) (LD 6:18) will not maintain the reproductive system of a hamster, because light will only coincide with the subjective night of the cycle. The inverse photoperiod (i.e., LD 18:6) would maintain the reproductive system because 6 hours of light would coincide with the photosensitive/inducible part of the cycle. Many studies across different taxa have produced results consistent with Bünning's hypothesis. In principle, a few seconds of light per day, appropriately timed to coincide with an individual's photosensitive/inducible phase, would evoke a "long-day" response.

There have been several tests of Bünning's hypothesis in vertebrates. One type of test of circadian involvement in photoperiodic time measurement makes use of the resonance paradigm (Nanda & Hamner 1958; Elliott 1976). In resonance studies, groups of animals are maintained on photoperiod cycles that couple a fixed-light phase (e.g., 6 or 8 hours) with various durations of darkness resulting in LD cycles of varying lengths. For example, hamsters can be maintained in LD 6:18, LD 6:24, LD 6:30, LD 6:42, and LD 6:54 photoperiods. If hamsters use an hourglass timer for photoperiodic time measurement, then animals maintained in each of these photoperiods should have regressed reproductive systems; the duration of light in each instance is less than 12 hours/day. Alternatively, if hamsters use a circadian timer for photoperiodic time measurement, then only animals housed in the LD 6:18 and LD 6:42 photoperiods should have regressed reproductive systems, because only during these photoregimens is light-restricted to the putative photoinsensitive/noninducible phase. Animals exposed to LD 6:30 and LD 6:54 photoperiods should have reproductive systems that are comparable to long-day males because light would coincide with the subjective night every other day or every third day, respectively. The results of these types of studies are consistent with Bünning's hypothesis of photoperiodic time measurement (Figure 5.2).

Skeleton photoperiods have also been used to test Bünning's hypothesis. In skeleton photoperiods, the full light cycle is usually replaced by appropriately timed light pulses. For example, the reproductive system of male hamsters maintained in LD 16:8 photoperiods could be sustained in LD 2:22 photoperiods if 1 hour of light occurred from 8:00 to 9:00 hours and another hour of light occurred at 23:00 to 24:00 hours. This skeleton photoperiod would provide a pulse of light in the morning to entrain the cycle of

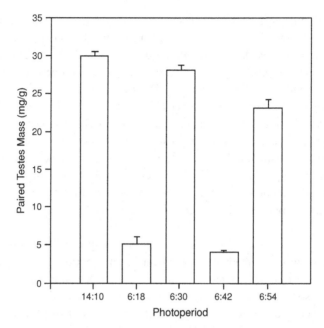

Fig. 5.2. *Testicular development in response to resonance photoperiods* shows that the absolute ratio of light to dark does not mediate photoperiodic responsiveness in hamsters. Mean (±SEM) paired testes mass corrected for body mass (mg/g) of hamsters housed either in LD 14:10 (long-day control), LD 6:18, LD 6:30, LD 6:42, or LD 6:54 photoperiods. Note that these photoperiods provided 6 hours of light every 24, 36, 48, or 60 hours, and that testicular function regressed only when the photocycle was 24 or 48 hours. (Reproduced with permission from Elliott 1976.)

photo-noninducible/photoinducible phases and a second pulse of light near the middle of the photoinducible phase. Thus, an appropriately timed, very short photoperiod of LD 2:22 could "trick" the reproductive system into a long-day response. Of course, it is possible for the individual to entrain in such a way that the 1 hour pulse of light between 23:00 to 24:00 hours would entrain the cycle of photo-noninducible/photoinducible phases and the second pulse 8 hours later would coincide with the photo-noninducible phase resulting in short-day responses. Thus, predictions about specific skeleton photoperiods depend on entrainment patterns; monitoring locomotor activity or other circadian cycles is important in explaining photoperiodic responsiveness to skeleton photoperiods.

Internal Coincidence Model

This model of photoperiodic time measurement assumes the existence of two or more internal oscillators that change in their phase relationship during the annual change in day length (Pittendrigh & Minis 1964). For example, if the peak blood concentration of glucocorticoids was coupled to a "dawn" oscillator and the peak blood concentration of prolactin was coupled to a "dusk" oscillator, the difference in the timing of the peak values of these two hormones might range from 15 to 9 hours throughout the year among temperate zone animals. According to the internal coincidence model, a photoperiodic response should be observed when certain internal phase relations are attained. Although the existence of internal coincidence processes in vertebrate photoperiodic time measurement systems has not been firmly established (but see Hyde & Underwood 1993), it continues to be an attractive hypothesis because this model is based on entrainment theory and does not rely on a special photoinducible oscillator.

5.1.3. Entrainment

The process of synchronization of endogenous biological rhythms to a periodic cue in the environment is called entrainment. Light is an important entraining agent for circadian rhythms, including the rhythm of photoresponsiveness/inducibility-photononresponsiveness/noninducibility. An understanding of phase-response curves is useful to understand the process of entrainment. Phase-response curves are graphic representations of the differential effects that light has on the timing of biological rhythms; these temporal effects depend on the phase relationship of the light to the circadian organization of the individual in question (Figure 5.3). Exposure to light at different times throughout the day does not result in uniform responses of free-running biological rhythms. For example, if hamsters are maintained in constant dark conditions, then they begin to free-run, the period of the onset of their daily locomotor activity rhythms deviating slightly from 24 hours each day.

The circadian period can be divided into a subjective day and subjective night. Because hamsters are nocturnal, the beginning of activity usually coincides with the onset of its subjective night, whereas the rest period begins at the onset of the animal's subjective day. For diurnal animals, the onset of activity coincides with the beginning of the animal's subjective day, and the rest period begins at the onset of the animal's subjective night. If free-running hamsters that are housed in constant dark conditions are exposed to a 1-hour pulse of light at any time during the middle of the subjective day, the time of onset of the next bout of locomotor activity is unaffected; that is, for a hamster with a free-running period of 24.2 hours, wheel-running still begins 24 hours and 12 minutes after the onset of the previous bout of activity. However, if the 1-hour pulse of light is given early in the subjective night (or late in the

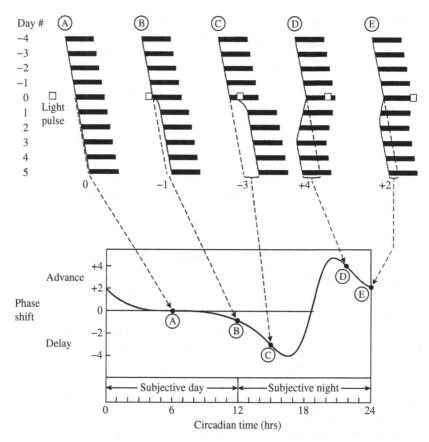

Fig. 5.3. *Phase-response curves* are graphic representations of the differential effects that light has on the timing of biological rhythms. A free-running nocturnal individual maintained in constant dark conditions was exposed to a 1-hour light pulse at various times during the subjective day and night. (A) When light was given in the middle of the subjective day, the subsequent activity onset was unaffected; the middle of the subjective day is thus called the "dead zone." (B) A light pulse at the end of the subjective day phase-delayed activity the next day by about 1 hour. (C) With a light pulse at the start of the subjective night, a substantial phase delay (3 hours) was observed. (D) Light given later during the subjective night caused a substantial advance (4 hours) of activity onset the next day. (E) Finally, a light pulse at the start of the subjective day caused a 2-hour phase advance in activity the next day. The effect of light on the endogenous rhythm is involved in synchronizing circadian rhythms to exactly 24 hours each day. (After Moore-Ede et al. 1982.)

subjective day), then the hamster does not begin its activity 24.2 hours later, but rather delays its activity onset for 1–4 hours, depending on when exactly the pulse of light occurs. If the 1-hour pulse of light is given late in the subjective night (or early in the subjective day), then again the hamster does not begin its activity 24.2 hours later, but rather advances its activity onset by 1–4 hours, depending on exactly when the pulse of light occurs. Thus, the largest phase delays can be produced early during the subjective night, when a nocturnal animal has just awakened and a diurnal animal has just retired, and the largest phase advance can be produced late during the subjective night. Light appears to have little effect on the circadian system during the subjective day of either nocturnal or diurnal animals. Although these phase responses to light are best observed in free-running animals, these same effects of light also operate on individuals that are entrained to a light-dark cycle. The function of an animal's daily cycle of responsiveness to light is to reset the biological rhythms to exactly 24 hours. Phase response curves have been generated for many species, including humans, and it is also apparent that many species measure day length. However, the extent to which some species, including humans, couple reproductive function to day length remains unspecified.

The extent to which immune function is coupled to day length has been studied only in a few species. These studies typically only examine differences in immune function and susceptibility to infection among animals housed in either short or long days. The extent to which entrainment or skeleton photoperiod affect immunity has not been examined. Of the few studies conducted, hamsters, are commonly used. Species such as laboratory strains of rats and mice, that do not display reproductive responsiveness to photoperiod, however, retain immunological responsiveness to photoperiod, suggesting that the effects of photoperiod on immunity may be independent of reproductive responsiveness to photoperiod. Because the effects of photoperiod on immunity may be sex steroid hormone-independent, melatonin remains a likely mediator of photoperiodic effects on immunity and susceptibility to infection.

5.2. Pineal Gland and Photoperiodism

The pineal gland, or *epiphysis cerebri*, remains one of the most mysterious and misunderstood organs in the body. The pineal gland was originally thought by the Greek physician, Herophilus, to be a valve regulating the flow of spirits between the ventricles of the brain. Much later, the French philosopher Rene Descartes described the pineal body as the "seat of the soul" due, in part, to its unpaired nature in the brain, and this notion was at the center of dualist beliefs between mind and brain. Between Descartes' time and the present, the pineal gland has also been referred to as the "third eye," as well as a calcified vestigial structure with no apparent function (reviewed in Arendt 1995). It was only in the late 1950s that the indolamine melatonin, the pineal gland's primary biological secretion, was first isolated in pigs (Arendt 1995). This discovery opened the door to a flurry of exciting research in the latter portion of the twentieth century on the role of the pineal gland and specifically melatonin, on many aspects of physiology and behavior. In addition to the now extensive evidence for the function of melatonin in regulating sleep patterns, as well as circadian and seasonal rhythms in a wide range of vertebrate species, including humans, recent evidence has focused on the potential role of melatonin in immune function, aging, and disease (Bubenik et al. 1998).

The pineal gland is a small (about the size of a pea in humans), unpaired structure that extends from the diencephalon of the brain via the pineal stalk (Figure 5.4) (Arendt 1995). In most nonmammalian species, the pineal gland serves as a direct photoreceptor, conveying ambient day length information directly to the brain. In mammals, however, the pineal gland is primarily a secretory organ with no known photoreceptive capability. The pineal gland receives direct sympathetic innervation from the superior cervical ganglion, although the ultimate regulation of the gland is via the suprachiasmatic nucleus (SCN), the "biological clock" in mammals. Within the pineal gland, melatonin (N-acetyl-5-methoxytryptamine) is synthesized from the neurotransmitter serotonin via a two-step biochemical process. First, serotonin is converted into N-acetylserotonin via the enzyme N-acetyltransferase (NAT).

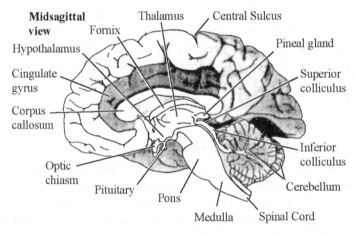

Fig. 5.4. *The "pine cone-shaped" pineal gland is about the size of a pea in humans.* The pineal is indicated in this midsagittal (midline) view.

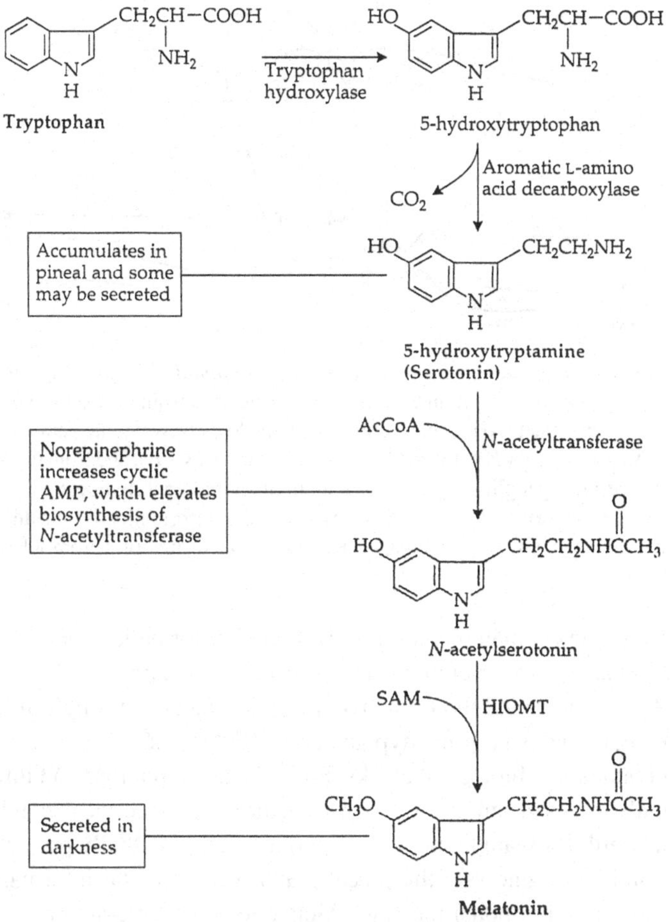

Fig. 5.5. *Melatonin synthesis pathway.* Serotonin is formed from tryptophan. During the light hours of the day, tryptophan is converted in a two-step reaction to serotonin. In darkness (shaded area), increased norepinephrine secretion causes an increase in N-acetyltransferase, the first of two enzymes that convert serotonin to melatonin. SAM, S-adenosylmethionine; HIOMT, hydroxyindole-O-methyltransferase.

N-acetylserotonin is then converted to melatonin via the enzyme hydroxy-O-methyltransferase (HIOMT) (Figure 5.5) (Klein 1974). Melatonin is produced and released from the pineal gland almost exclusively during the nighttime, with day light suppressing melatonin secretion, resulting in a cyclic pattern of melatonin secretion.

The pineal gland, and its hormone melatonin, mediate photoperiodic time measurement in mammals. Pinealectomy blocks responsiveness to

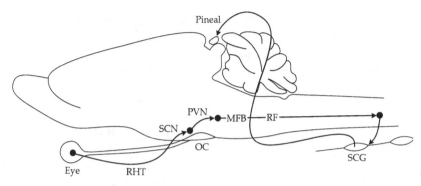

Fig. 5.6. *Photoperiodic time measurement requires an intact circuit among the pineal gland, SCN, and the eyes.* Environmental light enters the SCN from the lateral eyes via the retinohypothalamic tract (RHT). Projections from the SCN travel to the paraventricular nuclei (PVN), from the PVN to the medial forebrain bundle (MFB), and from the MFB to the superior cervical ganglion (SCG). Postganglionic noradrenergic fibers project back into the brain and innervate the pineal gland, where neural information is transduced into a hormonal message. OC = optic chiasm; RF = reticular formation (From Klein et al. 1983.)

photoperiod in every mammalian species studied. Information about environmental light arrives to the brain via the lateral eyes in mammals. A nonvisual neuronal pathway called the retinohypothalamic tract carries light information directly to the SCN of the hypothalamus (Figure 5.6). The SCN are the primary mammalian biological clocks. From the SCN, photoperiod information actually leaves the brain and synapses in the superior cervical ganglion in the spinal cord. Postganglionic noradrenergic fibers eventually project back into the brain and innervate the pineal gland, where neural information is transduced into a hormonal message. Melatonin, an indole-amine hormone secreted rhythmically by the pineal, can induce photoperiodic responses when administered in a number of ways. Nevertheless, particular aspects of the responses observed following constant release implants, daily injections, or daily infusions of melatonin have led to some controversy regarding the significance of various temporal characteristics of melatonin secretion. Melatonin is normally secreted in a circadian fashion, with an extended peak occurring at night and basal secretion during the day. The duration of this nocturnal "peak" varies inversely with day length in several species, including humans. There has been significant controversy regarding the relative importance of the phase versus the duration of the nocturnal elevation of melatonin secretion from the pineal gland in the mediation of this gland on photoperiodic responsiveness. Results obtained in sheep and in Siberian hamsters strongly favor the overriding importance of the duration of the melatonin peak. When pinealectomized hamsters or sheep were infused with melatonin on a daily basis, the types of responses elicited depended on the duration, but not on the phase, of the daily infusion (Carter & Goldman 1983; Bittman & Karsch 1984). For example, when male hamsters received melatonin for 6 hours or less each day, they exhibited functional testes (a long-day response); when melatonin was administered in daily infusions of an 8-hour duration or more, the animals inhibited reproductive function (a short-day response). More recent data have been obtained from similar studies in several other mammalian species; in each instance, the results support the conclusions reached in the studies using hamsters and sheep, viz. that the duration of melatonin is the critical physiological parameter providing photoperiod information.

5.3. Photoperiod, Melatonin, and Immune Function

5.3.1. Photoperiod and Immune Function

Some of the earliest experimental studies implicating melatonin in changes in immune function involved the functional suppression of melatonin concentrations via experimental manipulations of the photoperiod. As previously described, the duration of melatonin secretion is extended during the night and at its nadir during daylight; furthermore, light suppresses melatonin release by the pineal gland. Thus, a "functional pinealectomy" can be induced by exposing experimental animals to constant light to inhibit melatonin synthesis. Laboratory rats maintained in constant light for 4 weeks demonstrate decreased thymic mass by more than 50%, compared with long-day values (Mahmoud et al. 1994). When animals are maintained in constant dark, however, thymic mass is increased by 300%, with the majority of the increase seen in the lymphatic tissue of the thymic medulla, as well as in the total number of thymocytes (Mahmoud et al. 1994). It is important to note that laboratory rats are not traditionally considered photoperiodic animals (i.e., they do not undergo gonadal regression or changes in body mass in short days) and constant dark does not alter steroid hormone concentrations in this species (c.f., Nelson et al. 1994; Heideman & Sylvester 1997). Thus, the effects of constant lighting conditions on thymic mass and presumably thymic function are likely due to a direct effect of melatonin on immune function.

Although the effect is not as dramatic as constant dark, maintaining animals in short photoperiods (e.g., LD 8:16) also prolongs the duration of melatonin secretion relative to long-day animals (e.g., LD 16:8) and can affect immune function. Thus, changes in immune function that occur in response to maintaining animals in short days are likely related, either directly or indirectly, to elevated melatonin secretion. Along these lines, there is increasing evidence of short-day enhancement of immune function in a variety of rodent species (reviewed in Nelson & Demas 1996). For example, one of the earliest studies to report photoperiodic effects on immunity demonstrates that rats maintained in short days have a moderate increase in splenic mass, compared with long-day housed animals (Wurtman & Weisel 1969). This effect is fairly robust, and short-day increases in splenic mass have since been reported in deer mice (*Peromyscus maniculatus*) (Vriend & Lauber 1973; Demas & Nelson 1996), and Syrian hamsters (Brainard et al. 1985; Vaughan et al. 1987). In addition, total splenic lymphocyte numbers and macrophage counts are significantly elevated in short-day housed Syrian hamsters, compared with long-day animals (Brainard et al. 1985). It is important to note that photoperiod does not affect all aspects of immune function in these studies; humoral immunity, as assessed by serum antibody concentrations, is unaffected by changes in day length in deer mice or Syrian hamsters. Photoperiodic effects on immunity also extend to changes in total cell counts. For example, lymphocyte numbers and total white blood cell counts are elevated in deer mice housed in short days, compared with long-day housed animals (Blom et al. 1994). This effect does not seem to be due to nonspecific changes in general hematocrit values, because other nonimmune cell types are unaffected by photoperiod (Blom et al. 1994). Measures of innate immunity (i.e., neutrophil counts) were also unaffected by photoperiodic manipulations (Blom et al. 1994). Maintenance of animals in short days can enhance cell-mediated immune function as well; deer mice housed in short days display enhanced splenocyte proliferation to the T-cell mitogen concanavalin A (Con A), compared with long day animals (Demas et al. 1996, 1997b; Demas & Nelson 1998a, b).

Although the studies discussed previously in rats suggest that enhancement of immune function can occur in species that are reproductively unresponsive to photoperiod, more recent data suggest that reproductive responsiveness is indeed important for the immune-enhancing effects of short days in some species. For example, cell-mediated immune function was examined in two subspecies of deer mice maintained in long and short days. Deer mice originating from high latitudes (*Peromyscus maniculatus bairdii*) typically respond to short day lengths by regressing their reproductive systems and undergoing several other winter adaptations (e.g. body mass, pelage changes). In contrast, *Peromyscus maniculatus luteus* originating from low latitudes typically are unresponsive to short days, with the majority of animals from this subspecies maintaining their reproductive systems in short days. Both subspecies were maintained in short days for 10 weeks, and then splenocyte proliferation was examined using the T-cell mitogen Con A (Demas et al. 1996). As expected, most of the *P. m. bairdii* responded to short days by regressing their reproductive systems, whereas very few of the *P. m. luteus* responded in kind. Interestingly, splenocyte proliferation was higher in short-day *P. m. bairdii*, compared with long day housed animals; no such increase occurred, however, in short-day housed *P. m. luteus* (Demas et al. 1996). Because lack of responsiveness to photoperiod is believed to be due to reduced sensitivity to the short-day melatonin signal rather than any general change in the pattern of melatonin secretion in this species, the effect of exogenous melatonin was also examined in these two subspecies. Similar to the results of the photoperiodic manipulation, exogenous melatonin treatment triggered gonadal regression in most of *P. m. bairdii*, but none of the *P. m. luteus* underwent gonadal regression. Furthermore, melatonin-treated *P. m. bairdii* also demonstrated enhanced immune function relative to untreated mice, whereas melatonin-treated *P. m. luteus* did not display any enhancement of immune function, compared with the untreated control animals (Demas et al. 1996). Surprisingly, however, the degree of enhanced immune function did not correlate with the extent of gonadal regression; the only important parameter was whether animals responded or not. These results suggest that, in at least some species, responsiveness to short days and likely the short-day pattern of melatonin secretion is important in eliciting changes in immune function. In other words, reproductive and immune responsiveness to day length, and likely melatonin, appear to be uncoupled in some species (Nelson & Drazen 1999). Collectively, these results clearly have important implications for melatonin treatment in humans, who are generally not considered responsive to day length cues (Bronson 1995).

Melatonin is only one of many hormones that are influenced by day length. Exposure to short day lengths changes the pattern of melatonin secretion; long nights evoke increased duration of melatonin secretion that inhibits reproductive function in many rodent species. Inhibition of reproductive function causes severe reductions in several hormones, including sex steroids, gonadotropins, GnRH, and prolactin. Importantly, both estrogens and androgens can affect immune function (see Chapter 7). Testosterone, the primary androgenic hormone secreted by males of most species, typically suppresses immune function (reviewed in Grossman 1984; Konstadoulakis et al. 1995; Klein 2000a, b). Thus, increased immune function in rodents housed in short days or treated with melatonin may be caused by reduced circulating testosterone concentrations; in other words, there may be a "release" from the immunosuppressive effects of androgens, rather than a direct effect of day length on immune function. In contrast to testosterone, estradiol, the primary estrogen secreted by females, generally enhances immune function (Grossman 1984; Klein 2000a, b). If photoperiodic changes in immune function are due to fluctuations in estrogen, then female rodents housed in short days, that have low estrogen concentrations, should exhibit reduced immune function compared with long-day animals. Alternatively, if the effects of short days or melatonin on immune function are independent of changes in gonadal steroid hormones, then immune enhancement should occur in both males and females housed in short days, regardless of circulating concentrations of gonadal steroids. Recent evidence supports the latter hypothesis. Specifically, both male and female deer mice housed in short days, display enhanced cellular immune function, compared with long-day housed animals, regardless of gonadal status (Demas & Nelson 1998a, b). Gonadectomized male and female animals that lack circulating concentrations of testosterone and estradiol, respectively, display similar enhancement of cellular immune function, compared with intact animals. Additionally, exogenous hormone replacement does not change this effect (Demas & Nelson 1998a, b). Furthermore, the photoperiodic effects on immune function do not appear to be due to changes in glucocorticoid concentrations, because corticosterone concentrations do not differ between short- and long-day mice (Demas & Nelson 1998a, b). Thus, it appears that the immunoenhancing effects of short days are probably caused by direct effects of melatonin, although sex steroids can directly affect lymphocyte proliferation.

In a recent study, Siberian hamsters were gonadally intact, gonadectomized, or gonadectomized but receiving steroid hormone replacement therapy (Bilbo & Nelson 2001a, b). Females exhibited the expected increase in antibody production over males, independent of hormone treatment condition, whereas gonadectomized male and female hamsters decreased lymphocyte proliferation in response to the T-cell mitogen, phytohemagglutinin (PHA), compared with intact animals, and this effect was reversed in gonadectomized hamsters following testosterone and estradiol treatment, respectively. In a related experiment, testosterone, dihydrotestosterone, and estradiol all enhanced cell-mediated immunity *in vitro*, suggesting that sex steroid hormones may be enhancing immune function through direct actions on immune cells. In response to an acute mitogen challenge of lipopolysaccharide (LPS), hamsters significantly suppressed lymphocyte proliferation to PHA in intact males but not females, suggesting that males may be less reactive to a subsequent mitogenic challenge than females. Contrary to evidence in many species – such as rats, mice, and humans – these data suggest that sex steroid hormones enhance immunity in both male and female Siberian hamsters (Bilbo & Nelson 2001a, b).

At an ultimate level of analysis, why is it adaptive for immune function to be enhanced in short "winter-like" days? As stated in Chapter 1, our working hypothesis is that short-day enhancement of immune function has evolved as an adaptive response to counter the immunosuppressive effects of environmental stressors present during the winter (e.g., low ambient temperatures, reduced food availability) (Nelson & Demas 1996). To test this hypothesis, the interactive effects of multiple environmental factors on immune function were initially examined in male deer mice (Demas & Nelson 1996). Specifically, the effects of day length and ambient temperature on immune serum IgG concentrations were examined. As expected, animals housed in short days had higher serum IgG concentrations, compared with long-day housed mice. Additionally, mice housed in low ambient temperatures had increased corticosterone concentrations and decreased IgG concentrations, compared with animals housed at thermoneutrality (Demas & Nelson 1996). Interestingly, the short-day enhancement of immune function counteracted the immunosuppressive effects of low ambient temperatures such that short-day mice kept in low temperatures have similar IgG concentrations, compared with mice kept in long days at mild ambient temperatures. These data lend support to the idea that short-day enhancement of immunity might have evolved to counter environmental stressors. It is important to note, however, that basal (i.e., unstimulated) levels of IgG were measured in the previous study; the effects of environmental variables on specific antibody concentrations were not examined. To test immune function more directly, a recent study examined photoperiodic changes in lymphocyte proliferation to Con A and the interactions with both ambient temperatures and food availability (Demas & Nelson 1998). Similar to the findings reported previously, deer mice housed in short days had higher lymphocyte proliferation compared with long-day housed animals. Both low ambient temperatures and reduced food availability suppressed immune function in long-day animals. Maintenance in short days could counteract the immunosuppressive effects of either low temperatures or reduced food availability, but not the effects of both factors (Demas & Nelson 1998). Thus, it appears that enhanced immune function in short days can counteract some, but not all, of the effects of winter stress-induced immune suppression. Because there are many environmental stressors during winter, these results may explain why immune function is commonly suppressed in winter in field studies, while enhanced immune function is the norm for short-day housed animals in controlled laboratory settings.

Changes in day length not only alter immune function of adult animals, but can also impact on the immune systems *in utero*. Photoperiodic information can be transmitted from the mother to the pups *in utero* and can subsequently affect postnatal immune function. For example, deer mice gestated in short days, but raised in long days, display enhanced immune function, compared with animals gestated in long day, regardless of the photoperiod under which they were subsequently raised (Nelson & Blom 1994). Additionally, lymphocyte numbers, white blood cell counts, and neutrophil numbers are higher in deer mice gestated in short days, compared with mice gestated in long days. Although the exact function of *in utero* transfer of photoperiodic effects on immunity remains unknown, it is likely that this phenomenon represents an adaptation to enhance survival of offspring born late in the breeding season (Nelson & Demas 1996). Recent research in Siberian hamsters suggests that short days selectively enhance some immune parameters while suppressing other, less critical immune parameters. For example, natural killer cell cytotoxic activity, as well as spontaneous blastogenesis, are enhanced in short-day hamsters, compared with hamsters maintained in long days (Yellon et al. 1999). By contrast, both phagocytosis and oxidative burst activity by granulocytes and monocytes are suppressed in short days. Specific immune parameters may be selectively enhanced in short days to allow animals to successfully cope with immune challenges prevalent during winter. In a similar study of Siberian hamsters, short-day exposure suppressed lymphocyte proliferation, a measure similar to spontaneous blastogenesis (Prendergast & Nelson 2001). To determine whether melatonin influences cell-mediated immunity through a direct action on lymphocyte proliferation, *in vitro* responsiveness to mitogens and melatonin was assessed in systemic and splenic lymphocytes from adult female Siberian hamsters housed in either long or short days for 13 weeks. Short days provoked reproductive regression and reduced lymphocyte proliferation. Physiological concentrations of melatonin (50 pg/ml) inhibited *in vitro* proliferation of circulating lymphocytes, whereas higher concentrations (>500 pg/ml) were required to inhibit proliferation of splenic lymphocytes. This may reflect the lack of melatonin 1b receptor subtypes in this species. Immunomodulatory effects of melatonin were restricted to lymphocytes from long-day hamsters; *in vitro* melatonin had no effect on circulating or splenic lymphocytes from females in short-days (Prendergast & Nelson 2001). Responsiveness in short-day lymphocytes may be maximized by the already expanded nightly pattern of melatonin secretion in short days. These data support the hypothesis that melatonin acts directly on lymphocytes from long-day hamsters to suppress blastogenesis. The difference between the two studies (Yellon et al. 1999; Prendergast & Nelson 2001) include the time of short-day exposure (6 vs. 13 weeks), sex (male vs. female), and incubation times (96 vs. ~48 hours), and these variables might have affected the final results.

Changes in photoperiod can have significant influences on immune function. Exposure to pathogens or to inflammatory stimuli such as LPS results in activation of the immune system, and the onset of several physiological and behavioral responses that are collectively termed the acute phase response. These responses are ubiquitous among vertebrates, and include fever, iron-withholding, increased slow-wave sleep, reductions in food and water intake, and reduced interest in parental, sexual, and other social interactions (Hart 1988; Avitsur & Yirmiya 1999). Rather than being nonspecific manifestations of illness, these responses are initiated and maintained by the host via endogenous proinflammatory cytokines [e.g., interleukin (IL)-1], and are often critical to survival (Kluger et al. 1998). During times of energetic shortage, however, it may be detrimental for individuals to prolong energetically demanding symptoms, such as fever and anorexia. During winter, animals often must cope with reduced food availability coincident with low ambient temperatures and increased energetic needs. Seasonal energy shortages are predictable, and individuals may adjust their immune responses prior to winter by using day length to anticipate these energetically demanding conditions. If the expression of sickness behaviors is constrained by energy availability, then one might expect that cytokine production, fever, and anorexia would be attenuated in infected Siberian hamsters housed under simulated winter photoperiods.

Siberian hamsters were housed in either short- (LD 10:14) or long-(LD 14:10) day lengths for 10 weeks, and assessed for cytokine production, anorexia, and fever after injections of LPS (Bilbo et al. 2001). Short days attenuated activation of the immune response to LPS in Siberian hamsters, by decreasing the production of IL-6 and IL-1β, and diminishing the durations of fever and anorexia (Figure 5.7). Short-day exposure in Siberian hamsters also decreased the ingestion of dietary iron, a nutrient vital to bacterial replication (Bilbo et al. 2001). Taken together, short-day lengths attenuate the symptoms of infection, known as sickness behaviors (Hart 1988), presumably to optimize energy expenditures and survival outcome.

Experimental manipulation of photoperiod has been an extremely useful model for assessing the role of natural changes in the pattern of melatonin secretion on immune function in seasonally breeding animals. In almost all cases, photoperiodic changes in immunity may be due to coincident changes in the pattern of melatonin secretion by the pineal gland. Although this is a

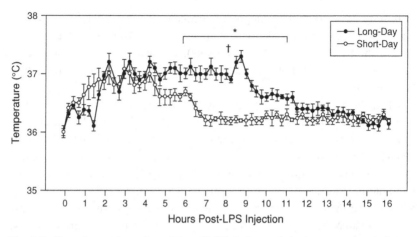

Fig. 5.7. *Short days reduces fever.* Mean (±SEM) change in body temperature in long-day and short-day hamsters following intraperitoneal injections of 50 μg of LPS. Fever duration is attenuated in short-day hamsters. Fever latency and amplitude do not differ as a function of day length. *$p < 0.05$. †Corresponds to when animals were handled. (From Bilbo et al. 2001.)

plausible hypothesis, studies examining photoperiodic changes in immunity are not a substitute for studies that examine the effects of melatonin on immune function.

5.3.2. Melatonin and Immune Function

In addition to its role in mammalian photoperiodism, the pineal gland and melatonin also play a role in immune function (reviewed in Maestroni 1993; Nelson & Demas 1996). Removal of the source of endogenous melatonin (i.e., pinealectomy) can alter immune function in many animal species. As described previously, melatonin is secreted in a circadian pattern, with the highest blood concentrations observed during the nighttime, and the nadir in melatonin blood concentrations observed during the light portion of the day. Functional pinealectomy via pharmacological inhibition of pineal gland function with the β-adrenergic antagonist propranolol (which inhibits melatonin synthesis by pinealocytes) or by p-chlorophenylalanine (which inhibits serotonin, the precursor to melatonin in the pineal gland) results in atrophy of lymphatic tissue, as well impairment in somatic growth and antibody production (Maestroni & Pierpoli 1981). Exogenous melatonin administration restores normal immune functioning in these drug-treated animals (Maestroni et al. 1987).

Immune function, in common with many other physiological and behavioral processes, undergoes circadian changes. These changes appear to be adaptive responses that presumably serve to optimize the immune response to defend against pathogens, the prevalence of which may fluctuate throughout a 24-hour period (Haus & Smolensky 1999). Along these lines, there is considerable evidence that specific immune responses vary across a 24-hour period. For example, immune responsiveness, as well as circulating lymphocytes, fluctuate according to a 24-hour rhythm (Fernandes et al. 1976; Abo et al. 1980). Additional evidence suggests that the 24-hour fluctuations in immune parameters display phase relationships that differ from the pattern of circulating glucocorticoid concentrations (Eskola et al. 1976; Skwarło-Sonita et al. 1983, 1987; Lévi et al. 1992). Thus, despite the well-established immunosuppressive effects of glucocorticoids on immune function, the circadian fluctuations in immunity appear to be independent of glucocorticoid secretion (Skwarło-Sonita 1996). Although the circadian fluctuations in immune cell function are well-established, the biological significance of these fluctuations remains unspecified. The first evidence of a functional relationship between pineal melatonin and immunity was the demonstration of a disorganization of thymic tissue after pinealectomy (Csaba et al. 1977). Several studies have since demonstrated more direct links between pinealectomy and immune function. For example, neonatal pinealectomy impairs murine antibody-dependent cellular cytotoxicity (ADCC) in mice (Vermeulen et al. 1993). Additionally, the increases in splenic mass of male Syrian hamsters housed in short days are reversed after pinealectomy (Vaughan et al. 1987). Pinealectomy of Wistar rats at 6 weeks of age impairs immunity; antibody production against bovine serum albumin is reduced, and there is a delay in skin graft rejection (Jankovic et al. 1970). Additionally, several studies illustrate that pinealectomy reduces humoral immunity (Kuci et al. 1983), natural killer (NK) cell activity, IL-2 production (Del Gobbo et al. 1989), and collagen II-induced autoimmunity (Hansson et al. 1992) in mice. These data are especially intriguing because the pineal glands of the inbred strains of mice used in these studies (e.g., BALB/c, C57BL6J) reportedly fail to synthesize functional melatonin (Ebihara et al. 1986; Goto et al. 1989). Recent studies, however, have demonstrated the existence of very low melatonin concentrations in these mouse strains. Alternatively, immunological effects may be due to other nonmelatonin pineal products, such as arginine vasotocin (Maestroni & Conti 1991).

The pineal gland receives input from the peripheral immune system (Skwarło-Sonita 1996). To examine the differential contributions of direct sympathetic innervation of immune target tissue, as well as the influences of adrenal medullary catecholamines, to photoperiodic changes in immune function, male Siberian hamsters received either bilateral adrenal demedullations or sham surgeries and were maintained in long days (LD 16:8) or short days (LD 8:16) (Demas et al. 2001). In a related experiment, animals received either surgical denervation of the spleen or sham surgeries and were then housed in long or short days. Short-day animals displayed reduced humoral immunity, compared with long-day animals. Adrenal demedullations reduced immune function, but only in long-day hamsters. In contrast, splenic denervation reduced humoral immunity, but only in short-day animals. Splenic norepinephrine content was increased in short days and by adrenal demedullations. Norepinephrine content was markedly reduced in denervated hamsters, compared with sham-operated animals. Collectively, these results suggest that the sympathoadrenal system is associated with photoperiodic changes in immune function (Demas et al. 2001).

Much of the research has focused on the role of cytokines in mediating the photoperiodic effects on immune function. For example, interferon (IFN)-γ can enhance melatonin production when added to cultured rat pinealocytes (Withyachumnarnkul et al. 1990) or reverse the enhancement in NAT activity after administration of a β-adrenergic agonist (Withyachumnarnkul et al. 1991). In addition, administration of recombinant human IL-1β can decrease circulating melatonin concentrations in rats in a dose-dependent fashion (Mucha et al. 1994). Furthermore, this effect can be ameliorated by administration of anti-human IL-1β receptor antibody (Mucha et al. 1994). Lastly, the glycoproteins granulocyte colony-stimulating factor and granulocyte-macrophage colony-stimulating factor, which are secreted by the immune system in response to activation of specific cells, can increase melatonin synthesis by the pineal gland both *in vitro* and *in vivo* (Zylinska et al. 1995). These intriguing data, along with the relatively wide distribution of receptors for specific cytokines throughout the mammalian brain, suggest that cytokines play a critical role in a bidirectional communication between the pineal gland and the immune system (Cardinali et al. 2000).

Melatonin and Immunoregulation

The role of melatonin in mediating specific immune responses was first demonstrated by eliminating endogenous melatonin both functionally and pharmacologically. For example, maintenance of mice under constant light suppresses melatonin secretion and this model can be used to examine the effects of the absence of normal melatonin secretion on immune function. When mice are bred continuously under this lighting regimen, mice grow poorly after about three generations and display an impaired ability to mount antibody responses to T-cell-dependent antigens. These mice also show a depletion of T-cells in the cortex of the thymus gland, as well as atrophy of the white pulp of the spleen (Maestroni & Pierpaoli 1981). There are many other physiological perturbations that can occur in constant light, including disturbances in sleep and food intake. Disruptions in these other physiological processes, rather than the suppression of melatonin, may provide alternative explanations for reduced immunity in constant light. The use of pharmacological melatonin inhibitors, either propranolol or *p*-chlorophnylalanin, has revealed that suppressed melatonin itself is the likely cause of reduced immune functioning. For example, administration of these pharmacological agents reduces the primary antibody response to sheep red blood cells (SRBCs) and also leads to reductions in autologous mixed lymphocyte reactions; these reductions in immune function can be completely reversed by evening administration of melatonin (Maestroni et al. 1986). Additionally, both the primary antibody response to T-cell-dependent antigens, as well as mixed lymphocyte reaction, are reduced by evening treatment with propranolol in mice. Notably, these effects do not occur after morning treatment with propranolol (Maestroni et al. 1986, 1987). Thus, a circadian sensitivity to the immunoenhancing properties of melatonin may exist. In hamsters, injections of propranolol decreases spleen mass and reduces T-cell blastogenesis stimulated by Con A (Champney & McMurray 1991). Exogenous melatonin treatment reverses the effects of propranolol on splenic mass and blastogenesis (Champney & McMurray 1991).

In summary, the absence of melatonin, at least during a specific time of the day, can adversely affect immune function in several rodent species. These studies, however, have focused on immunological impairments observed in the absence of normal melatonin. Although it is likely that melatonin is responsible for the immune enhancement, these studies only test this hypothesis indirectly.

Many studies have examined the effects of exogenous melatonin administration on immune function in pineal intact animals (reviewed in Maestroni 1993). For example, exogenous melatonin (0.01–10 mg/kg) can enhance both the primary and secondary responses to SRBCs in mice, but only when the melatonin injections occurred in the late afternoon. Melatonin given at this time also increased the secondary, but not the primary, response by cytotoxic T-cells in response to infection with vaccinia virus (e.g., cowpox) (Maestroni et al. 1988). High doses of melatonin ($\sim$200 mg/kg), however, can actually suppress these immunological responses (Maestroni et al. 1988). Melatonin treatment can also restore impaired T-helper cell activity in immunocompromised mice (Caroleo et al. 1992). Antigen presentation by splenic macrophages to T-cells is enhanced by melatonin treatment (Pioli et al.

1993). Melatonin administration to normal and immunocompromised mice significantly increases antibody responses, as well as enhances antigen presentation by macrophages (Bubenik et al. 1998). Melatonin can partially inhibit cyclic AMP production in human lymphocytes and reduce their proliferative ability, but only at large, pharmacological doses (Rafii-El-Idissi et al. 1995). Exogenous melatonin treatment in normal animals can increase the number of splenic plaque-forming cells, as well as enhance both the primary and secondary antibody responses to SRBCs (Maestroni et al. 1987). Melatonin has been reported to both enhance (Caroleo et al. 1992) and suppress (Del Gobbo et al. 1989) NK cell activity. Melatonin can also enhance IL-2 production (Del Gobbo et al. 1989; Caroleo et al. 1992; Sze et al. 1993), ADCC (Giordano & Palmero, 1991), lymphocyte blastogenesis (Franschini et al. 1990; Champney & McMurray 1991; Sze et al. 1993), as well as increase the T-helper/T-Cytotoxic ratio (i.e., $CD4^+/CD8^+$ ratio) (Lissoni et al. 1987). Lastly, melatonin enhances mitogen-induced T-cell blood lymphocyte and T-cell and B-cell splenocyte proliferation in male broiler chickens (Kliger et al. 2000).

All of these studies suffer from the limitation that it is difficult to parse out the influences of melatonin from the other hormonal effects *in vivo*. A recent study sought to tease apart the direct and indirect effects of melatonin on one aspect of immune function by examining the influence of *in vitro* melatonin on splenocyte proliferation in female prairie voles held in long (LD 16:8) or short (LD 8:16) days (Drazen et al. 2000) (Figure 5.8). Female prairie voles have undetectable sex steroid hormones if not induced into estrus by male chemosensory cues (Carter et al. 1980). Splenocyte proliferation in response to the T-cell mitogen Con A was enhanced by the addition of melatonin *in vitro*, compared with cultures receiving no melatonin. Body mass increased in short-day housed prairie voles, indicating that the animals were responsive to photoperiod. However, photoperiod did not affect splenocyte proliferation in the present study. These results support the hypothesis that melatonin exerts a direct effect on splenocyte proliferation, potentially via high-affinity melatonin receptors localized on splenocytes. The findings also indicate that, irrespective of photoperiod, melatonin exerts direct effects on splenocytes to enhance immune function. Similar results were obtained in male prairie voles (Kriegsfeld et al. 2001).

Melatonin, added to *in vitro* cell cultures of murine spleen cells, can enhance splenocyte proliferation to Con A, a response that can be reduced with the addition of luzindole, a melatonin 1a (MT1) receptor antagonist (Drazen et al. 2001a). Luzindole is not a pure MT1 receptor antagonist, but its affinity for the MT1 receptor is ~25 times higher than for the

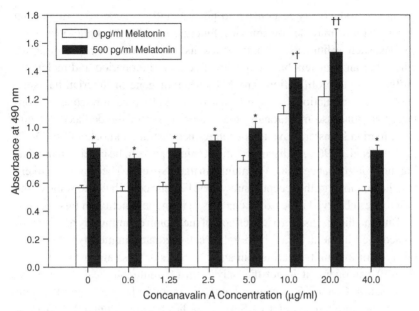

Fig. 5.8. *In vitro enhancement of splenocyte proliferation by melatonin.* This graph shows the mean ($\pm$SEM) proliferation values (represented as absorbance units) from female prairie voles in response to stimulation with Con A. Cross-hatched bars represent proliferative values of individuals that received 500 pg/ml melatonin in each assay well. Note that proliferative values are elevated after *in vitro* melatonin treatment at all Con A concentrations. (Reprinted from the *Journal of Pineal Research*, vol. 28, D. L. Drazen, S. L. Klein, S. M. Yellon, and R. J. Nelson, "In vitro melatonin enhances splenocyte proliferation in female prairie voles (*Microtus ochrogaster*)," pp. 34–40, copyright © 2000.)

melatonin 1b (MT2) receptor (Dubocovich et al. 1998). Because of its possible effects on both melatonin receptors, another study was conducted to examine the contribution of the specific melatonin receptor subtypes, MT1 and MT2, in mediating melatonin-induced enhancement of cell-mediated and humoral immune function in mice (Drazen and Nelson 2001). Melatonin enhanced both splenocyte proliferation and anti-KLH IgG concentrations in both wild-type and mice lacking a functional gene for melatonin 1a receptor (MT1$-/-$), suggesting that the MT1 receptor does not mediate these responses. In addition, luzindole attenuated melatonin-induced enhancement of splenocyte proliferation in both wild-type and MT1$-/-$ mice. Taken together, these results implicate the MT2 receptor subtype in mediating melatonin-induced enhancement of immune function and provide the first evidence for a specific melatonin receptor subtype involved in melatonin-induced immune enhancement (Drazen et al. 2001a; Drazen and Nelson 2001). Furthermore, these mice are not reproductively responsive to photoperiod or melatonin, suggesting that melatonin may act on immune function of other "nonphotoperiodic" species, such as humans. Also, these results may provide an explanation for why Siberian hamsters have opposite effects of photoperiod and melatonin on lymphocyte proliferation. The MT2 receptor gene in Siberian hamsters encodes a nonfunctional receptor; that is, the MT2 receptor sequence contains two nonsense mutations in the coding region of the deduced protein, and Siberian hamsters have thus been termed "nature's knockout" for MT2 (Weaver et al. 1996). As discussed, melatonin appears to be crucial in mediating short-day enhancement of immunity in many species. Therefore, it is likely that the mutation of the receptor subtype MT2 gene contributes to some of the differences observed in seasonal changes in immune function in this species.

Thus far, the discussion of the effects of melatonin on immunity has focused on acquired immunity (i.e., humoral or cell-mediated immunity). Although this type of immunity is critical in an organism's defense against potentially harmful pathogens, it is not the only form of immunological defense. As discussed in Chapter 2, innate or natural immunity also plays an important role in immunological defense. Several studies have examined the effect of melatonin on innate immunity (reviewed in Maestroni 1993). The effects of oral melatonin administration on innate immunity have been examined in humans. Unlike the studies described in rodents, oral melatonin had no effect on NK cell cytotoxicity against herpes-simplex infection (Maestroni & Conti 1991). Whether the difference between these findings and those reported in rodents represents differences in immune responsiveness to melatonin or rather simply differences in the dosage and route of administration of the hormone remains unspecified.

The effects of melatonin on cytokine production have also been examined. For example, both IFN-γ and IL-2 concentrations can be influenced by melatonin (Angeli et al. 1988; Del Gobbo et al. 1989; Caroleo et al. 1992). These Th1 cytokines are secreted by activated T-lymphocytes and stimulate NK cell activity. Exogenous melatonin can stimulate IFN-γ production from activated lymphocytes in humans (Muscettola et al. 1992). Interestingly, IFN-γ can also increase the synthesis of melatonin in the pineal gland (Withyachumnarnkul et al. 1990). In contrast, IL-2 administration can abolish the melatonin rhythm in humans (Lissoni et al. 1988). Melatonin can also affect other cytokines such as tumor necrosis factor (TNF)-α. For example, melatonin administration inhibits the production of both IFN-γ and TNF-α in cultures of human peripheral blood lymphocytes when incubated with PHA; melatonin alone, however, does not affect these cytokines (DiStefano & Paulescu 1994). The dose of melatonin is an important factor in immunological studies. Low melatonin doses stimulates mitogen-induced INF-γ release by peripheral blood lymphocytes; in contrast, high doses inhibited release (Muscettola et al. 1994). Inhibition of INF-γ by melatonin subsequently stimulates blood mononuclear leukocytes to synthesize melatonin, suggesting the existence of a feedback loop (Arzt et al. 1988; Finocchiaro et al. 1991; Skwarlo-Sonita 1996). Melatonin activates human Th1 lymphocytes by increasing the production of IL-2 and IFN-γ *in vitro*; Th2 cells do not appear to be responsive to melatonin because IL-4 is not affected by melatonin (Garcia-Maurino et al. 1997).

Melatonin-Glucocorticoid Interaction and Immune Function

Although several studies indicate direct effects of melatonin on immune function, it remains possible that the immunoenhancing effects of melatonin represents a physiological reversal of the stress-induced secretion of glucocorticoids (e.g., corticosterone, cortisol). Specifically, melatonin may increase the general sensitivity of the hypothalamo-pituitary-adrenocortical axis to negative feedback (Demas et al. 1997). Consistent with this hypothesis, melatonin administration inhibits the reduction of antibody production induced by either restraint stress or treatment with glucocorticoids (Maestroni et al. 1988). In addition, maintaining deer mice in short days, which functionally increases endogenous melatonin production, antagonizes the stress-induced secretion of corticosterone after administration of the metabolic stressor 2-deoxy-D-glucose and also increases splenocyte proliferation to Con A (Demas et al. 1997a, b). Similar, antistress effects have been reported after administration of a melatonin agonist in Syrian hamsters (Bolinger et al. 1996). Melatonin treatment can also completely block restraint stress-induced mortality after infection with sublethal doses of encephalomyocarditis virus in inbred mice (Maestroni et al. 1988). Additionally, melatonin prevents the reduction in thymic mass induced by increased corticosterone secretion in mice infected with the vaccinia virus. These results are consistent with the findings that exogenous melatonin reduces viremia and postpones disease onset and even death in mice infected with Semliki Forest virus (Ben-Nathan et al. 1995). Melatonin also plays an important protective role against adrenergic stress. Chronic treatment of rats with epinephrine or norepinephrine suppresses lymphocyte proliferation in the presence of a β-adrenergic antagonist; this effect appears to be due to specific activation of the α_2-adrenergic receptor subtype (Felsner et al. 1992). Blockade of α_2-receptors with chlonadine also leads to immunosuppression, but only with simultaneous blockade of the β-adrenergic receptors (Felsner et al. 1995). Thus, α_2- and β-adrenergic receptors may serve antagonistic roles in mediating immune function. Exogenous melatonin treatment blocks the immunosuppressive effects of both catecholamine and chlonadine administration in rats (Liebmann et al. 1996). Melatonin may protect lymphocyte effector functions from α_2-adrenergic suppression (Liebmann et al. 1996). These results may explain circadian as well as seasonal rhythms in immune function that appear to follow closely changes in melatonin secretion (reviewed in Nelson & Demas 1996). Consistent with this idea is the observation that exogenous melatonin can enhance antibody-dependent cellular cytotoxicity (ADCC), but only during the summer (Giordano & Palermo 1991).

Direct Actions of Melatonin on Immune Function

Consistent with the effects of melatonin and pinealectomy on immune function and disease, receptors for melatonin have been isolated on specific immune cells (reviewed in Calvo et al. 1995; Nelson & Demas 1996). For example, melatonin receptors have been identified on circulating lymphocytes (Figure 5.9) (Pang & Pang 1992; Poon & Pang 1992; Liu & Pang 1993; Pang et al. 1993; Calvo et al. 1995), thymocytes, and splenocytes (Lopez-Gonzales et al. 1992; Martin-Cacao et al. 1993; Rafii-El-Idrissi et al. 1995) in mice. Melatonin binding sites have also been localized in the spleens of guinea pigs, ducks, chickens, and pigeons, as well as the thymus and bursa of Fabricius in birds (Calvo et al. 1995). In almost all cases reported, the K_d values for these receptors are between 10–100 pM; this range is similar to the values found

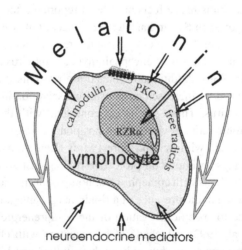

Fig. 5.9. *Possible mechanism for a direct effect of melatonin on lymphocytes via nuclear melatonin receptors.* Small arrows indicate interactions at the cellular level, whereas the large arrow indicates indirect effects. RZR, retinoid orphan receptor; PKC, protein kinase C. Liebmann et al. 1997.

in rat and hamster brains (Calvo et al. 1995). The low K_d values reported for melatonin binding on immune targets suggest the presence of high-affinity receptors at these sites. Thus, these receptors appear to serve a physiological role rather than merely being an artifact of nonspecific binding.

Specific melatonin receptors have been localized on circulating lymphocytes in humans using a radioactively tagged ligand, [^{125}I]melatonin (Lopez-Gonzalez et al. 1992). There is also evidence for low-affinity melatonin binding sites on human granulocytes (Lopez-Gonzalez et al. 1992). These findings suggest that melatonin may exert a direct effect on specific immune cells. Melatonin is a lipid-soluble molecule that is readily available to immune tissue. Recent evidence suggests that melatonin can interact with immune tissue in a wide range of mammalian and avian species. The localization of high-affinity melatonin binding sites on lymphoid tissue suggests an important physiological role of melatonin in the regulation of immune function. Although very little is known about the adaptive significance of melatonin receptors on peripheral lymphoid tissue, it is likely that the presence of these receptors provides an adaptive advantage by allowing for a potentially rapid increase in immune function during times when fitness may be compromised (e.g., winter, migration, malnutrition).

Melatonin and Endogenous Opioids

An increasing number of studies have focused on the role of endogenous opioids in mediating the effects of melatonin on immune function (Figure 5.10) (reviewed in Maestroni & Conti 1991). The role of the endogenous opioids was first suspected by the finding that administration of the specific μ-opioid receptor antagonist naltrexone reversed the effects of melatonin on stress-induced immunosuppression in a dose-dependent fashion (Maestroni et al. 1988, 1989). Consistent with this hypothesis, receptors for two classes of endogenous opioids, endorphins and enkephalins, have been isolated in lymphocytes (reviewed in Wybran 1985). In addition, immune cells appear capable of secreting endogenous opioids (Smith & Blalock 1981). To examine the role of the endogenous opioid system in regulating the effects of melatonin on immune function, various opioid antagonists have been examined for their ability to counteract the antistress actions of melatonin on immunity. The results of these studies suggest that antagonists (e.g., naltrexone) that act primarily on μ-opioid receptors can attenuate the immune enhancing antistress properties of melatonin, whereas antagonists (e.g., ICI 174,864) that act primarily on δ-receptors do not appear to have any such effects (Maestroni et al. 1987, 1988). Thus, the role of endogenous opioids appears to be due primarily to actions of melatonin on μ-receptors. Both naltrexone and ICI

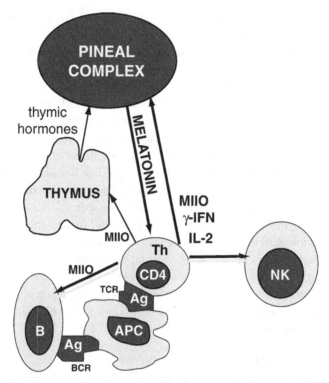

Fig. 5.10. *Interactions between melatonin, the endogenous opioid system and immunity.* (INF, γ-interferon, IL-2, interleukin-2, Th, T-helper cell; TCR, T-cell receptor; BCR, B-cell receptor; APC, antigen-presenting cell; Ag, antigen; B, B-cell; NK, natural killer cell; MIIO, melatonin-induced immuno opioids. (Reprinted from the *Journal of Pineal Research*, vol. 14, G. J. M. Maestroni, "The immunoneuroendocrine role of melatonin," pp. 1–10, copyright © 1993, with permission of Munksgaard International Publishers Ltd., Copenhagen, Denmark.

174,864 have moderately high affinities for κ-receptors, however, so the role of the κ-receptor subtype in mediating immune changes cannot be ruled out.

Several studies illustrate that endogenous opioids can directly affect immune function and may mediate photoperiodic effects on immunity. Administration of endogenous opioids, including β-endorphin and dynorphin, mimics the immune enhancing and antistress effects of melatonin (Maestroni & Conti 1989). In addition, there is a strong circadian component to the effectiveness of opioids on immunity, with evening treatment being more effective than morning administration (Maestroni & Conti 1989). To examine whether melatonin acts by stimulating the release of endogenous opioids, researchers incubated spleen cell cultures with melatonin (Maestroni & Conti 1991). Melatonin is able to activate T-helper cells to release endogenous opioids, which enhance immune function and counteract stress-induced immunosuppression. Further-more, melatonin is also effective in displacing naloxone binding in mouse brains (Maestroni & Conti 1991). These results, coupled with the evidence for direct effects of melatonin on immune function, suggest that melatonin may mediate changes in immunity through several mechanistic pathways.

5.4. Melatonin, Free Radicals, and Aging

In addition to the effects of the pineal gland and melatonin on immune func-tion, there is growing evidence that the pineal gland acts as a "homeostatic headquarters," integrating and coordinating an organism's ability to cope with environmental factors (i.e., light-dark cycle, temperature, stress, pathogens) (Maestroni et al. 1986). In general, the process of aging can be considered as a progressive reduction in the ability of the body to cope with environ-mental factors. Thus, researchers hypothesized that the pineal gland may play an important role in aging. The first evidence for this role of the pineal gland is a marked correlation between aging and reductions in pineal func-tion (see Zeitzer et al. 1999). The circadian rhythm in melatonin secretion dampens with age, and this reduction in rhythmicity has been hypothesized to lead to dysregulation of neuroendocrine and immune functions (Maestroni et al. 1986). It is possible that this dysregulation leads to suppressed immune function and renders the organisms susceptible to a variety of opportunis-tic pathogens. Consistent with this idea, restoring the melatonin rhythm by administration of exogenous melatonin prolongs the lifespan of male mice (Bubenik et al. 1998). Mice given $10\text{-}\mu$ g/ml melatonin daily in the drinking water prolonged life span by ~20%, compared with control animals. Ad-ditionally, the general appearance and body conditions of melatonin-treated animals were reported to be better than control mice. It has been suggested that indoleamine hormones, specifically melatonin, are involved in the regula-tion of electron transfer and act as potent endogenous free-radical scavengers, detoxifying reactive radical intermediates (Reiter et al. 1993; Tan et al. 1993). For example, dietary restriction, which reduces the generation of free radi-cals, preserves pineal structure and function in old age (Reiter 1992; Reiter et al. 1993). Chemical agents that inhibit the accumulation of free radicals also prevent age-associated decline in nocturnal pineal and blood melatonin concentrations in rodents (Scaccianoce et al. 1991).

5.5. Melatonin and Cancer

In addition to the effects on specific immune cell subtypes, the earliest indications of the immunomodulatory capabilities of melatonin stem from research demonstrating dramatic antitumor effects in several mammalian species (reviewed in Liebmann et al. 1997). For example, treatment of humans suffering from cancer with pineal extracts appeared to slow the growth of specific cancers and also improved the quality of life of these patients (Hofstatter 1952). Additionally, pinealectomy in rats accelerated the growth of experimentally administered tumors (Lapin 1976). Since these early studies, a large number of studies have focused on the role of melatonin in mediating the proliferation of both normal and neoplastic cells, as well as tumor growth and development in both human and nonhuman animals (reviewed in Blask 1984). The defining feature of cancer is the uncontrolled division of normal cells leading to abnormal growth (Blask 1984).

Extensive research has demonstrated that the pineal gland is capable of regulating cell proliferation in a variety of tissues. For example, pinealectomy in male rats at an early age increases mitotic cell division in the spleen, small intestine, anterior pituitary, and liver (Bindoni 1971). Additionally, normal rats that are blinded experimentally, and thus do not experience the inhibitory input of light on the pineal, fail to display the normal increase in pituitary weight seen in unmanipulated animals. Taken together, these results suggest that the pineal gland provides a "brake" on cellular proliferation, and removal of this brake leads to increased cell proliferation.

The studies described here suggest a role of the pineal gland and melatonin on normal cell proliferation. Melatonin also has been reported to affect abnormal, or neoplastic, cell proliferation both *in vitro* and *in vivo* (reviewed in Blask 1984). For example, partially purified sheep extracts can completely inhibit mitotic cell division of KB cells, neoplastic cells derived from a specific human carcinoma, when added to cell cultures (Bindoni et al. 1976). Melatonin, however, failed to inhibit mitosis in HeLa cell cultures, cells derived from a form of human cervical cancer (Fitzgerald & Veal 1976). Furthermore, the addition of melatonin in the cell cultures prevented the mitotic arrest that can be induced by adding colchicine to the cultures. Clearly, the role of the pineal gland depends on the type of neoplastic growth, as has been suggested previously (Blask 1984). *In vivo* studies also suggest a role of the pineal gland in regulating neoplastic growth. For example, pinealectomy increases the mitotic index and cell growth in mice with Ehrlich's tumors (Billiteri & Bindoni 1969). Additionally, injections of melatonin reduces cell division in male mice with transplanted hepatomas (Vasil'ev 1979). Collectively, these results provide support that the pineal gland, and specifically melatonin, inhibit neoplastic cell proliferation both *in vitro* and *in vivo*.

There is substantial evidence that the pineal gland can affect the growth of several types of tumors. For example, neonatal pinealectomy caused a large increase in Yoshida tumor growth in rats, while also substantially decreasing the survival time for animals with these tumors (Lapin 1974). Additionally, tumor metastases in the pancreas increased in pinealectomized animals compared with control rats. Interestingly, daily melatonin injections in intact rats just prior to lights off did not alter tumor growth. Administration of NAT, however, resulted in ~50% reductions in tumor size, whereas serotonin, the precursor to melatonin, also reduced tumor growth, but only when administered in the morning (Lapin & Frowein 1981). Administration of P-chlorophenylalanine, which inhibits serotonin synthesis, or S-adenosyl-homocystein, which inhibits HIOMT, both increased tumor growth. Both compounds lead to decreases in circulating melatonin. Exogenous melatonin treatment can also alter tumor growth. For example, daily injections of melatonin increased tumor size in Swiss albino mice with Ehrlich's tumors when given in the morning, but decreased tumor growth when given just prior to lights off (Bartsch & Bartsch 1981). Melatonin administration during midday had no effects on tumor growth in mice.

Photoperiod also affects the growth of an experimentally induced colon cancer in mice (Waldrop et al. 1989). Using a mouse colon adenocarcinoma cell line that produces tumors in a dose-dependent manner when injected subcutaneously, significantly greater tumor area, weight, and group mortality were found in mice exposed to LD 12:12 photoperiods, compared with either short (LD 6:18) or long (LD 18:6) days; difluoromethylornithine (DFMO) was a more effective inhibitor of tumor growth under the short photoperiod, compared with intermediate or long days (Waldrop et al. 1989).

The role of the pineal gland and exogenous melatonin has been investigated in experimental leukemias. For example, pinealectomy appears to increase the rate of leukemia in thymectomized rats treated with reserpine compared with sham pinealectomized animals (Lapin 1978). Additionally, exogenous melatonin treatment substantially reduces leukemias in BALB/c mice transplanted with LSTRA leukemia cells. Specifically, both the number of tumors and the extent of tumor growth in those animals with tumors were reduced by melatonin treatment (Buswell 1975).

Melatonin administration can affect the prevalence and growth of chemically induced tumors. For example, one of the most common models of tumor growth in rodents is DMBA-induced mammary tumors. 9,10-dimethyl-1,2-benzanthracene (DMBA) is a chemical carcinogen that causes a high prevalence of mammary tumor in a variety of rodent species. In one early study, pinealectomy alone did not increase DMBA-induced tumor growth in Wistar rats; however, pinealectomized rats treated with reserpine, which depletes catacholamines, showed a 100% increase in the incidence of tumors, compared with control rats (Lapin 1978). The development of DMBA-induced tumors was also assessed at several time points after prepubertal pinealectomy in female Sprague-Dawley rats (Tamarkin et al. 1981). Only ~20% of the sham-pinealectomized rats had mammary tumors 240 days after DMBA treatment. Almost 90% of pinealectomized rats demonstrated mammary tumors at this same time point. Additionally, daily afternoon administration of melatonin decreased the incidence of DMBA-induced mammary tumors in intact rats by 75%, compared with saline-treated rats (Tamarkin et al. 1981). Melatonin did not alter tumor incidence in pinealectomized rats, however. Other researchers have found that twice-weekly injections of melatonin in the late afternoon decreases mammary tumor incidence by 20–30% in both sham and pinealectomized rats maintained in a LD 12:12 photoperiod or constant light (Aubert et al. 1980). These data suggest that the pineal gland plays an important role in the development of DMBA-induced mammary tumors. The role of the pineal gland in mammary tumors has also been studied using manipulations of photoperiod to alter pineal melatonin secretion. For example, rats subjected to constant light, which inhibits the normal nightly increase in melatonin secretion, increases DMBA-induced mammary tumors in rats (Hamilton 1969; Kothari et al. 1982). More recently, the incidence of DMBA-induced mammary tumor was reduced by short day lengths. Specifically, 90% of deer mice housed in long days and treated with DMBA displayed mammary tumors (Blom et al. 1994). Conversely, none of the DMBA-treated deer mice displayed mammary tumors when housed in short days for 8 weeks. Furthermore, the inhibition of mammary tumors in short-day animals was mimicked completely by implanting long-day mice with Silastic capsules filled with melatonin to simulate a short-day pattern of secretion of this hormone (Blom et al. 1994).

An underlying circannual cycle of susceptibility to DMBA-induced mammary carcinogenesis may exist in female Sprague-Dawley rats housed in constant environmental conditions (photoperiod, temperature, air humidity, food) (Loscher et al. 1997). DMBA was administered orally, and rats were palpated weekly for the presence of mammary tumors. Autopsies were conducted at 13 weeks, and the number and size of mammary tumors were recorded. When the experiment was performed twice within 2 years during spring/summer, the tumor incidence was 56 and 61%. When the same experiment was performed in autumn, however, a significantly lower tumor incidence (34%) and tumor burden were obtained. When the experiment was started during winter, tumor incidence was similar to the spring/summer groups, but tumor burden was lower (Loscher et al. 1997). Data indicate a seasonal, and perhaps circannual, rhythm in the development and growth of DMBA-induced breast cancer in Sprague-Dawley rats. These results may represent the seasonal variation in pineal melatonin production and immune function previously reported in rodents under constant environmental conditions.

5.5.1. Melatonin and Disease: Clinical Implications

Melatonin has been used in both clinical and experimental settings as an immunotherapeutic agent in the treatment of several diseases including acquired-immunodeficiency-syndrome (AIDS) and cancer (Blask 1984; Maestroni 1993). For example, oral administration of melatonin in the evening increased peripheral blood mononuclear cell number, as well as response to the T-cell mitogen PHA in AIDS patients (Maestroni 1993). A growing body of literature implicates pineal melatonin as a regulator of oncostatic processes (Blask 1984; Maestroni 1993). NK cells attack malignant or virus-infected cells while sparing normal cells (Ortaldo & Heberman 1984). IL-2 can potentate NK cell activity. Recent therapies have centered around combining exogenous melatonin treatment with IL-2 immunotherapy in the treatment of cancer (Maestroni 1993). Research has demonstrated that melatonin treatment can potentiate the anticancer actions of IL-2 in humans (Maestroni & Conti 1993). Exogenous treatment of pulmonary metastases with either melatonin/IL-2 or melatonin alone had a significant antitumor effect, reducing tumor size (Maestroni & Conti 1993). Combined melatonin/IL-2 immunotherapy has also been examined in metastatic renal cancer, nonsmall cell lung cancer, colorectal cancer gastric carcinomas, and mammary carcinomas (Maestroni 1993; Lissoni et al. 1995). Preliminary results suggest that melatonin potentiates the immunotherapeutic effects of IL-2 within the context of these types of cancers (Maestroni 1993). TNF is another immunological agent that exerts cytotoxic activity on human cancer cells (Haranaka & Satomi 1981). Like the results reported for IL-2, oral administration of melatonin to cancer patients also appeared to modulate the cytotoxic effects of TNF (Brackowski et al. 1994). Additional research combining melatonin with administration of human lymphoblastoid INF suggests that melatonin may enhance the ability of INF to activate NK cell activity and inhibit tumor growth in patients with renal cell carcinoma (Neri 1994). Collectively, the results of these clinical studies suggest that the use of exogenous melatonin as part of a multifactorial treatment regime is promising. Much of this research is in the early stages, and the long-term benefits of melatonin treatment on cancer remain unknown. Although the use of melatonin to reduce the spread of cancer is indeed promising, much more research, both basic and applied, must be conducted on the long-term effects of melatonin administration in humans.

5.6. Summary and Conclusions

Taken as a whole, data presented in this chapter provide compelling evidence
that melatonin, the primary indolamine secreted from the pineal gland in virtually all species, possesses a wide range of immune-enhancing, antistress,
and disease-preventing properties. For example, melatonin can enhance both
innate and acquired immune function, increase the numbers and ratios of
many immune cells types, and affect cytokine production and release. Additionally, melatonin can help prevent proliferation of some forms of cancer and
reduce the symptoms of diseases such as AIDS. Clearly, many of these findings have potentially important clinical applications, for example, in how we
treat patients suffering from AIDS, various forms of cancer, and autoimmune
diseases. Despite these potentially exciting benefits, we must take a cautious,
scientific approach in evaluating the data on melatonin and immunity, because
it is easy to get "caught up in the melatonin hype" (e.g., Reppert & Weaver
1995). As discussed earlier, many of the findings demonstrating enhanced
immune function after treatment with melatonin have been performed in seasonally breeding species. Recall that melatonin is the primary physiological
cue for determining the time of year in these species; therefore, individuals
of these species undergo a suite of physiological and biochemical changes
in response to melatonin, including changes in body mass and metabolism,
reproductive function, as well as other morphological changes that allow an
animal to survive in its environment. Some of the immune enhancing effects
of melatonin may be due to these other changes and not due to a direct effect
of melatonin on the immune system. Alternatively, if immune enhancement
by melatonin is indeed direct, such enhancement may not be seen in species
that are generally unresponsive to melatonin. Given that humans are not generally considered to be a seasonally breeding species (cf. Bronson 1995), it
is quite possible that many of the immune-enhancing properties of melatonin
may not occur in humans. Obviously, further research is necessary to evaluate
this possibility. It is important to note that, because most of the research on
the immune-enhancing effects of melatonin is relatively recent, there are no
long-term studies on potential adverse effects of prolonged melatonin treatment. Melatonin is a hormone, and hormones can have very powerful effects
on many aspects of our physiology and biochemistry. Thus, widespread use
of melatonin to boost immune function in otherwise healthy individuals is
not warranted at this time. Only continued research in this rapidly growing
and exciting area of research will determine the full benefits of melatonin in
health and disease prevention.

6

Energetics and Immune Function

> In winter and spring the bowels are naturally the hottest, and the
> sleep most prolonged; at these seasons, then, the most sustenance
> is to be administered; for as the belly has then most innate heat, it
> stands in need of most food.
>
> Hippocrates, ca. 400 B.C.
> *Aphorisms*

6.1. Introduction

Animals require a relatively steady supply of energy to sustain biological functions. Although these energy demands are somewhat continual, both daily and seasonal fluctuations in energy requirements occur as physiological and behavioral activities wax and wane. Animals typically do not feed continuously to supply constant energy to cells; rather, adaptations have evolved to store and conserve energy. Eventually, however, the chronic need for energy depletes stores and requires animals to obtain food. In most habitats, significant daily and seasonal fluctuations in food availability occur. For example, outside of the tropics, food availability is generally low during winter, whereas thermogenic energy demands are typically high. Consequently, energy intake and energy expenditures are never perfectly balanced because of fluctuations in both energy requirements and energy availability.

Most animals restrict breeding to specific times of the year when food is abundant and survival and reproductive success are most likely. Inhibition of winter breeding is one of the central components of a suite of energy-saving adaptations evolved among nontropical animals (Goldman & Nelson 1993). In addition to the seasonal onset and termination of reproductive function among nontropical vertebrates, it is also well-established that endocrine, metabolic, growth, neural, activity, and thermogenic processes undergo seasonal changes to enhance winter energy conservation (Bronson 1989; Bronson & Heideman 1994; Moffatt et al. 1993). Mechanisms in virtually every physiological and

151

behavioral system have evolved to cope with this winter energetic "bottle-neck"; presumably, animals possessing these energy-conserving adaptations enhance their survival, and ultimately increase their reproductive success, compared with conspecific individuals that do not (Bronson & Heideman 1994).

As described previously, immune function has traditionally been assumed to remain relatively constant across the seasons (see Sheldon & Verhulst 1996; Nordling et al. 1998; Lochmiller & Deerenberg 2000). Recent evidence, however, suggests that immune function varies seasonally (Lochmiller et al. 1994; Nelson & Demas 1996; Sinclair & Lochmiller 2000; Chapter 4). Maintaining optimal immune function is energetically expensive; the cascades of dividing immune cells, the onset and maintenance of inflamation and fever, and the production of humoral immune factors all require substantial energy (Ardawi & Newsholme 1985; Grimble 1994; Maier et al. 1994; Demas et al. 1997a, b; Spurlock 1997). Mounting an immune response likely requires resources that could otherwise be allocated to other functions (To aid the reader, Fig. 4.1 is repeated as Figure 6.1; Figure 6.1) (Sheldon & Verhulst 1996; Lochmiller & Deerenberg 2000). Thus, it is reasonable to consider immune function from an ultimate perspective of energetic trade-offs. Investment in current offspring compromises investment in future offspring. Recent studies on the proximate level of this issue suggest that individuals "optimize" immune function, and allocate energetic resources among the costs of immune function and other maintenance or reproductive functions (e.g., Festa-Bianchet 1989; Gustafsson et al. 1994; Råberg et al. 1998). Seasonal changes in immune function are consistent with this energetic perspective and are likely driven by seasonal changes in energy requirements and availability (Nelson & Demas 1996). According to this hypothesis, animals inhabiting environments where energy needs and demands continuously fluctuate maintain the highest level of immune function that is energetically possible (without evoking autoimmune diseases) given the constraints of other survival needs, growth, and reproduction (Sinclair & Lochmiller 2000).

The observation that immune function is generally compromised during specific energetically demanding times, such as during breeding activities, winter, migration, or molting, is consistent with the hypothesis that immune function is optimized (Zuk 1990; John 1994). For example, pregnant or lactating mammals often display compromised immune function (reviewed in Szekeres-Bartho et al. 1990; McCruden & Stimson 1991; Drazen 2001). Traditionally, the medical hypothesis for immunosuppression during pregnancy has been to protect the fetus from being attacked as foreign tissue by the maternal immune system (McCruden & Stimson 1991). However, reduced

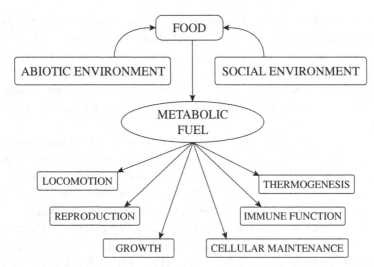

Fig. 6.1. Competing behavioral demands that require substantial metabolic energy. When demands are higher than the energy immediately available, individuals engage in trade-offs among various energy demands to reduce total energy needs. During energetic shortages, locomotion, reproduction, and growth are curtailed with little effect on survival. As energy becomes increasingly scarce, survival may be compromised because of reduced energy allocation to thermogenesis and immune function. Extrinsic factors, such as social environment, are hypothesized to influence the partitioning of metabolic reserves to these various behavioral and physiological processes. (Adapted from Wade and Schneider 1992. Reprinted from *Reviews of Reproduction*, vol. 4, S. L. Klein & R. J. Nelson, "Influence of social factors on immune function and reproduction," pp. 168–78, copyright © 1999.)

immune function during pregnancy may (also) represent an energy-savings adaptation; this perspective may not be obvious in recent studies because most laboratory and clinical studies have been conducted on individuals with *ad libitum* food availability. Few women currently living in North America or Western Europe are energetically challenged during pregnancy. However, examination of the literature on pregnant women's health in the United States and Western Europe prior to 1920 or current studies of pregnant women in nonindustrialized nations reveal an elevated incidence of many diseases and immune dysfunctions during pregnancy, particularly during the third trimester (Sakamoto-Momiyama 1977).

The mechanisms that mediate seasonal changes in other physiological, morphological, and behavioral processes also appear to mediate the seasonal changes in immunity (see Chapters 7 and 8). Individuals use photoperiodic information to initiate or terminate specific seasonal adaptations, including reproduction, to maintain a positive energy balance (reviewed in

Bartness & Goldman 1989; Heldmaier et al. 1989; Saafela & Reiter 1994). Although many domesticated species (including humans) appear to have escaped photoperiodic constraints on reproduction, these species continue to display rhythms of pineal melatonin secretion comparable to photoperiodic species (Reiter 1998). Thus, the effects of photoperiod on immune function may remain extant in humans (e.g., Nelson & Demas 1996; Nelson et al. 1995).

To summarize our working hypothesis, we propose that individuals maintain the highest level of immune function that is energetically possible within the constraints of other survival needs, growth, and reproduction in habitats in which energy requirements and availability often fluctuate. The observation that immune function is generally compromised during specific energetically demanding times, such as winter, breeding (including pregnancy and lactation), migration, or molting is consistent with the hypothesis that immune function is energetically optimized (Zuk 1990; John 1994; Lochmiller & Deerenberg 2000; Sinclair & Lochmiller 2000). In this chapter, we also propose the hypothesis that metabolic fuels and the hormones associated with their metabolism serve as the proximate mechanism mediating the fluctuation in immune function and disease prevalence. Similar to seasonal reproductive function (Schneider & Wade 1999), availability of metabolic fuels are presumably monitored, and a physiological decision is made to allocate resources to maximize immune function. It appears that the hormones associated with stress are critical in this physiological decision-making process.

6.2. Stress and Immune Function

The risk of infection and death is highest when insufficient energy reserves are available to sustain immunity. Stress can compromise immune function (for reviews, see Dunn 1989; O'Leary 1990; Ader & Cohen 1993). Prolonged or severe food shortages may evoke secretion of glucocorticoid hormones (Nakano et al. 1987; Jose & Good 1973); glucocorticosteroids actively compromise several aspects of immune function (Kelley 1989; Munck & Guyre 1986; Maier et al. 1994; see Chapter 7). Environmental or biotic conditions perceived as stressful, such as reduced food availability, low ambient temperatures, overcrowding, lack of shelter, or increased predator pressure can recur seasonally leading to seasonal fluctuations in immune function among individuals, and seasonal changes in population-wide disease and death rates (Lochmiller et al. 1994; Silverin 1998; Nelson & Klein 1999). A dynamic relationship exists between longevity and reproductive output (Stearns 1976); all other factors being equal, longer-lived individuals produce more offspring and are more fit than individuals that die prematurely. Thus, suppressed immune function impairs survival and, ultimately, fitness. This is the proximate mechanism underlying the ultimate hypothesis that investment in current offspring occurs at the expense of investment in future offspring.

Laboratory studies have established that stress inhibits immune function (Keller et al. 1983; Laudenslager et al. 1983; reviewed in Ader et al. 2001). As mentioned "stress" has been a notoriously ethereal concept in biology. The term "stress" has been conflated to include both the stressor and the physiological stress response (Sapolsky 1992). In many cases, the implicit definition of a stressor is anything that increases glucocorticoid secretion; glucocorticoids suppress certain aspects of immune function (Kelley 1991; Munck & Guyre 1986; Maier et al. 1994). Individual differences in coping with similar stressors reflect differences in perception and processing of information about stressful situations (Sapolsky 1998), but may also represent differences in energy availability or efficiency in energy use. Traditionally, there has been little concern about the functional explanation for why individuals exposed to stressors evolved mechanisms to compromise immune function (but, see Sapolsky 1992, 1998). In other words, what is the adaptive advantage to compromised immune function (Råberg et al. 1998; Svensson et al. 1998)? To address this issue, it is important to propose an operational definition of "stressor" and change levels of analysis. Stressors disturb homeostasis. Because restoring homeostasis requires more energy than maintaining homeostasis (Sapolsky 1992), exposure to a stressor increases energy demands on individuals. During slight homeostatic perturbations, adrenalin mobilizes energy and prepares the individual for "fight or flight" activities. Animals have evolved to function within an optimal range of conditions (Figure 6.2). During long-term perturbations of homeostasis (e.g., when temperatures are low or food availability becomes scarce), glucocorticoids are released and energy is mobilized to restore homeostatic equilibrium. Immune function may be reduced during reinstatement of homeostasis because both processes are energetically costly. As noted previously, there is a trade-off between "maximal" immune function and "perfect" homeostatic balance. Steroid hormones appear to be one of the physiological mediators of this trade-off.

Immune function is often compromised during the breeding season (John 1994). The odds of close social interactions and the risk of infection both increase at this time. Presumably, socially polygynous animals are at higher risk for communicable diseases and infections than monogamous animals. As described previously, there is a trade-off between maximal immune and reproductive function. Immune function is generally suppressed among male, compared with female conspecifics. The physiological mechanisms

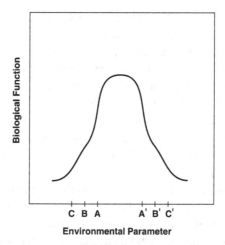

Fig. 6.2. Individuals have evolved to perform best under optimal conditions. Under suboptimal conditions, reproductive function is impaired. Animals might breed when conditions are between A and A'. As conditions deteriorate, however, immune function is impaired and survival is endangered (e.g., between C and C').

compromising male immune function during breeding are likely mediated by blood androgen concentrations (Alexander & Stimson 1988; Olsen & Kovacs 1996; Klein 2000a). Males of monogamous species generally have lower androgen concentrations than individuals of polygynous species (Wingfield 1994; Klein 2000b).

The glucocorticoids (e.g., cortisol or corticosterone) represent one well-known class of hormonal modulators of immune function (Jefferies 1991; Dhabhar et al. 1993). Glucocorticoids have typically been studied in the context of "stress," and generally, glucocorticoids have been considered to mediate the immunocompromising effects of stress (see Chapter 4). From a regulatory perspective, glucocorticoids are released when energy must be mobilized from glycogen and adipose stores to fuel the body. Glucocorticoids are commonly called "stress hormones" because they are released in response to conditions that jolt individuals out of equilibrium. However, several other hormones are affected during stress and can affect immune function (Figure 6.3).

Many extrinsic factors can stimulate the release of glucocorticoids that serve to increase circulating glucose concentrations in the blood. Virtually all energetic challenges, including cold water swim or other low temperature exposure, food restriction, restraint, and social isolation, evoke glucocorticoid secretion. Many other unexpected or frightening factors that activate

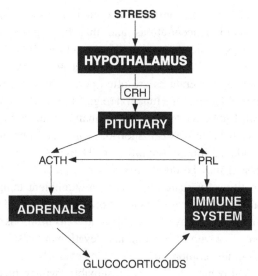

Fig. 6.3. Hypothalamic-pituitary-adrenocortical axis. Stressors evoke the production and release of corticotrophic-releasing hormone (CRH) from the hypothalamus, which stimulates adrenocorticotropic hormone (ACTH) secretion from the anterior pituitary, which stimulates the secretion of glucocorticoids from the adrenal cortex. All three of these hormones can suppress immune function. Prolactin (PRL) can partially reverse the inhibitory effects of the hypothalamic-pituitary-adrenocortical hormones on immune function. (From Davis 1998.)

the sympathetic system, including electric shock, predator attack, or loud noise, increase cardiac and respiratory output, which are energetically expensive. From a psychological perspective, unpredictable or uncontrollable evocative stimuli induce adrenalin and glucocorticoid secretion that activates sympathetic responses, such as increased respiration, blood pressure, and heart rate, with a concomitant elevation in energy consumption. Accordingly, we view an extrinsic "stressor" as *any event that causes an individual to increase energy consumption above baseline.* The adaptive response to such an event is to release glucocorticoids in order to bring energy out of storage (Sapolsky 1998). In nature, this metabolic mechanism may not represent a "stress-coping" response; rather, this response may be appropriately considered part of a dynamic system that mobilizes energy and operates to keep the individual functioning at optimal levels (Sapolsky 1998). In addition to stimulating glucose availability, glucocorticoids also function to suppress unnecessary physiological energy costs. Thus, chronic glucocorticoid secretion suppresses reproduction and immune function – processes that are enormous energy sinks (Figure 6.1).

Previous studies have demonstrated that environmental stressors elevate blood glucocorticoid concentrations and that high glucocorticoid concentrations suppress immune function (Hauger et al. 1988; Claman 1972; Besedovsky & del Rey 1991; Black 1994; Ader & Cohen 1993). For example, low ambient temperatures are often perceived as stressful and can potentially compromise immune function (e.g., Claman 1972; Monjan 1981; MacMurray et al. 1983). Winter survival in small animals is hypothesized to require a positive balance between short-day enhanced immune status and glucocorticoid-induced immunosuppression (Demas & Nelson 1996). This immunosuppression may be due to many factors, including overcrowding, increased competition for scarce resources, low ambient temperatures, reduced food availability, increased predator pressure, or lack of shelter. Each of these potential stressors may cause high blood concentrations of glucocorticoids. Winter breeding with its concomitant elevation in sex steroid hormones may also cause immunocompromise (e.g., Tang et al. 1984; Lochmiller et al. 1994; Sinclair & Lochmiller 2000). Presumably, winter breeding occurs when other environmental stressors, such as temperature and food availability, are not severe. The balance of enhanced immune function (i.e., to the point where autoimmune disease becomes a danger) against stressor-induced immunosuppression (i.e., to the point where opportunistic pathogens and parasites overwhelm the host) must be met for animals to survive and become reproductively successful.

Thus, many aspects of the lives of individuals are energetically challenging. Many animals have evolved timing mechanisms to parse out energetically demanding activities to different times of year. For example, birds engage in predictable cycles of migration, molting, breeding, parenting, molting, and over winter migration. These energetically expensive activities do not overlap. As noted previously, both extrinsic and intrinsic factors are energetically challenging, including winter, pregnancy, lactation, and migration. According to our hypothesis, immune function is predicted to be low during energetically demanding activities.

6.2.1. Seasonal Changes in Stress Responses

One issue that arises with studies of wild-caught animals is the experience of being captured generally causes a rapid surge in epinephrine and glucocorticoids. These changes in stress hormone concentrations affect other hormones. For example, if blood samples are obtained immediately, then reproductive hormone concentrations are generally much higher than if blood samples are obtained after a few hours postcapture (e.g., Licht et al. 1983; Mendonca &

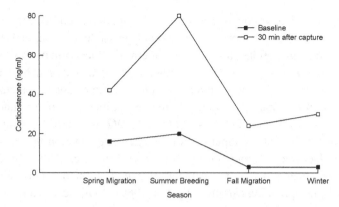

Fig. 6.4. Seasonal changes in stress response. Corticosterone concentrations were measured in blood samples obtained from white-crowned sparrows immediately or 30 minutes after capture. Corticosterone release was "blunted" during the fall and winter. (Reprinted from *Comparative Biochemistry and Physiology C*, vol. 116, L. M. Romero, M. Ramenofsky & J. C. Wingfield, "Season and migration alters the corticosterone response to capture and handling in an artic migrant, the white-crowned sparrow (*Zonotrichia leucophrys gambelii*)," pp. 171–7, copyright ©1997, with permission from Elsevier Science.)

Licht 1986; Orchinik et al. 1988; Mahmoud & Licht 1997; Tyrrell & Cree 1998). Obviously, the interval of time between capture and the blood sample is important when trying to determine hormone concentrations in wild-caught animals.

Redpolls (*Acanthis flammea*) are birds that live in the challenging arctic conditions of Alaska. Both males and females showed the typical marked increase in blood corticosterone concentrations when captured for about 1 hour, indicating that they responded similarly to many other vertebrates (Wingfield et al. 1994). However, the amount of the elevation in postcapture corticosterone varied across the year (Figure 6.4). Corticosterone concentrations were generally blunted during January and maximal during June when birds were breeding. It appeared that birds that had the highest fat stores secreted the lowest amounts of glucocorticoids postcapture (Wingfield et al. 1994). Similar effects have been reported for other birds living at high latitudes, including the white-crowned sparrows (*Zonotrichia leucophrys gambelii*), snow buntings (*Plectrophenax nivalis*), Lapland longspurs (*Calcarius lapponicus*), and willow tits (*Parus montanus*) (Romero et al. 1997, 1998a–d; Silverin 1997). The seasonal modulation of corticosterone secretion appears to be regulated at multiple sites within the hypothalamo-pituitary-adrenocortical (HPA) axis; during the winter (or during the energetically expensive, feather-replacing molting process), the HPA axis appears relatively insensitive to the effects of exogenous adrenocorticotropic hormone in stimulating corticosterone production (Romero et al. 1998a–d). It is certainly possible that mammals living in boreal or arctic regions show similar patterns of glucocorticoid responses.

If changes in energy availability alter homeostasis, then it seem reasonable to expect immune function should be compromised in response to metabolic stress. As an example of metabolic stressors inhibiting immune function, the chemical compound 2-deoxy-D-glucose (2-DG) was used to manipulate energy availability at the input end of the energetic equation in seasonally breeding deer mice (*Peromyscus maniculatus*) (Demas et al. 1997b). 2-DG is a glucose analog that inhibits cellular utilization of glucose, thus inducing a state of glucoprivation (Smith & Epstein 1969). 2-DG acts as a metabolic stressor by increasing serum corticosterone concentrations (Lysle et al. 1988), and 2-DG glucoprivation induces an anestrous state in female Syrian hamsters (*Mesocricetus auratus*) (Miller et al. 1994), and torpor in female Siberian hamsters (*Phodopus sungorus sungorus*) (Dark et al. 1994). 2-DG-induced metabolic stress also affects immune function; 2-DG administration inhibits murine splenic T-lymphocyte proliferation in a dose-dependent manner in laboratory strains of rats (*Rattus norvegicus*) (Lysle et al. 1988) and mice (*M. musculus*) (Miller et al. 1994).

Environmental and biotic influences on immune function are certainly complex, but not empirically intractable. Low ambient temperatures, reduced food availability, and other energetically demanding winter conditions compromise immune function, and these factors can be manipulated and controlled in the laboratory. Because many stressors are seasonally recurrent in nature, individuals of some species may have evolved mechanisms to anticipate and counteract these recurring threats to maximize immunity. Short photoperiods have been hypothesized to serve as a cue used by animals to enhance immune function in advance of energy-compromising conditions (Nelson & Demas 1996; Yellon et al. 1999). This enhancement appears to be mediated by the pineal hormone, melatonin. Increased duration of melatonin treatment (mimicking long nights) enhances immune function, either directly or indirectly by affecting the secretion of steroid hormones and prolactin (Goldman & Nelson 1993).

Exposure to short-day lengths in a laboratory buffered deer mice against glucoprivation stress (i.e., 2-DG treatment). Long-day deer mice injected with 2-DG had elevated corticosterone concentrations, compared with long-day mice injected with saline (Figure 6.5); corticosterone concentrations were not significantly elevated in short-day mice injected with 2-DG. 2-DG-treated long-day mice displayed reduced splenocyte proliferation to concanavalin A

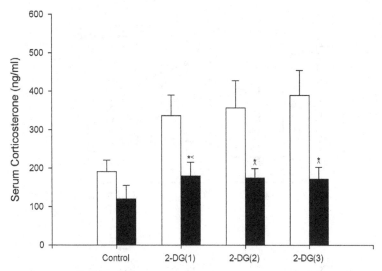

Fig. 6.5. Mean (±SEM) serum corticosterone (ng/ml) concentrations in deer mice housed in long days (open bars) or short days (solid bars). Statistically significant differences between means are indicated with asterisks. Numbers in parentheses correspond to the number of daily 2-deoxy-D-glucose (2-DG) injections received. (Reprinted from the *American Journal of Physiology*, vol. 272, G. E. Demas, A. C. DeVries, & R. J. Nelson, "Effects of photoperiod and 2-deoxy-D-glucose-induced metabolic stress on immune fuction in female deer mice (*Peromyscus maniculatus*)," pp. R1762–7, copyright ©1997b, with permission from the American Physiological Society.)

(ConA), compared with saline-injected mice. Splenocyte proliferation did not differ among short-day deer mice, regardless of experimental treatment; i.e., all short-day animals exhibited enhanced immune function. Overall, short-day mice treated with 2-DG displayed higher splenocyte proliferation than long-day mice treated with 2-DG (Figure 6.6) (Demas et al. 1997b).

Melatonin also appears to have an important protective role in adrenergic stress. Chronic treatment of rats with adrenaline or noradrenaline suppresses the proliferative response of both B- and T-cells to mitogenic stimulation in peripheral blood lymphocytes in the presence of a β-adrenergic antagonist (Felsner et al. 1992); this effect appears to be due to specific activation of the α_2-adrenergic receptor subtype. Blockade of α_2-receptors with chlonadine also leads to imunosuppression, but only with simultaneous blockade of the β-adrenergic receptors (Felsner et al. 1995). Thus, it appears that α_2- and β-adrenergic receptors serve antagonistic roles in mediating immune function. Interestingly, exogenous melatonin treatment blocks the

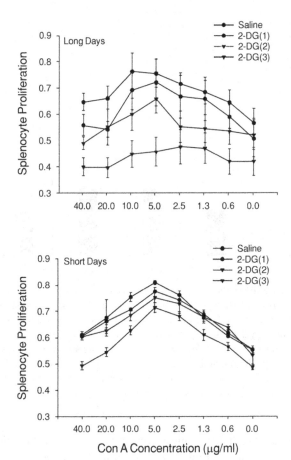

Fig. 6.6. Mean (±SEM) splenocyte proliferation to concanavalin A (Con A) (represented as absorbance units) of deer mice housed in long (LD 16:8) (upper panel) or short (LD 8:16) (lower panel) days. Experimental mice received daily injections of 2-deoxy-D-glucose (2-DG) across 1, 2, or 3 consecutive days. Control mice in each photoperiod received daily intraperitoneal injections of sterile 0.9% saline across three consecutive days. Statistically significant differences between means are indicated by an asterisk. (Reprinted from the *American Journal of Physiology*, vol. 272, G. E. Demas, A. C. DeVries, & R. J. Nelson, "Effects of photoperiod and 2-deoxy-D-glucose-induced metabolic stress on inmune funtion in female deer mice (*Peromyscus maniculatus*)," pp. R1762–7, copyright © 1997b, with permission from the American Physiological Society.)

immunosuppressive effects of either catecholamine or chlonadine administration in rats (Liebmann et al. 1996). It has been hypothesized that melatonin protects lymphocyte functions from α_2-adrenergic suppression and that suppressed immune function is the result of inhibited melatonin release

(Liebmann et al. 1996). These results may explain circadian as well as seasonal rhythms in immune function that appear to follow closely changes in melatonin secretion (reviewed in Nelson & Demas 1996). Consistent with this idea, previous findings have shown that exogenous melatonin can enhance antibody-dependent cellular cytotoxicity, but only during the summer (Giordano & Palermo 1991).

These data are consistent with the hypothesis that short days buffer against metabolic stress. Reduced corticosterone concentrations in animals maintained on short days or treated with melatonin are likely due to improved metabolic function (Saafela & Reiter 1994). Accordingly, improved immune function in short days represents one component of numerous winter-coping adaptations that may be mediated by melatonin.

6.3. Energy Shortages and Immune Function

To be successful, energy input must be greater than or equal to energy output. If immune function requires substantial energy, then any perturbation at either end of the energy balance equation should influence the immune system. In general, studies have established that either nutritional restriction or intense caloric expenditure compromises immune function. Many of these studies have been conducted within the context of understanding the relationship between illness and poor production or growth of agricultural market animals. Depending on the state of the animals in question, energy and other nutrients will be (1) diverted from growth in young, infected individuals; (2) diverted from reproductive function in sexually mature individuals; or (3) deflected away from fetal or placental development in pregnant, infected individuals (Johnson 1997). Studies of the effects of restricted nutritional input and elevated energetic output on immunity are reviewed here. In addition to simple caloric input, the effects of protein or lipid intake, as well as micronutritional intake on immune function, will be presented.

6.3.1. Limited Caloric Intake and Immune Function

Several studies of domesticated animals have reported that reduced food intake is associated with compromised immune function (reviewed in Spurlock 1997). For example, proinflammatory cytokines – particularly interleukin (IL)-1, IL-6, and tumor necrosis factor (TNF) – interrupt anabolic processes and alter nutrient uptake and utilization (e.g., Grimble 1994; Johnson 1997). These cytokines may also cause the release of glucocorticoids, catecholeamines, and prostaglandins, which can all modulate metabolism and immune function. Although proinflammatory cytokines are critical in the responses to infection, excess cytokines can damage tissue and even lead to death. For instance, excess or inappropriate secretion of cytokines has been linked to morbidity in a number of conditions, including malaria, sepsis, meningitis, inflammatory bowel disease, and rheumatoid arthritis (Beutler & Cerami 1988; Waage et al. 1989; Kelley 1990). Consequently, a balanced release of cytokines must be achieved for maximal effectiveness in protecting the host from any infection. One characteristic of infection is wasting of peripheral tissues (Grimble 1994). Wasting results from two processes: (1) the loss of appetite and reduced ingestion of new nutrients, and (2) the loss of tissue fats, protein, and trace nutrients that are used in activating the immune system and protecting the host during infection. Indeed, tumor necrosis factor was also known as cachectin because of its involvement in cachectis (wasting) that is often observed in advanced cancer or AIDS patients.

Reduced caloric intake impairs immune function in domesticated animals and humans (e.g., Kramer et al. 1997; Spurlock 1997; Klasing 1998; Lin et al. 1998). For example, two separate groups of U.S. Army Rangers in training lost 15.6% and 12.6%, respectively, of their body mass during 2 months of intense training (Kramer et al. 1997). Their T-lymphocyte responses to *in vitro* stimulation by phytohemagglutinin (PHA) and tetanus toxoid were reduced. Individuals who received a 15% boost in daily food intake increased their T-lymphocyte counts by >20% within 9 days. This study was motivated by an elevated incidence of pneumococcal pneumonia among Ranger Corps trainees during training.

Body mass loss may be the critical parameter that mirrors compromised immune function. Young animals continue to grow under a dietary regime of 50% restriction, although adults typically lose about 25% of their body mass if food intake is restricted by this amount (Franklin et al. 1983). Young mice receiving a 50% reduction in *ad libitum* food intake displayed a reduced rate of thymic cell division (Franklin et al. 1983). In adults, the same level of caloric restriction resulted in a marked depletion of thymocytes, suggesting that the body mass loss triggered the regression of thymocytes.

Energy restriction (50%) decreases the number of circulating natural killer (NK) cells and immunoglobulin levels in obese women (Kelley et al. 1994). Although the health status of these women was not jeopardized, the authors warned that immune function and health could be compromised with more severe dietary restrictions or even after moderate restriction in susceptible individuals (Kelley et al. 1994). In another study of obese women, a moderate energy restriction regime (4.19–5.44 mJ or 1,200–1,300 kcal/day) resulted in decreased immune parameters, including several measures of T-cell, B-cell, monocyte, and granulocyte function. Taken together, these studies do not support the contention that obesity is associated with mild-to-moderate decreases in the function of certain aspects of immune function (Nieman et al. 1996), although the idea of extended longevity can be achieved by restricted caloric intake [and consequential reduction of reactive oxygen species-mediated damage to biological macromolecules (Tian et al. 1995)].

Immune responses were assessed in over- and underfed cattle (Fiske & Adams 1985). Underfed cattle were provided 0.5% of their body weight for 87 days, followed by 0.3% of their body weight for another 55 days to induce significant weight loss, whereas overfed cattle were fed *ad libitum* to maximize weight gain. A control group was fed 1.5% of their body mass to maintain initial body mass. Energy-restricted (i.e., underfed) cattle had significantly reduced levels of plasma protein and circulating lymphocytes. Furthermore, cattle eating low-protein diets had impaired lymphoproliferative responses to pokeweed mitogen, compared with cows with *ad libitum* food intake. Although cell-mediated immune function was altered, humoral immune function was similar among the groups. Specifically, neither IgG nor IgM levels differed significantly among any group (Fiske & Adams 1985).

In contrast with the results of severely malnourished individuals, patients with anorexia nervosa exhibited normal T-lymphocyte populations and *enhanced* mitogen responsiveness to PHA and Con A (Golla et al. 1981). Supplemental feeding of these patients served to return the proliferative lymphocyte response to control values. Energy restriction decreases the number of circulating NK cells and immunoglobulin concentrations in obese women (Kelley et al. 1994).

6.3.2. Limited Protein Intake and Immune Function

Immune function is impaired by low protein diets (Jennings et al. 1992; Lin et al. 1998). For example, protein-energy malnutrition (PEM) in baboons caused marked deficiencies in immune function prior to the onset of any fever (Qazzaz et al. 1981). For example, burned children healed faster when provided with protein supplements (whey protein) (Alexander et al. 1980). Infection rates are generally higher among malnourished children (e.g., Chandra 1979, 1993). Protein supplements also improved immune function in malnourished elderly patients with impaired cytokine production (Keenan

et al. 1982). Supplemental arginine feeding in humans enhances wound healing and lymphocyte responses (Kirk et al. 1993).

PEM decreases cellular immunity. In one study, malnourished children in
Bolivia were hospitalized for 60 days while receiving a special four-step diet
(Chevalier et al. 1998). Immune function was monitored by daily clinical ex-
aminations and weekly sonographic examination of the thymus. Nutritional
recovery (i.e., 90% of median reference weight for height) was achieved after 1 month when these children would normally be discharged. However,
immunological recovery was not achieved until after 2 months of supplemental feeding. Because these children lived in impoverished conditions and
had high exposure to pathogens, it is possible that discharges based on body
weight are typically premature, and frequent relapses might reflect persistent
immunocompromise (Chevalier et al. 1998).

Similarly, cell-mediated immunity was depressed in malnourished children
from rural Bangladesh, and associated with an elevated incidence of upper
respiratory infections (Zaman et al. 1997). PEM has been linked to compromised cell-mediated immunity, especially reduced monocyte phagocytosis,
complement components, and production of both IL-1 and IL-2 in young
(5–20 months of age) patients treated in Egypt (Lofty et al. 1998). The
authors concluded that impairment of the development of specific immune
responses predisposed these patients to increased susceptibility to infection
(Lofty et al. 1998). Several other studies on humans with PEM have made the
same conclusion that protein malnutrition severely inhibits immune function
(e.g., Salimonu et al. 1983; Garre et al. 1987; Fakhir et al. 1989; Chandra
1991, 1992, 1997; Delafuente 1991; Redmond et al. 1991; Ozkan et al.
1993).

Similar findings on the effects of protein malnutrition on immune function have been reported in meat-producing animals, such as cows (Doherty
et al. 1996). For example, antibody titers were assessed in food-restricted
or unrestricted pregnant beef cows immunized with tetanus toxoid, keyhole
limpet hemocyanin (KLH), and chicken red blood cells (Olson & Bull 1986).
Protein-restricted cows had lower tetanus toxoid antibody titers than unrestricted cows; however, anti-chicken red blood cells and KLH antibodies were
not affected by food restriction. Similarly, immunocompetence of neonatal Holstein bulls fed for either maximal growth or PEM was determined
in vivo and *in vitro* by assessing IL-2 and lymphocyte proliferation in response to Con A, respectively (Griebel et al. 1987). After 2 weeks of PEM,
calves had significantly lower body masses, IL-2 activity, and lymphocyte
proliferation, compared with age-matched controls that were fed *ad libitum*.

Also, immune function was assessed in dairy cows in relation to energy balance and metabolic status (Ropstad et al. 1989). Two experimental groups received high protein either in high- or low-energy diets. Similarly, two groups received low protein in high- or low-energy diets. There was a significant positive relationship between energy balance and lymphocyte response to mitogenic challenge. Cows fed high-energy diets showed greater proliferative responses than cows maintained on low-energy diets. However, metabolic status, as assessed by plasma concentrations of fatty acids and glucose, did not correlate with immune function (Ropstad et al. 1989).

The effects of PEM on immune function was also assessed in rats fed isocaloric diets containing 0.5%, 5%, and 18% casein (Shimura et al. 1983). Growth was retarded in rats fed protein-insufficient diets (i.e., 0.5% and 5% casein). Furthermore, PEM reduced serum IgG and IgA levels, as well as antibody responses to the T-cell-dependent antigen, dinitrophenylated bovine γ-globulin.

The influence of PEM on the cellular immune response was also assessed in mice (Nejad & Mohagheghpour 1975). Male and female Swiss mice were either provided a diet deficient in protein and total energy or fed *ad libitum*. Delayed-type hypersensitivity (DTH) (Chapter 2) to 2,4-dinitro-chlorobenzene (DNCB) was assessed by placing the chemical on the ear and measuring ear thickness. A vigilant immune system would be indicated by a robust DTH response (i.e., swollen ear in response to DNCB). Malnourished mice displayed impaired skin sensitivity to DNCB, compared with control-fed mice, indicating impaired immune function (Nejad & Mohagheghpour 1975). PEM reduced phagocytotic capabilities of macrophages in mice despite increased oxygen production (Teshima et al. 1995).

The effects of varying degrees of dietary protein on *in vivo* and *in vitro* cellular immunity has also been assessed by assessing DTH in bobwhite chicks (*Colinus virginianus*) (Lochmiller et al. 1993). Immune function was assessed by intradermal injections of PHA; immune function was suppressed in birds fed a low-protein diet. Both splenic and bursa masses were reduced in birds receiving low-protein diets. However, lymphocyte proliferation in response to several mitogens (e.g., Con A, pokeweed mitogen) did not differ among birds (Lochmiller et al. 1993).

PEM is associated with defective macrophage phagocytic function. For example, phagocytotic activity of macrophages in PEM mice was suppressed, despite increased respiratory burst (Teshima et al. 1995). However, the role of restricted dietary protein intake on antigen presentation (AP) is unclear

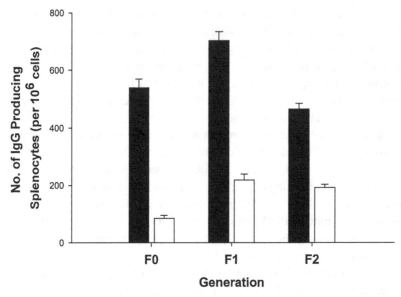

Fig. 6.7. Mean (±SEM) number of splenocytes from food-restricted female rats (F_0) (open bars), compared with control animals fed *ad libitum* (solid bars). Females were mated with males with *ad libitum* access to food. The number of splenocytes from the resultant F_1 and subsequent F_2 offspring were examined at 6 weeks of age; these offspring continued to show suppressed immune function two generations removed from the energy restriction. (From Chandra 1975.)

(Redmond et al. 1995). One study examined AP in peritoneal macrophages of mice that were fed either a low or moderate protein-rich diet for 8 weeks (Redmond et al. 1995). Both AP and splenocyte responses were maintained in mice with PEM.

Female rats were food restricted by 35%, compared with control animals fed *ad libitum*. IgG antibody-producing splenocytes were reduced by approximately six-fold (Chandra 1975). Females were mated with *ad libitum*-fed males, and the splenocytes of the resultant F_1 and subsequent F_2 offspring were examined at 6 weeks of age; these offspring continued to show suppressed immune function two generations removed from the energy restriction (Chandra 1975) (Figure 6.7).

In summary, it seems that low-protein intake can adversely affect immune function. Numerous clinical studies have correlated PEM with compromised immune function (Keusch 1982; Latshaw 1991). PEM appears to affect cell-mediated immune function more than cellular immune function. Because protein availability can change seasonally, even among herbivores, PEM may play an important role in the seasonal modulation of immune function and disease prevalence.

6.3.3. Limited Fat Intake and Immune Function

Caloric density of fats is nearly twice that of carbohydrates or protein. Nonetheless, high intake of dietary fat is usually correlated with *impaired* immune function. For example, the effects of nutrient density and dietary energy source on immune function were studied in weanling pigs (Van Heugten et al. 1996). Pigs were fed a standard laboratory diet *ad libitum* or pair-fed diets consisting of a 70% laboratory diet and 30% of either starch or isocaloric amounts of lard. Half of the animals were then challenged with lipopolysaccharide (LPS). Both cellular (lymphocyte proliferation in response to PHA or pokeweed mitogen) and humoral (anti-sheep red blood cells (SRBCs) production) immunities were assessed. Injection (immunization) with LPS reduced feed intake and weight gain, but did not affect energy utilization. Increasing energy density did not alter the depressed performance in LPS-treated pigs. The addition of fat improved efficiency of energy conservation, but depressed humoral immunity (Van Heugten et al. 1996).

Similarly, the effects of lipids on the immune system were investigated in adult C57/Bl6 mice. Animals were fed for 10 weeks on 1 of 4 diets: high or low polyunsaturated fatty acids, high saturated fatty acids, or standard laboratory chow (Crevel et al. 1992). High fat diets significantly reduced delayed hypersensitivity response after sensitization with tuberculin. Further IgM antibody formation against LPS was transiently reduced in high-fat fed mice, but IgG response to SRBC and IgE response to ovalbumin were unaffected. These results also demonstrate reduced immunocompetence induced by high-fat diets (Crevel et al. 1992). Thus, caloric energy is not the critical factor in the energetic modulation of immune function; rather, the quality of the metabolic fuel ingested (or lacking) seems to be the important feature in the support of immune function.

Because high-fat food availability varies on a seasonal basis, it is possible that the endocrine correlates to fat consumption might be involved in the seasonal modulation of immune function. To test that hypothesis, photoperiodic alterations in reproductive function and leptin concentrations were examined to determine any relationship to photoperiod-modulated changes in immune function (Drazen et al. 2001a, b). Siberian hamsters (*Phodopus sungorus sungorus*) were housed in either long- (LD 16:8) or short- (LD 8:16) day lengths. After 9 weeks in these photoperiodic conditions, blood samples were collected to assess leptin concentrations. One week later, animals were injected with KLH to evaluate humoral immunity. Body mass, body fat content, and serum leptin concentrations were correlated with reproductive responsiveness to photoperiod; short-day animals with regressed gonads exhibited a reduction in these measures, whereas short-day non-responders resembled long-day animals. In contrast, immune function was influenced by photoperiod but not reproductive status. Taken together, these data suggest that humoral immune function in Siberian hamsters is independent of photoperiod-mediated changes in leptin concentrations (Drazen et al. 2001a, b).

Treatment of mice with leptin improved phagocytosis and the expression of proinflammatory cytokines (Finck et al. 1998); these observations are consistent with earlier reports that mice with genetic defects in either leptin production or leptin receptor development display reduced macrophage phagocytosis and decreased production of proinflammatory cytokines (reviewed in Loffreda et al. 1998). The role of other metabolic hormones, such as orexin (hypocretin), neuropeptide Y, α-melanocyte stimulating hormone, or urocortin on immune function, has not been reported, but we anticipate that these regulatory metabolic pathways will also affect immunity.

Fat oxidation provides fuel for many tissues and, importantly, decreased glucose use and oxidation in muscle are signaled by fat oxidation byproducts. It is possible that fat fuels provide information for immunomodulation, but few studies address this issue (reviewed in Newsholme et al. 1993).

6.3.4. Limited Micronutrients and Immune Function

Micronutrients are critical for proper immune function. In many cases, the effects of caloric malnutrition may (also) reflect the effects of missing micronutrients on immune function. For example, zinc is often an important cofactor in immune reactions, and intake is often low during PEM or caloric malnutrition (Chandra 1997). Mice fed a zinc-deficient diet had reduced antibody production, and decreased mitogen and antigen-induced T-cell proliferation during the challenge part infection (Shi et al. 1997). Interferon (IFN)-γ production was also reduced in zinc-deficient mice (Shi et al. 1997). Furthermore, zinc-deficient mice had reduced numbers of spleen cells, and when infected with the parasitic nematode, *Helignosomoides polygyrus*, zinc-deficient mice displayed higher worm burdens and reduced IL-4 and IL-5 production than mice fed a zinc-sufficient diet (Shi et al. 1998). It is possible that many of the immunosuppressive effects of PEM might be attributable to zinc alone (Good & Lorenz 1992); if true, then administration of zinc to children with

PEM should restore cellular immunity to these individuals. In addition to zinc, other micronutrients that have been shown to support immune function include the trace elements, copper, selenium, and iron (Grimble 1994). Furthermore, vitamins A, E, D, B_6, and folic acid have been reported to enhance cytokine production (Grimble 1994; Chandra 1997).

Replacement of missing micronutrients can restore immunity, but there is little evidence that additional intake of micronutrients enhances immune function. For example, wound healing is impaired in individuals lacking vitamin A, C, or zinc because of enhanced epithelialization, collagen synthesis, and cell proliferation, respectively (Trujillo 1993). Because the availability of micronutrients to animals living in the wild vary on a seasonal basis, it is possible that fluctuations in these substances may be important for seasonal patterns of immune function and disease. To our knowledge, this possibility has not been addressed experimentally.

6.3.5. Prolonged Exercise and Immune Function

A strong link has been made between strenuous exercise and compromised immune function (Table 6.1) (Shephard & Shek 1995; Shephard et al. 1998). As noted previously, even a temporary reduction in immune function can increase the likelihood of infection, septic reactions to tissue injury, and potentially certain types of cancer (e.g., Nieman et al. 1989, 1990; Brenner et al. 1994; Shephard & Shek 1998). Athletes are often plagued by infection during training or competition (Shephard & Shek 1995). However, moderate exercise training can counteract the usual suppression of NK cell activity in obese women (Scanga et al. 1998).

Individuals who are most likely to suffer the effects of intense physical activity on immune function are athletes and military personnel. It is often

Table 6.1. *Prolonged Exercise and Immune Responses.*

Circulating NK cell count and activity	↓
Mitogen-induced lymphocyte proliferation	↓
Immunoglobulin concentrations	↓
Macrophage function	↓
Proinflammatory cytokine secretion	↓
Cutaneous response to antigen	↓
Downregulation of IL receptors	↑
Susceptibility to viral infections	↑

NK, natural killer; IL, interleukin.
From Shephard et al. 1998.

difficult to study these individuals because the intense physical activity often interacts with other stressors, including the positive and challenging aspects of the competition for athletes or the stresses of battle (real or simulated), sleep deprivation, and exposure to the elements (Shepard et al. 1998). Laboratory studies of strenuous exercise, either a single intense bout or a period of repetitive moderate activities, in which extraneous variables are controlled, have demonstrated that heavy exercise inhibits: (1) NK cell count (Shek et al. 1995), (2) lymphocyte proliferation in response to mitogenic stimulation (Verde et al. 1992a, b), (3) circulating immunoglobulin concentrations (Baum & Liesen 1997), and (4) DTH (Baum & Liesen 1997).

Although the immune response may only be compromised for hours, rather than days, intense exercise can still have important clinical consequences (Shepard et al. 1998). For example, data obtained from the United States military special warfare trainees indicates that the rate of respiratory illnesses during the 25 weeks of grueling training was over twice the overall United States incidence rate (Linenger et al. 1993). Individuals taking part in a United States Army Ranger training course exhibit a progressive increase in the rate of infections, despite heavy antibiotic usage, as the course progresses (Martinez-Lopez et al. 1993); furthermore, a subset of these Ranger students display decreased lymphocyte proliferation in response to mitogens (Martinez-Lopez et al. 1993). Again, immune function was compromised for only a few hours postexercise.

Other studies have demonstrated prolonged suppression of immune function after strenuous exercise. For example, in one study of military personnel, NK cell activity remained less than 50% of baseline 1 week after a single bout of prolonged exercise (65% of aerobic power for 120 minutes) (Shek et al. 1995). In all reported cases, immune deficits were reported after strenuous exertion (reviewed in Shephard & Shek 1995; Shephard et al. 1998). This has important clinical consequences, especially for individuals who may be required to engage in strenuous exercise under a number of conditions associated with negative energy balance, including sleep deprivation, psychological stressors, high or low ambient temperatures, or hypogravity conditions (see Shephard et al. 1998).

Taken together, either shortages on the input side or excesses on the expenditure side of the energy balance equation result in impaired immune function. Compromised immune function is common among individuals engaged in strenuous exercise, especially under conditions of negative energy balance, sleep deprivation, and psychological or physical stressors (Shephard et al. 1998). Compromised immune function often impairs both physical and cognitive performance by increasing susceptibility to opportunistic infection.

In the following section, the effects of specific perturbations to energy balance will be described.

6.4. Energetic Challenges to Immune Function

Animals must maintain a relatively constant flow of energy (i.e., energy intake $\geq$ energy expended) to the body, despite potentially large fluctuations in energetic availability in their environment (Wade & Schneider 1992). As noted previously, mounting an immune response likely requires resources that could otherwise be allocated to other biological functions (Sheldon & Verhulst 1996), and immune function should be "optimized" so that individuals can tolerate small infections if the energetic costs of mounting an immune response outweigh the benefits (Behnke et al. 1992). The following section will describe immune function during various energetically challenging events, including pregnancy, lactation, and migration.

6.4.1. Indirect Studies of Energetics and Infection

Within the past two decades, there has been a growing interest in the effects of pathogens and pathogenic infection on host life history strategies, population dynamics, as well as health and disease in both human and nonhuman animals (Hamilton & Zuk 1982; Sheldon & Verhulst 1996). Parasite infection is typically associated with reductions in growth, reproductive success, and survival of their infected hosts (Boonstra et al. 1980; Forbes & Baker 1991). From the pathogens' perspective, their hosts merely reflect the resources needed for the benefit of their own growth and reproductive success (Ewald 1996). Faced with a large variety of potential pathogens that attempt to exploit their hosts, many species of animals have evolved adaptive defenses to deter potential pathogenic infection and to minimize the potential harm of such infection (Forbes 1993; Sheldon & Verhulst 1996). Among diverse host defenses – such as autogrooming, protection through some physical barrier (e.g., fur or skin), or ingestion of antiparasitic foods – activation of the immune system is probably the most common defense. However, mounting an immune response, like any other physiological process, has an energetic cost. Increased immunity requires utilizing resources that could otherwise be allocated to other biological functions (Sheldon & Verhulst 1996). In fact, it has been suggested that immune function is likely "optimized" so that individuals would be able to tolerate small, less costly parasitic infections if the energetic costs of mounting an immune response against this parasite outweighs the benefits of preventing parasitic infection in the first place (Sheldon & Verhulst 1996;

Møller et al. 1998). The energetic costs of pathogenic infection or the elevation in immune function necessary to fend off these infections have been quantified only very recently.

Although energy availability can affect immune function, surprisingly few studies have directly assessed the energetic costs of mounting an immune response. Presumably, the initiation of an immune response, which involves activation of cytokines, induction of fever, and production of antibodies, requires a significant amount of energy. The majority of studies of energy use and immune function usually employ an indirect measure of energy use, such as growth or productivity (Spurlock 1997). For instance, the metabolic cost of mounting an antibody response was examined in pullets (Henken & Brandsma 1982). Adult pullets were injected with antigen SRBCs and energy consumption, protein, and fat gain were also assessed. During days 1–5 of the immune response, SRBC pullets retained more energy, deposited more fat, and had a lower maintenance requirement of metabolizable energy, compared with control pullets. However, on days 6–10, SRBC pullets deposited significantly less fat and significantly more protein. SRBC immunized animals consumed more food, but gained less weight than control pullets. These findings suggest that antibody production involves considerable energy consumption by shifting metabolism from energy storage in the form of body fat toward mobilization of energy.

Activation of immune function directly suppresses growth hormone (GH) secretion (Spurlock 1997). Endotoxin or LPS treatment mimics an infection without making the animal ill. Treatment with endotoxin or LPS results in reduced pulsatile release and low GH concentrations in rats and cattle (Elsasser et al. 1988; Kenison et al. 1991; Fan et al. 1994, 1995; Peisen et al. 1995), but not sheep (Coleman et al. 1993). IL-1β appears to mediate the effect of endotoxin on GH (Peisen et al. 1995). LPS treatment also redirects energy partitioning and suppresses growth in mice (Laugero & Moberg 2000). These studies provide a direct link between activation of immune function and inhibition of growth.

Insulin-like growth factor-I (IGF-I) induction in liver (and possibly muscle) is partially responsible for the anabolic actions of GH (Spurlock 1997). Immunological challenges result in low blood IGF-I concentrations in several species (Elsasser 1988; Prickett et al. 1992; Fan et al. 1994, 1996; Hathaway et al. 1996). IGF-binding proteins (IGF-BPs) are central in the regulation of IGF-I activity (Kelley et al. 1996). Several reports have indicated that IGF-BP concentrations increase during an immunological challenge. For example, IGF-BPs are increased in response to endotoxin or IL-1β treatment or parasite infection (Elsasser et al. 1988; Prickett et al. 1992; Fan et al.

Table 6.2. *Effects of Energy Deficit
on Human Immune Function.*

$CD4^+$ and $CD8^+$ counts	↓
NK cell counts	↓
B-cell counts	↓
Lymphocyte proliferation	↓
Cytokine production	↓
Susceptibility to infections	↑

NK, natural killer.
From Shephard et al. 1998.

1994, 1995). Taken together, immune challenges change anabolic processes by evoking adjustments in GH, IGF, and IGF-BP, and permit energy to be shifted to the energetically demanding immune system during activation (Ardawi & Newsholme 1985). The effects of an energy deficit on immune function are shown in Table 6.2.

Much of the early literature suggesting that infection is energetically demanding on host individuals derives from studies of bird species. For example, in great tits (*Parus major*) the prevalence of natural infections with the blood protozoan *Leucocytozoon* spp. increases with enlarged broods (Norris et al. 1994). Although other possibilities have not been ruled out, the increased infection rates are likely due to the increased energetic demands placed on the host in maintaining a large brood. That is, the greater the number in the brood, the greater the energy demand and, thus, the greater the potential reduction in immune function and subsequent increase in disease susceptibility. In support of this hypothesis, energy turnover in Eurasian kestrels (*Falco tinnunculus*), as assessed by the double-labeled water method, confirmed that increases in brood size alters daily energy expenditure (Jonsson et al. 1996).

The increase in parasite infection with increased brood size can be explained by the supposition that producing a larger brood requires additional energy; increased energy use reduces the total pool of available energy and may reduce the amount of energy allocated to immune function. Compromised immune function may leave an animal susceptible to parasitic infection. However, there is also an increase in parental care with increasing brood size because of the increased number of offspring that require care. The increase in energetic investment with increased parental care may also explain the reductions in immune function. To test this hypothesis, the association between parental investment and the prevalence of malaria was examined in great tits (*Parus major*) (Richner et al. 1995). Brood size was experimentally manipulated (thus controlling for the energy required for brood production), and both feeding effort and the prevalence of the hemoparasite *Plasmodium* were assessed in blood smears. Males that experienced large broods had significantly higher rates of food provisioning and had more than double the rate of parasite infection, compared with small-brood males. Thus, the increased energetic expenditure allocated toward food provisioning appears to increase susceptibility to parasite infection, presumably through reduced immune function. These results support the idea of a trade-off between reproductive effort (e.g., energy) and parasitic infection.

Although the results discussed thus far suggest an important relationship between energetics and parasite infection, the critical assumption that mounting an immune response is energetically expensive was not directly tested in these studies. Few data have been collected that speak to this question. However, some preliminary evidence supports the idea that parasitic infection and subsequent activation of the host immune responses has an energetic cost. For example, the effects of the parasite *Eimeria acervulina* on energy metabolism were assessed in broiler chickens (Takhar & Farrell 1979). Chickens were infected by injections of *E. acervulina*, and energy use was assessed by monitoring changes in body mass and food intake, as well as metabolizable energy and heat production using respiration chambers. As a result of parasite infection, chickens gained less weight, displayed increased heat production, and reduced metabolic use of their diets than uninfected chickens. These results suggest that infection with *E. acervulina* places significant energetic demands on the host animal. The energetics of immunity has also been assessed in the European starling (*Sturnus vulgarus*). Male and female nonbreeding European starlings were fed *ad libitum* commercial dog and cat food (Connors & Nickol 1991). Birds were then infected by placing a drop of water containing *Plagiorhynchus cylindraceus*, a relatively nonpathogenic worm species (i.e. does not cause serious illness in the host), in their gullets while control birds received water alone. Food consumption and urine and feces excretions were monitored, and host energy use and oxygen consumption was determined via bomb calorimetry. Energy use was divided into Ingested Energies (IE) (total energy consumed), Excretory Energies (EE) (energy excreted in urine and feces), and Metabolizable Energies (ME) (total amount of food energy available to an animal for use). Males infected with *P. cylindraceus* had significantly higher IE (i.e., they ingested more energy) than uninfected control males. Infected males also had higher EE and ME, but lower oxygen consumption, compared with control males. Female birds, however, did not differ on any parameter. These results suggest that parasitic infection (even nonpathogenic ones) affects total energy flow, increasing total energy demand, at least in male birds. These results are consistent with many studies showing impaired immune function in males, compared with females (Chapter 4).

There is a small but growing literature on the energetic costs of parasitic infection in mammals. Several studies have assessed the effects of energetic restrictions on prevalence and spread of parasitic infection in mammals. For example, the effects of fasting versus high-energy diets on IFNγ production and virus replication was assessed in calves (d'Offay & Rosenquist 1988). IFN levels were highest in nonfasted calves given the normal diet; however IFNγ was higher in fasted calves given the high-energy diet, compared with fasted calves given the normal diet. Energy intake may have influenced virus replication and excretion; fasted calves tended to secrete more virus than nonfasted calves. Similar results have been found for pigs. For example, worm burden was assessed in young pigs infected with the intestinal parasite *Ascaris suum* and maintained on either a high- or low-protein diet (Stephenson et al. 1980). Intestinal worm burden was higher in pigs maintained on the low protein diet, compared with high-protein diet animals. Gut weight was also significantly heavier in high protein-infected animals, compared with low protein-infected animals, which correlated with the number of worm harbored. Furthermore, laboratory rats infected with the intestinal parasite *Nippostrongylus brasiliensis* showed marked decreases in body mass, compared with uninfected control rats (Ovington 1985). Taken alone, these results suggest increased energy consumption by infected rats. However, food consumption was also reduced in infected rats. Whether reductions body mass was due to increased energy use or reduced food consumption cannot be teased apart in this study.

In a direct test of metabolic costs of mounting an immune response, house mice were injected with either saline or KLH (Demas et al. 1997a). Oxygen consumption and metabolic heat production were monitored periodically throughout anti-KLH antibody production using sensitive, indirect calorimetry. KLH-injected mice mounted significant immune responses, consumed more oxygen, and produced more metabolic heat than mice treated with saline (Figure 6.8). These results suggest that mounting an immune response requires significant energy and therefore requires using resources that could otherwise be allocated to other physiological processes (Demas et al. 1997a). Another study of the effects of mounting an antibody response was conducted in blue tits (*Parus caeruleus*) (Svensson et al. 1998). The authors reported an 8–13% change in basal metabolic rate, and dismissed this as insufficient to account for the reduction in immune response observed in low temperature-stressed individuals (Svensson et al. 1998). A 10% energy savings is sufficient to maintain numerous energy-conserving adaptations, including torpor, and may indeed support the adaptive trade-off between thermoregulation and immune function reported earlier.

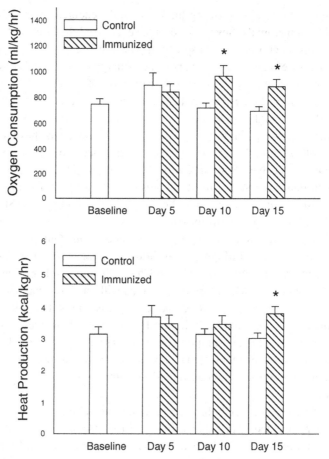

Fig. 6.8. (Top panel) Mean (±SEM) oxygen consumption (ml/kg) 5, 10, and 15 days after male house mice were injected subcutaneously with 150 μg of keyhole limpet hemocyanin suspended in 0.1 ml of sterile saline. (Bottom panel) Mean (±SEM) metabolic heat production 5, 10, and 15 days after male house mice were injected subcutaneously with 150 μg of keyhole limpet hemocyanin suspended in 0.1 ml of sterile saline. Significant differences ($p < 0.05$) are indicated by the asterisk. (Reprinted from *American Journal of Physiology*, vol. 273, G. E. Demas, V. Chefer, M. C. Talan & R. J. Nelson, "Metabolic costs of an antigen-stimulated immune respone in adult and aged C57B1/6J mice." pp. R1631–7, copyright ©1997a, with permission from the American Physiological Society.)

Food restriction and body mass loss are associated with deficits in reproductive function (Wade & Schneider 1992). Reproductive attempts by animals in poor condition can lead to low parental return on reproductive investment or even death of the mother and offspring. Ovarian hormones evoke changes in energy intake, partitioning, and utilization, but reproductive function appears to be more dependent on the availability of oxidizable metabolic fuels than body mass, body fat content, or hormone concentrations, per se (Schneider & Wade 1999). For example, simultaneous pharmacological blockade of both glycolysis or fatty acid oxidation inhibited estrous cycles in hamsters (Schneider & Wade 1989). If either one of the metabolic pathways was available, then estrous cycles continued. Pharmacological blockade of the fatty acid oxidation pathway had no effect on immune function, but inhibition of glycolysis compromises immune function (Demas et al. 1997a, b; Drazen et al. 2001).

When lymphocytes are activated by an appropriate immune stimulus, they go from a "resting" state into a state of high biochemical activity (Ardawi & Newsholme 1985). In addition to rapid cell division, production of antibodies, lymphotoxins, chemotactic factors, and mitogenic agents occurs. Although the primary energy source for lymphocytes appears to be ATP from glucose metabolism, a significant amount of lymphocyte energy is provided by glutamine (Cheung & Morris 1984; Ardawi & Newsholme 1985).

6.4.2. Temperature and Immune Function

Given the assumption that immune function requires substantial energy, immune function should be compromised in animals maintained in low, compared with high temperatures. To test this hypothesis, the interaction between photoperiod and temperature was examined on IgG concentrations and splenic mass in male deer mice (*Peromyscus maniculatus*) (Demas & Nelson 1996). Animals were maintained in LD 16:8 or LD 8:16 photoperiods and either in 20°C or 8°C temperatures. Serum IgG levels were elevated in short-day mice maintained at normal room temperature (i.e., 20°C), compared with long-day animals housed at either 20°C or 8°C (Figure 6.9). Long-day deer mice kept at 8°C temperatures had reduced IgG concentrations, compared with long-day mice maintained at 20°C, whereas mice exposed to short days and low temperatures had IgG concentrations comparable with long-day mice maintained at 20°C. In other words, short days elevated IgG concentrations over long days. Low temperatures caused a significant reduction in IgG concentraions. The net effect of short-day enhancement and low-temperature reduction of IgG levels is no appreciable difference from baseline (i.e., long-day mice kept at 20°C) (Demas & Nelson 1996). This adaptive system may help animals cope with seasonal stressors and ultimately increase reproductive fitness. To enhance immune function in anticipation of demanding winter conditions, animals must initiate these adaptations well in advance of the demanding

Energetics and Immune Function

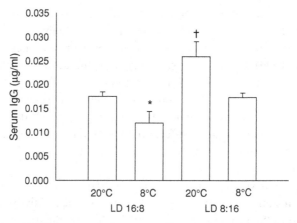

Fig. 6.9. Mean (±SEM) serum IgG concentrations (μg/ml) of male deer mice (*Peromyscus maniculatus*) housed in long (LD 16:8) or short (LD 8:16) days and either mild (20°C) or low (8°C) temperatures. The results from this study suggest that short days lead to enhancements of immune function in anticipation of winter, whereas stressors such as low temperatures lead to a reduction in immune function. When animals are exposed to short-day lengths in combination with low temperatures, the result is a "balance" between short-day enhancement of immune function and a "stress-induced" decrement in immune function. †Significantly greater than all other groups ($p < 0.05$). *Significantly less than all other groups ($p < 0.05$). (From Demas & Nelson 1996).

conditions. Again, the most reliable environmental cue for time of year is the annual pattern of changing photoperiod.

Acute low-temperature exposure of blue tits (*Parus caeruleus*) caused a reduction in antibody production against novel proteins (Svensson et al. 1998). Energy requirements increased between 8–13% of the basal metabolic rate after injection of the nonreplicating antigen (diphtheria-tetanus vaccine) (Svensson et al. 1998). Presumably, higher energy costs would have accrued for a replicating antigen, such as a typical bacterial or viral infection might evoke.

Some mammals contend with severe seasonal food scarcity and low ambient temperatures by undergoing hibernation. Hibernators typically regulate core body temperature at 1°–2°C above ambient temperature. In common with other hibernating mammals, ground squirrels do not hibernate continuously, but arouse at approximately 7-day intervals, during which body temperature increases to normothermic values of 37°C for ≈16 hours; thereafter, they return to hibernation and sustain low body temperatures until the next arousal. A hibernation season encompasses many such arousals that ultimately consume 60–80% of a squirrel's winter energy budget. Ubiquitous among hibernators, periodic arousals are assumed to be adaptive, although their functional significance is unknown and disputed. Based on morphological data, it appears that host-defense mechanisms are downregulated during the hibernation season and preclude normal immune responses. In support of this hypothesis, the acute-phase response to LPS is arrested during ground squirrel hibernation and fully restored upon arousal to normothermia (Prendergast et al. 2001). Treatment of normothermic squirrels with LPS resulted in a 1°–1.5°C fever that was sustained for more than 8 hours. LPS was without effect in hibernating squirrels, neither inducing fever nor provoking arousal (Prendergast et al. 2001). LPS-treated squirrels did, however, develop a fever days later, when they aroused from hibernation, and the duration of this periodic arousal was increased six-fold above values generated by control treatments. Intracerebroventricular infusions of prostaglandin E_2 provoked arousal from hibernation and induced fever, suggesting that neural signaling pathways that mediate febrile responses are functional during hibernation, and that periodic arousal functions to activate immune function (Prendergast et al. 2001).

6.4.3. Social Factors and Immune Function

Importantly, social conditions also dramatically influence immune function (Rabin & Salvin 1987; Rabin et al. 1987; Karp et al. 1993; Barnard et al. 1994; Klein & Nelson 1999), and a large part of this effect may reflect the energetic savings of huddling behavior. Social isolation is a common laboratory method of evoking a glucocorticoid response among laboratory mice and rats. Traditionally, the stress of social isolation has been assigned to lack of contact comfort. However, the isolation response is not invariable among naturally selected species. The stress response to isolation often changes as a function of reproductive condition, which probably reflects the tendency of rodent social organization to change from highly territorial during the breeding season to the formation of communal groups during the winter (McShea 1990). The contribution of changes in social organization in the mediation of seasonal fluctuations of immune function has not been well examined; but, in the field, social huddling may conserve sufficient energy to enhance immune function individuals, compared with isolated animals. Furthermore, the role of the social system (i.e., socially monogamous versus polygynous) in seasonal fluctuations in immune function remains unspecified. The physiological processes by which social factors are transduced to enhance or compromise immune function are also unknown, but we hypothesize that glucocorticoids are involved. Immune function is not a static process, but an optimized energetically expensive physiological process that is affected by numerous environmental and biotic forces that affect energy availability, consumption, and utilization.

We would predict that immune function would be compromised in individuals that failed to engage in huddling behavior. Social factors, such as

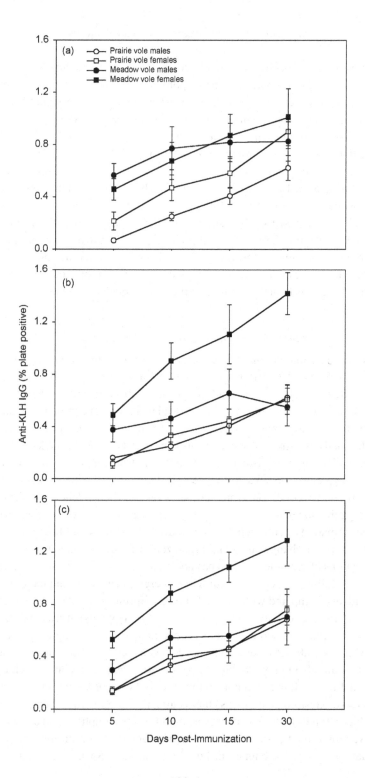

182

male–male interactions or exposure to receptive mates can influence the timing and extent of seasonal reproductive quiescence. Social factors also influence immune function. For example, compared with house mice (*Mus musculus*) housed five per cage, individually housed mice show higher proliferative responses to the T-cell mitogen, Con A (Rabin et al. 1987), increased primary and secondary antibody responses (Rabin & Salvin 1987; Karp et al. 1993), more cytokine production (Rabin et al. 1987), and greater resistance to infection (Plaut et al. 1969; Rabin & Salvin 1987). The impact of social factors on immune function has not been fully examined in an ecological context. Recently, the effect of social interactions on immune function was examined in polygynous meadow voles (*Microtus pennsylvanicus*) and monogamous prairie voles (*Microtus ochrogaster*).

Among individually housed animals there were no apparent species or sex differences in antibody production against KLH (Klein & Nelson 1999). Pairing animals with either a same sex or opposite sex conspecific for 28 days unmasked both sex and species differences in humoral immunity (Klein & Nelson 1999) (Figure 6.10). Among animals housed in either same-sex or mixed-sex pairs, sex differences were only apparent among the polygynous meadow voles, in which females had higher antibody responses than conspecific males. Sex differences were not observed among monogamous prairie voles. Overall, meadow voles displayed higher antibody responses than prairie voles in both same-sex and mixed-sex pairs. These data suggest that sex and species differences become apparent only in the context of the social environment in which breeding occurs; and because social organization changes seasonally, seasonal changes in immune function may also reflect changes in social relations across the year.

6.4.4. Pregnancy and Immune Function

Because of the increased metabolic costs, we would predict that immune function should be compromised in pregnant, compared with nonpregnant individuals. A large clinical and animal literature suggest that immune function is compromised during pregnancy. The prevailing hypothesis to explain compromised immune function during pregnancy is the necessity to protect the genetically incompatible fetus from rejection by the maternal immune

---

Fig. 6.10. Serum anti-keyhole limpet hemocyanin (KLH) IgG responses in male and female meadow and prairie voles housed individually (a), in same-sex pairs (b), or in mixed-sex pairs (c) for 28 consecutive days prior to immunization with KLH. Blood samples were collected 5, 10, 15, and 30 days later. (From Klein & Nelson 1999.)

system. Essentially, the fetus may be considered an intrauterine fetal allograft tissue transplant because of genetically dissimilar paternal antigens (Fischer et al. 1996).

Marked maternal immunosuppression has been reported during pregnancy in nonhuman mammals, especially around the time of parturition in ewes, dogs, and cows (Wells et al. 1977; Burrells et al. 1978; Yamamoto et al. 1980; Lloyd et al. 1983). In a recent study, splenocyte proliferation to Con A was inhibited in pregnant Siberian hamsters housed in long-day lengths (Drazen 2001). Remarkably, neither pregnant nor lactating female hamsters housed in short days experienced immunocompromise, compared with short-day females that were not pregnant or lactating. The environmental stressors present during winter decrease energy availability by reducing food availability, while also increasing physiological energy expenditure through an increase in metabolic rate (Blank & Desjardins 1984, 1985; Castle & Wunder 1995). Previous studies have shown that short days initiate adaptations to deal with the energetic stressors of winter by making energy partitioning more efficient during this time. For example, in Siberian hamsters, the capacity for nonshivering thermogenesis is greatly increased when animals are housed in short days (Wiesinger et al. 1989).

Short-day exposure can also have dramatic effects on body mass; depending on the species, an individual will either reduce or increase body mass as a winter-coping strategy. Reduction of body mass decreases absolute energy requirements, reduces foraging time, and limits exposure to unfavorable climatic conditions (Iverson & Turner 1974; Dark & Zucker 1983; Bartness & Wade 1985). Autumnal fattening provides a convenient energy store, as well as insulation (Davis 1976; Bartness & Wade 1985; Bartness 1995). In addition, deer mice held on short days appear to be buffered against 2-DG-induced metabolic stress; 2-DG suppressed cell-mediated immunity in long-day mice, but not in short-day mice (Demas et al. 1997a, b). It is possible that the lack of pregnancy- or lactation-induced immunosuppression in short days is another short-day adaptation to protect an individual during the harsh conditions of winter, making energy partitioning more efficient. Components of immune function are reduced in short-day housed Siberian hamsters; so, it seems reasonable, from an energetic perspective, that immune function should not be further compromised, given that this scenario would likely lead to death.

Humoral immunity appears to be compromised during pregnancy. Longitudinal measurements of antibody titers during pregnancy have revealed significant reductions in IgG levels, slight, but significant reductions in IgA levels, but no changes in IgM or IgE levels (Amino et al. 1978; Ostensen et al. 1983; Falkoff 1987). Preexisting, antibody levels specific to either herpes simplex, measles, rubella, or influenza A were reduced during pregnancy, suggesting suppressed immune function (Baboonian & Griffiths 1983).

Cell-mediated immune function also appears to be suppressed during pregnancy in euthermic mammals. For example, the incidence of malignancies is increased during pregnancy (reviewed in Szekeres-Bartho et al. 1990). Studies have indicated that the dermal response to tuberculin exposure is reduced or unchanged (Montgomery et al. 1968) among pregnant women, and the survival of skin homografts or kidney transplants is prolonged in pregnant women (Rani & Mittal 1989; Fischer et al. 1996). Susceptibility to several types of infections increase during pregnancy (Pickard 1968; Krishnan et al. 1996a); epidemiological studies have indicated increased susceptibility to viral, bacterial, protozoal, and fungal infections, especially among pregnant women in non-Western societies or in industrialized societies 40–50 years ago (Roberts et al. 1996; Styrt & Sugarman 1991). These results might reflect poorer sanitation and medical care in these women; however, because women living in nonindustrialized countries have less somatic energetic reserves than women currently living in industrialized countries, these results are also consistent with our energetic hypothesis that immune function should be compromised during pregnancy.

6.4.5. Viral Infections

During the 1957 Asian influenza epidemic in New York City, 40% of the deaths of individuals under the age of 50 were pregnant women, most in their third trimester. The odds of developing clinical symptoms of polio were 60% higher among pregnant women from 1949 to 1955 (Siegel & Greenberg 1955). The incidence of viral encephalitis and paralytic poliomyelitis after the 1917–1918 influenza pandemic was higher among pregnant women (Styrt & Sugarman 1991). Similarly, the incidence of nonA and nonB hepatitis in Kashmir was increased 600% among pregnant women. The odds of mortality from hepatitis were elevated from 0 to over 40% for women in the third trimester in India and Iran, but not in Canada (Khuroo et al. 1981; D'Cruz et al. 1968; Borhanmanesh et al. 1973; Delage et al. 1986).

6.4.6. Bacterial Infections

Increased concentrations of Hanson's disease (leprosy) bacilli in skin smears were observed in 35% of women followed prospectively through pregnancy, compared with only 2% increases observed in women prior to conception (Duncan et al. 1982). The majority of aggravated symptoms occurred during the third trimester of pregnancy, and a similar worsening was observed during lactation. Pregnant women were also at increased risk for flare-ups of tuberculosis prior to the availability of antibiotics, and pregnant women and domesticated mammals continue to be at increased risk for listeriosis (Nieman & Lorber 1980). Pregnancy also may be a risk factor for *Gonococcus salpingitis* and *Staphylococcus aureus* (Roberts et al. 1996).

6.4.7. Parasitic Infections

Parastitemia with *Plasmodium falciparum* (malaria) is increased among pregnant women (Bray & Anderson 1979). Sixty-eight percent of women, aged 15–40, dying from amoebic colitis in Nigeria, were pregnant (Abioye 1973). Amebobiasis due to *Entameba histolytica* is more severe during pregnancy (Abioye 1973), and infections due to *E. histolytica, Giardia lamblia,* or *Trichomonas hominis* were associated with fetal growth retardation (Brabin & Brabin 1992). Pregnancy also might be a risk factor for helminthic infections, visceral leishmaniasis, and African trypanosomiasis (sleeping sickness) in women (Brabin & Brabin 1992).

Studies examining *Leishmania major* infection in mice revealed that pregnant female mice were more susceptible to infection (i.e., develop larger cutaneous wounds and had higher parasite burden) than nonpregnant females (Krishnan et al. 1996a). Assessment of immune function in these mice revealed that Th1 responses (i.e., IFN-γ and IgG2a) against *L. major* were suppressed during pregnancy, and Th2 responses (i.e., IL-4, IL-5, IL-10, and IgG1) were elevated (Krishnan et al. 1996a). Consequently, although Th1 responses are important for resistance against *L. major* infection, elevated Th1 immune responses compromise fetal survival, whereas heightened Th2 immune responses confer fetal protection (Krishnan et al. 1996b; Mosmann & Sad 1996). These data support the hypothesis that pregnancy and resistance against parasite infection are energetically incompatible. The endocrine mediators of these immunological changes during pregnancy, as well as the seasonal changes in energy utilization for mounting Th1 and Th2 immune responses, have not been examined.

6.4.8. Fungal Infections

A number of fungal infections have been shown to increase in prevalence during pregnancy, including those caused by *Candida* and *Chlamydia* (Roberts et al. 1996). Coccidioidomycosis dissemination is increased during pregnancy apparently in response to elevated estrogen concentrations (Styrt & Sugarman 1991). Some *Bacteriodes* strains appear to be able to use estrogens as a growth factor in place of vitamin K (Styrt & Sugarman 1991). Pathogenesis of other fungi [e.g., *Saccharomyces cerevisiae* (Bakers' yeast)] appear to be unrelated to estrogen availability (Styrt & Sugarman 1991).

6.4.9. Lactation and Immune Function

Lactation is energetically costly (Fairbairn 1977; Bronson 1989). Consequently, we predict that immune function should be compromised in lactating, compared with nonlactating individuals. In early lactating cows, immune function remained compromised (Ropstad et al. 1989); however, there were no significant relationships between immune responses and serum indicators of metabolic status. As described previously, lactating Siberian hamsters display low splenocyte proliferation, compared with nonpregnant or nonlactating females (Drazen 2001). Housing in short days ameliorated this immunosuppression associated with lactation (Drazen 2001). Treatment of cows with glucocorticoids for ketosis during early lactation further suppressed immune responses (Ropstad et al. 1989). Similar findings have been reported in ewes (Burrells et al. 1989), dogs (Lloyd et al. 1983), and women (Yamamoto et al. 1980; also see Birkeland & Kristoffersen 1980).

Lactating bighorn ewes (*Ovis canadensis*) have increased parasite infection in fecal samples, compared with nonlactating females (Festa-Bianchet 1989). Elevated parasitic infection likely reflects compromised immune function in lactating ewes, but this hypothesis remains to be tested directly. In fact, few studies have directly assessed the energetic cost of mounting an antibody response, although the initiation of an immune response (i.e., inflammation, activation of cytokines, induction of fever) presumably requires substantial energy. For example, every 1°C increase in body temperature requires a 7–13% increase in caloric energy production, depending on the species (Maier et al. 1994). Because basal metabolic rate triples or quadruples during lactation (Bronson 1989), any energy shortages will be inconsistent with full immune function.

6.4.10. Migration and Immune Function

Migration is very energetically costly. We would predict that immune function should be compromised in migrating, compared with nonmigrating or nonmigratory individuals. Thymic tissue regresses at the onset of the spring migration and breeding, then regrows at the end of each breeding season (mid-summer) in both male and female adult mallard ducks (*Anas platyrhynchos*); the thymus of mallards also regresses immediately before the autumnal migration (Höhn 1947). The physiological stress associated with migration and breeding was considered incompatible with full thymic size and function (Höhn 1947). Similar observations have been made for house sparrows (*Passer domesticus*) and robins (*Turdus migratorius*) (Höhn 1956).

The reduced relative splenic size of white-crowned sparrows (*Zonotrichia leucophrys gambelii* and *Zonotrichia leucophrys nuttalli*) at the beginning of the breeding season in western North America cannot be attributed to the "metabolic stress of migration," because both migratory and nonmigratory populations displayed an identical seasonal pattern of splenic development (Oakson 1953, 1956). Data on the two populations of white-crowned sparrows are consistent with the hypothesis that metabolic stress associated with breeding may impair immune function. The adaptive significance of increased spleen mass in other species before the autumnal migration has been suggested to reflect an enhancement of immune function, particularly of the young birds after hatching, in advance of winter (Fänge and Silverin 1985; Silverin 1999). One parsimonious, proximate explanation for the seasonal pattern of lymphatic organ mass among birds is that the high energetic needs associated with migration (or breeding) are incompatible with highly developed lymphatic tissue, but this possibility remains untested in the field.

The possibility that immune function is enhanced in birds during the winter, especially in birds that migrate to thermoneutral climates, to keep parasite activities minimal also remains open (Beaudoin et al. 1971; John 1994). For example, it has been hypothesized that *Plasmodium* infections remain latent during the winter, but parasitemia becomes evident during vernal migration or breeding (Beaudoin et al. 1971; Alexander & Stimson 1988). Increased prevalence or incidence of relapse of blood protozoa of *Plasmodium, Leucocytozoon, Haemoproteus*, and *Trypanosoma* species has been reported during spring and summer, coincident with migration or breeding, among many avian species (see John 1994 for review).

6.5. Energy Restriction and Autoimmunity

The effects of chronic energy restriction on the development of autoimmunity were assessed in male BXSB mice (Kubo et al. 1992a). BXSB mice exhibit a human, lupus-like disease with B-cell hyperplasia in peripheral lymphoid tissue that occurs more often in males than females. Male BXSB and C57BL/6 mice, serving as controls, were given 1 of 3 diets: (1) a fixed diet equal to 12 kcal/day/mouse, (2) a 40% reduction in the control diet initiated at 6 weeks of age, or (3) a 40% reduced diet starting at 4 months of age. Animals were killed between 3–5 months of age, and several aspects of immune function were assessed at the time of death. Chronic energy restriction inhibited development of autoimmunity in BXSB mice, regardless of the age at initial restriction. Energy restriction also prevented normal reductions in immune function associated with age. Specifically, IL-2 production, cell-mediated cytotoxic responses, and mixed lymphocyte reactivity were reduced in energy-restricted mice.

The effects of chronic caloric restriction and development of autoimmunity were assessed in NZB mice (Kubo et al. 1992b). NZB mice restricted to 50% *ad libitum* food intake levels at 6 weeks of age had longer life spans, a return to normal numbers of T-cell and B-cell populations, as well as normal responses to mitogenic stimulation, compared with unrestricted NZB mice. Similarly, the effects of undernutrition without malnutrition were assessed in both BXSB and MRL/I mice, two inbred strains associated with lymphoproliferative and autoimmune diseases (Ogura et al. 1990). Mice were fed high-carbohydrate, low-fat diets either *ad libitum* or at 60.2% restriction. Chronic undernutrition inhibited autoimmunity in both BXSB and MRL/I mice. Energy restriction inhibited the accumulation of Ly-1 B-lymphocytes throughout the lymphoid system almost to the level of normal, autoimmune-resistant C57/B16 mice.

The effects of energy restriction on immune function was assessed in immune thrombocytepenic pupora-prone (NZW $\times$ BXSB) mice (Mizutani et al. 1994). Mice were either given free access to food or received 32% restriction at 6, 14, 17, or 22 weeks of age. The onset of autoimmunity was prevented by energy restriction at 6 weeks of age. Platelet-associated IgG autoantibody levels and the number of splenic antiplatelet antibody-forming cells were reduced. Life span was also extended in mice restricted at 6 and 14 weeks of age, but not at later times.

6.6. Summary

Optimal immune function requires energy. The observation that immune function is generally compromised during specific energetically demanding times – such as during mating activities, pregnancy, lactation, winter, migration, or molting – is consistent with the hypothesis that immune function is optimized (Zuk 1990; John 1994). Taken together, it appears that perturbations on either side of the energy balance equation can compromise immune function. If the energy deficit is severe or if the animal is exposed to opportunistic pathogens during compromised immune function, then disease may ensue. In the following chapters, the hormonal mechanisms that mediate the shunting of energy between immune function and other anabolic or energy-expensive processes will be reviewed.

7

Hormonal Influence on Immune Function

> The human body has been designed to resist an infinite number of
> changes and attacks brought about by its environment. The secret
> of good health lies in successful adjustment to changing stresses on
> the body.
>
> Harry J. Johnson, 1968

7.1. Introduction

Over half a century ago, Walter B. Cannon noted the ability of higher vertebrates to maintain their internal *milieu* within very narrow operating limits,
a process termed homeostasis (Cannon 1939). This phenomenon has been
appreciated for decades with regard to the internal steady-state balance of
physiology and biochemistry. However, the immune system has typically
been considered to be a coordinated unit that acts independently of the rest
of the brain/body in response to pathogens, with the goal being to eliminate
foreign factors and return to the steady-state that existed prior to the detection
of the invading substances. There is now considerable evidence that both neural and endocrine factors act to maintain the immune system within "safe"
operating limits (i.e., to avoid overactive immune responses that may lead
to autoimmune disease or underactive immune responses that may lead to
ineffective removal of foreign pathogens). There is also an abundance of evidence that suggests that the immune system acts to maintain the neural and
endocrine systems within narrow operating limits.

The focus of this chapter is to evaluate endocrine–immune interactions,
and to determine the extent to which these interactions contribute to seasonal
patterns in immune function and disease. In order for the immune system to
be considered part of a homeostatic network along with the neuroendocrine
system, three criteria must be met: (1) there must be communication between
the immune system and endocrine glands (or central nervous system (CNS)
mechanisms that regulate the output from endocrine glands), (2) the flow of

information between the endocrine and immune systems must be bidirectional, and (3) the immune system must influence physiological adjustment to environmental change (Husband 1995). There is an enormous body of literature providing evidence for complex interactions between the endocrine and immune systems that satisfy these criteria. Changes in hormone concentrations represent some of the most salient seasonal changes in physiology. The extent to which seasonal changes in immune function reflect these endocrine adjustments has only recently become the subject of study. The goal of this chapter is not to provide an exhaustive review of all the available literature, but to provide a general overview of what is known regarding endocrine–immune interactions and to place this knowledge in an integrated conceptual framework.

It remains controversial whether or not direct synaptic connections exist between neurons and immune cells, although direct polysynaptic projections from the brain to the spleen have recently been identified (Drazen et al. 2001). Regardless of whether or not nerve cells make direct synaptic contact with immune cells, receptors for numerous neurotransmitters have been identified on lymphocytes, and these factors can influence lymphocyte function and migration (Ottaway & Husband 1992, 1994). This observation suggests that neurotransmitters normally act under physiological conditions to modulate immune function. Numerous receptors for cytokines have been localized in brain regions, including the hypothalamus, thalamus, hippocampus, cortex, and cerebellum (reviewed in Besedovsky & del Rey 1996); cytokine administration can have pronounced effects on brain electrical activity (e.g., Xia et al. 1999).

In common with reciprocal interactions between the nervous and immune systems, similar interactions exist between the endocrine and immune systems (for recent reviews, see Campbell & Scanes 1995; Daynes et al. 1995; Drača, 1995; Fabris et al. 1995; Besedovsky & del Rey 1996; McEwen et al. 1997; Nussdorfer & Mazzocchi 1998; Raber et al. 1998). A model for these endocrine–immune interactions is presented in Figure 7.1. Receptors for cytokines have been identified on endocrine glands, such as the pituitary, thyroid, pancreas, and gonads, suggesting that the immune system is capable of influencing endocrine function (reviewed in Besedovsky & del Rey 1996; Table 7.1). In addition, numerous studies indicate that hormone administration or removal of endocrine glands can have a marked impact on immune function; leading to either suppression or stimulation, depending on the hormone administered or removed and the dose (Table 7.2). Likewise, numerous receptors for hormones have been identified on immune cells and lymphoid tissue (Table 7.3), suggesting that hormones may act directly on immune cells

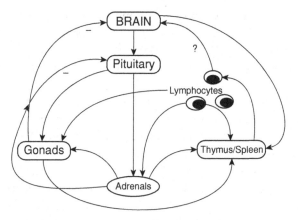

Fig. 7.1. A schematic depicting the major interactions among the nervous, endocrine, and immune systems. Other interactions between endocrine components and the immune system have been identified. However, this model is intended to provide a theoretical framework to understand some of the well-characterized interactions.

to influence immune function. The specific effects of endocrine factors on immune function, and immune factors on endocrine components, are discussed in detail.

7.1.1. Stress, the Hypothalamic-Pituitary Adrenal Axis, and Immune Function

Animals are exposed to several stressors that change seasonally, including competition for mates, reduced food and water availability, and extremes in fluctuating temperatures. Although these seasonal stressors are salient, very few studies have examined the effects of naturally occurring seasonal stressor on immune function and susceptibility to disease (see Demas & Nelson 1996, 1998a, b). In contrast, several laboratory studies have been conducted that illustrate that animals exposed to acute environmental stressors exhibit lymphoid tissue involution as part of an adaptation Hans Selye termed "General Adaptation Syndrome" (Selye 1936). This study also demonstrated that the HPA axis is involved in the response to stressors; stress-induced thymic involution was reduced in animals that had been adrenalectomized or hypophysectomized. It is now widely accepted that stress, whether psychological or somatic, can influence immune function and susceptibility to disease. The working definition for stress in this book is: a complex, dynamic condition in which disturbing forces (i.e., physical or psychological "stressors") disrupt or threaten homeostasis (Chrousos 1992; Sternberg et al. 1992; Wilder

Table 7.1. *Changes in Hormone Concentrations Following Cytokine Administration* in vivo.

Hormone	Cytokine	Route	Species	Result
ACTH	IL-1β	ip, icv, iv	Rat	↑
		ip	Mouse	↑
	IL-1α	ip, icv	Rat	↑
		iv	Rat	↑, ↔
	TNF-α	ip	Mouse, rat	↔
		iv	Rat, humans	↑
		icv	Rat	↔
	IL-2	iv	Human	↑
		ip	Rodent	↔
	IFN-γ	ip	Rodent	↔
		icv	Rat	↑
TSH	IL-1β	sc	Rat	↓
	IL-1	icv	Rat	↓
	TNF-α	iv, icv	Rat	↓
	INFγ	icv	Rat	↓
GH	IL-1β	icv	Rat	↓
		ip	Rat	↔
	IL-1	icv	Rat	↑
	TNF-α	iv	Human	↔
		iv	Bovine	↓
PRL	IL-1β	iv, icv	Rat	↔
	IL-1α	iv, icv	Rat	↔
	IL-1	icv	Rat	↑
	TNF-α	iv	Rat	↔
		iv	Human	↑
	INF-γ	icv	Rat	↔
LH	IL-1β	icv	Rat	↓
	IL-1α	icv	Rat	↓
		iv	Rat	↔
FSH	IL-1α	icv	Rat	↔

ip, intraperitoneal; icv, intracerebroventricular; iv, intravenous; sc, subcutaneous; ↑, increased; ↓, decreased; ↔, no significant change.
Source: Adapted from Scarborough 1990.

1995). In this sense, the general adaptation syndrome represents the body's attempt to counteract the effects of stressors to return to a state of homeostasis. The subject of the stress response is complex and controversial, and will not be further elucidated here. Reviews on this topic may be found elsewhere (e.g., Chrousos 1992; Chrousos & Gold 1992; Sternberg et al. 1992).

The stress response is regulated by a coordinated cascade of events involving the HPA axis and the sympathetic autonomic nervous system (Rose 1985; Antoni 1986;). After a stressor is perceived by the brain, corticotropin-releasing hormone (CRH) is released from the hypothalamus. In turn, CRH

Table 7.2. *Effects of Hormones and Neuropeptides on Immune Function.*

Hormone/ Releasing factor	Effect	Reference
Glucocorticoids	Decreased cytokine production	Kern et al. 1988; Culpepper & Lee 1985
	Reduced T-cell production	Daynes & Araneo 1989
CRH	Immunosuppression	Smith et al. 1992
GH	Stimulates cytokine production	Kelley 1989
GHRH	Stimulates proliferation, inhibits NK cell activity	Zelazowski et al. 1989
Androgens	Increased susceptibility to infection	Levine & Maddin 1962
	Ameliorates symptoms of autoimmune disease	Roubinian et al. 1977 Verheul et al. 1981
Estrogens	Exacerbates symptoms of autoimmune disease	Roubinian et al. 1978
	Depresses cell-mediated immunity	Albin et al. 1988, 1989 Kuhl et al. 1983; Terrazas et al. 1994
	Enhances response to T-cell antigens	Brick et al. 1985
GnRH	Reverse thymic atrophy	Marchetti et al. 1989
	Decreased absolute thymocyte numbers	Rao et al. 1993
	Decreased serum Ig concentrations	Jacobson et al. 1994
PRL	Enhances lymphocyte proliferation	Hartman et al. 1989
	Enhances several immune parameters	Berczi & Nagy 1991
TH	Promotes thymocyte maturation and differentiation	Johnson et al. 1992
TSH	Enhances antibody response	Blalock et al. 1985; Kruger & Blalock 1986
DHEA	Enhances immune parameters	Shealy 1995; Loria et al. 1996 Padgett et al. 1997
	Restored IL-2 production in aged animals	Weksler 1993; Daynes et al. 1993

acts on the anterior pituitary to cause the secretion of adrenocorticotropic hormone (ACTH). ACTH acts on the adrenal cortex to cause the secretion of the glucocorticoids. In humans, the primary glucocorticoid is cortisol, whereas the principal glucocorticoid in most rodents is corticosterone. In conjunction with the activation of the HPA axis, sympathetic terminals that directly innervate the adrenal medulla stimulate the release of epinephrine. Together, these two systems act to mediate the physiological and biochemical effects of stress.

Table 7.3. *Expression of Hormone Receptors on Leukocytes and Lymphoid Tissue.*

Receptor	Expressed on	Reference
Glucocorticoids	Lymphocytes	Lippman & Barr 1977
	Bursacytes	Sullivan & Wira 1979; Compton et al. 1990
	Thymocytes	Compton et al. 1987
	TEC	Dardenne et al. 1986
Gonadal steroids	TEC	Grossman et al. 1979; Sakabe et al. 1986
	Bursal epithelium	Sullivan & Wira 1979
	Thymocytes	Raveche et al. 1980
ACTH	Lymphocytes	Plaut 1987; Clarke & Bost 1989
TSH	Lymphocytes	Blalock 1985; Kruger et al. 1989
GH	Lymphocytes	Eshet et al. 1984; Kiess & Butenandt 1985
	Thymocytes	Arrenbrecht 1974
PRL	Lymphocytes	Russell et al. 1985; Hooghe et al. 1993
	TEC	Dardenne et al. 1989
Thyroid hormone	Lymphocytes	Csaba et al. 1977; Lemarchand-Beraud et al. 1977
	TEC	Villa-Verde et al. 1992
GnRH	Lymphocytes	Kruger et al. 1989; Costa et al. 1990
GHRH	Thymocytes	Guarcello et al. 1991
CRF	Lymphocytes	Smith et al. 1986
TRH	Lymphocytes	Kruger et al. 1989
DHEA	T-cells	Meikle et al. 1992

TEC, thymic epithelial cells.
Source: Adapted from Marsh & Scanes 1994.

Not only do stressors act to stimulate the HPA axis and the sympathetic nervous system, but numerous other systems in the CNS and periphery are affected by the stress response, and these secondary responses can also affect immune function. Thus, the effects of stress on immune function are complex and cannot be considered independent from secondary effects on other systems. For example, chronic stress inhibits numerous neuroendocrine factors, including growth hormone (Chrousos & Gold 1992), prolactin (Stephanou et al. 1992), thyroid-stimulating hormone (Chrousos & Gold 1992), gonadotropin-releasing hormone, and luteinizing hormone (Rivier & Rivest 1991). In turn, these hormones can affect immune function. Not only does activation of the HPA axis secondarily affect other systems, but reciprocal feedback from these systems also helps to regulate the HPA axis (e.g., Kamilaris et al. 1991; Spinedi et al. 1992). For the purposes of clarity, the effects of these factors will be discussed independently throughout subsequent sections of this chapter. The final section of this chapter will attempt to integrate the effects of the endocrine–immune interactions on seasonal changes in susceptibility to disease.

7.1.2. Reciprocal Interactions Between the HPA Axis and Immune System

HPA Influences on Immune Function

As previously mentioned, the immune system is maintained within safe-operating limits, in part, due to reciprocal interactions between the endocrine and immune systems. Hormones and neurotransmitters can influence immune function by binding directly to receptors on immune cells. For example, receptors for corticosteroids have been found on immune cells (Werb et al. 1978), and glucocorticoids are capable of affecting immune function. In general, cytokine production is reduced by glucocorticoid administration (e.g., Gillis et al. 1979; Snyder & Unanue 1982; reviewed in Munck & Guyre 1986). As will be shown, immune cells are capable of influencing HPA axis function. This bidirectional communication is depicted in Figure 7.2. The regulation of

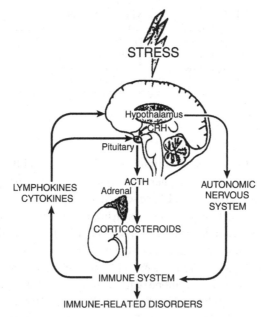

Fig. 7.2. Innervation of the immune system by the autonomic nervous system. Stressors (whether psychological or somatic) activate a cascade of endocrine events that leads to pronounced effects on the immune system. This model represents the interactions between various endocrine factors and releasing peptides/hormones and the immune system. (Reprinted from *Brain Research Reviews*, vol. 23, B. S. McEwen, C. A. Biron, K. W. Brunson, K. Bulloch, W. H. Chambers, F. S. Dhabhar, R. H. Goldfarb, R. P. Kitson, A. H. Miller, R. L. Spencer, & J. M. Weiss, "The role of adrenalcorticoids as modulators of immune function in health and disease: Neural, endocrine, and immune interactions," pp. 79–133, copyright © 1997, with permission from Elsevier Science.)

the HPA axis presumably acts to prevent overactivation of the immune system to avoid destruction of self tissue and cells.

Numerous studies have provided convincing evidence for glucocorticoid/ stress regulation of the immune response. Animals treated with an antigen (e.g., sheep red blood cells) have lower inflammatory responses than animals that have been adrenalectomized or treated with the glucocorticoid antagonist RU486 (Flower et al. 1986; Laue et al. 1988). Likewise, Lewis rats have an impaired ACTH and corticosterone response to stress and a high incidence of arthritis (Sternberg et al. 1989). Glucocorticoids can affect a number of cytokines (reviewed in Munck & Guyre 1986). For example, incubation of macrophage-activation factor, colony-stimulating factor, interferon-γ, or interleukin (IL)-2 with dexamethasone and the mitogen, concanavalin A (Con A), inhibits lymphocyte proliferation in a dose-dependent manner (Kelso & Munck 1984). Production of IL-1 and IL-3 is also blocked by glucocorticoid administration *in vitro* (Culpepper & Lee 1985; Kern et al. 1988). The predominant effect of glucocorticoid administration is reduced T-cell production (reviewed in Homo-Delarche et al. 1991). Likewise, administration of dexamethasone to animals decreases T-cell production of IL-2 and IFN-γ (Daynes & Araneo 1989). Finally, BALB/c mice are incapable of clearing an infection with the protozoan parasite, *Leishmania major*, and they eventually die. This is in contrast with other strains of mice (e.g., C57BL/6 and C3H) that typically clear the infection readily. The decreased capacity to clear this parasite is associated with a decrease in 11β-hydroxysteroid dehydrogenase, an enzyme responsible for converting corticosterone into an inactive keto form, in peripheral lymphoid tissues (i.e., spleen and lymph nodes) (Daynes et al. 1995).

Not only do glucocorticoids directly affect immune function, but upstream components of the HPA axis, such as ACTH and CRH, may also have direct effects on immune function, rather than by acting indirectly by increasing glucocorticoid production. For example, ACTH can bind with high affinity to human lymphocytes (Smith et al. 1987). In addition, ACTH can influence immune cell function directly (e.g., Alvarez-Mon et al. 1985). Similarly, CRH can act directly on rat splenocytes to modulate immune function *in vitro* (Audhya et al. 1991). These data suggest that hypothalamic and pituitary releasing factors can directly influence immune function rather that acting on the adrenals to increase glucocorticoid production.

Optimally, the HPA axis assists to maintain the immune system within very fine operating limits. However, HPA modulation of immune function can often have adverse consequences during periods of prolonged stress (e.g., unusually long winter or during extreme food shortages). There is a substantial body of literature addressing the effects of acute and chronic stress on immune function (reviewed in Leonard & Song 1996). Comparisons across human studies are particularly difficult because of the lack of a satisfactory operational definition of stress. Additionally, there are enormous individual differences in response to life stressors, making interpretation of results difficult and variability within studies large (Sapolsky 1998). Nonetheless, there appears to be a relationship between stress and immune function in several clinical studies. For example, divorced or separated men and women tend to have poorer immune function, compared with married men and women (Kiecolt-Glaser et al. 1987, 1988). Likewise, individuals experiencing posttraumatic stress disorder have fewer T-cells, B-cells, and natural killer (NK) cells than an unaffected population of individuals (Davidson & Baum 1986). Similarly, NK cell numbers and function are reduced by examination stress in college students (Kiecolt-Glaser et al. 1984; Glaser et al. 1986). Thus, not only can HPA output act to regulate immune function within normal operating parameters, but chronic (and acute) stress can also have an adverse effect on immune function.

Numerous animal models of stress have been used to study this adverse effect of stress on immune function. Studies of stress and depression have used manipulations such as olfactory bulbectomy, water or restraint stress, and learned-helplessness paradigms to mimic the physiological effects of stress and depression. These animal models have little ecological validity because exposure to artificial stressors may elicit a stress response that is uncharacteristic of responses to stressors encountered in nature. Laboratory studies of immune function in animals using ecologically relevant stressors have provided evidence for the relationship between exposure to stressors and seasonal changes in immune function (see Chapter 4). Using this approach, laboratory studies are able to evaluate the effects of stressors that animals typically experience during winter (e.g., reduced food availability, low ambient temperatures, increased predation pressure). These studies typically reveal that a single "natural" stressor can inhibit immune function, and the effects of additional stressors are often additive (Figure 7.3). This approach improves on previous animal models of stress and immune function because the stress response elicited by these "natural" stressors more likely reflects a stress reaction characteristic of what the animal would exhibit in nature. Thus, these types of animal studies may better parallel the human studies of stress and immune function.

Immune System Effects on the HPA Axis

There is reciprocal communication between the HPA axis and the immune system. Increased glucocorticoid concentrations are associated with activation

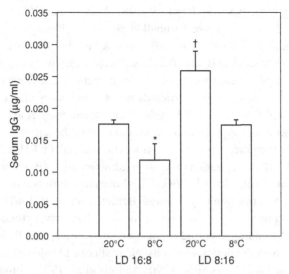

Fig. 7.3. Mean (±SEM) serum IgG concentrations (μg/ml) of male deer mice (*Peromyscus maniculatus*) housed in long (LD 16:8) or short (LD 8:16) days and either mild (20°C) or low (8°C) temperatures. The results from this study suggest that short days lead to enhancements of immune function in anticipation of winter, whereas stressors such as low temperatures lead to a reduction in immune function. When animals are exposed to short-day lengths in combination with low temperatures, the result is a "balance" between short-day enhancement of immune function and a "stress-induced" decrement in immune function. †Significantly greater than all other groups ($p < 0.05$). *Significantly less than all other groups ($p < 0.05$). (From Demas & Nelson 1996).

of the immune system by numerous antigens (e.g., Besedovsky et al. 1975; Saphier 1989). Apparently, this increase in HPA axis activity requires that the degree of activation of the immune system must reach a particular threshold; increased concentrations of corticosterone are only observed after an immunological challenge in Biozzi immunologically high-responding mice, but not Biozzi immunologically low-responding mice (Shek & Sabiston 1983). These results suggest that the immune system acts to regulate HPA function to help modulate its own activity.

The increase in glucocorticoids after injection with an antigen does not appear to be the result of the immune reaction or a nonspecific stress reaction in response to the injection. For instance, immune cells stimulated an *in vitro* increase ACTH and corticosterone *in vivo* when injected into animals (Besedovsky et al. 1981, 1985). Likewise, peripheral IL-1 and IL-6 administration or intracerebroventricular injections of tumor necrosis factor-α (TNF-α) lead to increased concentrations of ACTH and glucocorticoids

(Besedovsky et al. 1986; del Rey et al. 1987; Naitoh et al. 1988; reviewed in Besedovsky & del Rey 1996; Turnbull et al. 1997), although IL-1 is a most potent stimulator of the HPA axis (Besedovsky et al. 1991). Athymic nude mice lack both functional B- and T-cells; when these athymic mice are injected with IL-1, they also increase glucocorticoid concentrations. These results suggest that the increase in glucocorticoids is not a secondary effect of IL-1 on factors released from mature T-lymphocytes (Besedovsky et al. 1986). IL-1 appears to act at the hypothalamic level, rather than the anterior pituitary, to affect ACTH and glucocorticoid secretion; antagonists of CRH reduce the increase in ACTH concentrations after IL-1 administration (Berkenbosch et al. 1987; Sapolsky et al. 1987). Likewise, IL-1 does not stimulate ACTH release *in vitro* from anterior pituitary cultures (Berkenbosch et al. 1987).

In addition to acting at the level of the hypothalamus, cytokines can directly stimulate adrenal steroid production; IL-1, IL-2, and IL-6 stimulate glucocorticoid production from adrenal gland slices and cultured adrenal cells maintained *in vitro* (Salas et al. 1990; Andreis et al. 1991; Tominaga et al. 1991). Taken together, these data suggest that the immune system is regulated, in part, by cross-talk between the HPA axis and the immune system. When the immune system is activated, cytokines signal the hypothalamus to release CRH that, in turn, produces a decrement in the immune response to prevent overexpression of the immune reaction. The extent to which cytokines, especially proinflammatory cytokines such as IL-1, IL-6, and TNF-α, vary seasonally in nonhuman animals is currently unexplored. Presumably, because cytokine production varies seasonally (Maes et al. 1994; Mann et al. 2000a, b), cytokine effects on the CNS should vary seasonally as well.

Regulatory cross-talk between the HPA axis and immune system may serve to prevent the immune system from becoming overactive and damaging self tissue. For example, adrenalectomy dramatically increases the lethality of lipopolysaccharide (LPS) endotoxin injection in mice; this increase is mediated by overproduction of IL-1 and TNF (Bertini et al. 1988; Ramachandra et al. 1992). Likewise, circadian variation in several immune parameters is negatively correlated with adrenal hormone secretion (Angeli et al. 1992).

The immune system also may act locally within the brain to regulate HPA function. Receptors for IL-1 are located in the hypothalamus (reviewed in Besedovasky & del Rey 1996), and neurons in the hypothalamus produce IL-1 that may be responsible for increased CRH release in response to endotoxin. In addition, cytokines can affect pituitary secretion of ACTH and CRH induced ACTH release *in vitro* (reviewed in Besedovsky & del Rey 1996). Cytokines also have been localized in endocrine tissue, suggesting that immune cells may act locally through a humoral signal to influence endocrine function. IL-1β, IL-6, and TNF-α have been identified in the anterior pituitary, and their secretion or mRNA levels are increased in response to immune activation via lipopolysaccharide injections (Koenig et al. 1990; Spangelo et al. 1990, 1991; Spangelo et al. 1991; Gatti and Bartfai 1993; Schobitz et al. 1992). Peripheral cytokine receptors in endocrine glands also have been reported, suggesting that immune mediators may act in the periphery to modulate endocrine function. For example, cytokine receptors, including IL-1 and IL-6 receptors, have been localized in the adrenal gland (Cunningham et al. 1992; Gadient et al. 1995; Path et al. 1997). These data illustrate that cytokine receptors may modulate the effects of the immune system on glucocorticoid release and ACTH-mediated glucocorticoid release.

Not only does the immune system stimulate the release of ACTH and glucocorticoids as part of a regulatory response, but there is evidence that immune cells may produce endocrine products. For example, ACTH-like immunoreactivity has been found in lymphocytes (Smith & Blalock 1981). This locally produced ACTH is unlikely to modulate adrenal output of corticosterone; lymphocyte-derived ACTH is insufficient to stimulate adrenal steroidogenesis in hypophysectomized mice (Olsen et al. 1992). Therefore, lymphocyte-derived ACTH may act to regulate distinct, local populations of immune cells. Thus, not only does the immune system act to modulate function through feedback actions that affect HPA activity, but immune cells may also regulate their own function by producing small amounts of ACTH locally. Again, the seasonal effects of these interactions remain unspecified, but may contribute to seasonal changes in energy balance and regulation of the endocrine and immune systems.

7.1.3. Immune Function and the Hypothalamic-Pituitary Gonadal (HPG) Axis

The HPG axis begins in the hypothalamus with the neuropeptide known as gonadotropin-releasing hormone (GnRH) (Figure 7.3). In mammalian species, GnRH neurons form a loose continuum from the telencephalic diagonal band of Broca and more dorsal septal areas (i.e., medial and triangular septal nuclei), to the bed nucleus of the stria terminalis, and diencephalic areas (i.e., periventricular area, medial and lateral preoptic areas, anterior hypothalamus, and retrochiasmatic zone medial to the optic tract) (Silverman et al. 1994). The median eminence represents the final common GnRH neuronal pathway for regulation of anterior pituitary mediation of reproductive function (Silverman et al. 1994). The median eminence is located at the base of the hypothalamus and is continuous with the pituitary stalk.

Importantly, the median eminence contains a capillary plexus connected with the hypothalamo-pituitary portal system crucial for the regulation of anterior pituitary function.

The anterior pituitary contains neurons that synthesize, among other factors, two hormones important for the regulation of reproduction, namely luteinizing hormone (LH) and follicle-stimulating hormone (FSH). GnRH acts on the pituitary to cause the secretion of these two hormones (collectively called the gonadotropins). In turn, these hormones act on the gonads to regulate the production and secretion of the sex steroid hormones (e.g., testosterone and estrogen) (reviewed in Gorski 1979).

Most species exhibit seasonal patterns in reproductive behavior and physiology, including humans (Bronson & Heideman 1994; Bronson 1995). The most well-studied seasonally breeding species are long-day breeders such as small rodents. In anticipation of winter, or in laboratory-simulated winter conditions, most rodent species exhibit a dramatic decline in gonadotropin and gonadal steroid concentrations, eventually leading to regression of the reproductive system (reviewed in Bronson 1989; Bronson & Heideman 1994). The decline in the gonadotropins and gonadal steroids appears to be mediated by reductions in GnRH secretion (e.g., Turek et al. 1977; Kriegsfeld & Nelson 1999). Animals use the annual cycle of day length (photoperiod) to phase endogenous adaptations with changing seasons. Day length is transduced into a melatonin signal and the duration of nightly melatonin secretion signals the time of the year (Bartness et al. 1993; Chapter 5). Thus, long melatonin durations (short days) signal winter, whereas short melatonin durations (long days) signal spring and summer.

As with the HPA axis, numerous interactions exist between the HPG axis and the immune system. Importantly, the immune system acts to modulate sex steroid production by acting on different levels of the HPG axis. In turn, sex steroids also act to mediate immune function. Because sex steroid hormone concentrations vary between the breeding and nonbreeding seasons, sex steroid hormones are likely mediators of seasonal patterns of immune function and disease (see Chapters 3 and 8).

Sex Differences in Immune Function and Disease

The effects of gonadal steroids on immune function are best illustrated by the fact that there are profound sex differences in immune function. Sex differences in immune function are well established in vertebrates (Billingham 1986; Alexander & Stimson 1988; Schuurs & Verheul 1990). Males generally exhibit lower immune responses than female conspecifics (Billingham 1986; Schuurs & Verheul 1990; Zuk & McKean 1996). Specifically, humoral immune responses (i.e., antibody production by B-cells) are typically elevated in females, compared with males. Females of various species display higher IgM, IgG, and IgA concentrations than males, and are also better able to mount both primary and secondary antibody responses to antigenic challenge than males (Butterworth et al. 1967; Eidinger & Garrett 1972; Tartakovsky et al. 1981; Schuurs & Verheul, 1990). In mice (*Mus musculus*) infected with the parasite, *Giardia muris*, or the bacterium, *Corynebacterium kutscheri*, females have lower infection rates and higher antibody production than males, suggesting a functional advantage for elevated humoral immunity in females (Daniels & Belosevic 1994; Komukai et al. 1999).

Cell-mediated immune responses that involve primarily T-cells also differ between males and females. T-cells, in particular helper T-cells (Th cells), are functionally and phenotypically heterogeneous and can be differentiated based on the cytokines they release. Reliance on subsets of Th cells (i.e., Th1 or Th2 cells) to overcome infection differs between males and females with females exhibiting higher Th2 responses (i.e., higher IL-4, IL-5, IL-6, and IL-10 production) than males (Bijlsma et al. 1999). In response to lymphocyte stimulation with mitogens, females have higher proliferative responses than males (Krzych et al. 1981). Female rodents also have increased immunological intolerance to foreign substances; females reject skin grafts faster than males (Billingham 1986). Females of many species, however, more readily produce immune responses against self tissues and, therefore, are more likely to develop autoimmune diseases than males (Olsen & Kovacs 1996; Talal & Ansar Ahmed 1987). For example, the female-to-male ratio of incidence is 19:1 for autoimmune thyroiditis, 9:1 for systemic lupus erythematosus (SLE), and 3–4:1 for rheumatoid arthritis (reviewed in Homo-Delarche et al. 1991). Several autoimmune diseases tend to develop or flare during puberty in females. Higher immunocompetence often results in faster skin graft rejection in females, compared with males, suggesting that cell-mediated immune function is higher among females than males (Brent & Medawar 1966; Graff et al. 1969; see Inman 1978; Kalland 1980).

Males and females differ in susceptibility to various infections. For example, human males are more susceptible to dysentery, gonorrhea, meningitis, pneumonia, rabies, syphilis, tetanus, and certain types of cancers (e.g., lymphomas and leukemias) than females (Goble & Konopka 1973; Billingham 1986). Field studies of both mammals and birds illustrate that the prevalence (i.e., proportion of individuals infected) and intensity (i.e., severity of the infection) of parasite infections is often higher in males than females (Poulin 1996; Zuk & McKean 1996). Although clinical studies of humans and field studies of nonhuman animals are suggestive, several factors, including exposure rates, social behavior, habitat, and diet, cannot be held constant and could contribute to the observed differences in parasite infection. Laboratory studies of rodents, however, have demonstrated that, even in a controlled setting, males are more susceptible to infection than females, and this difference is related to sex steroid hormones (Alexander 1988; Sano et al. 1992; Daniels & Belosevic 1994; Tiuria et al. 1994; Klein et al. 1999, 2000).

During pregnancy, a number of immune changes occur. As noted in the previous chapter, the prevailing hypothesis is that immune function is suppressed presumably to prevent rejection of the fetus. For example, several components of cell-mediated immunity are depressed during pregnancy (Finn et al. 1972), whereas some measures of humoral immunity are elevated (Stimson & Hunter 1980). Likewise, a number of parasitic diseases may flare during pregnancy (reviewed in McCruden & Stimson 1991). The hormonal changes that occur during pregnancy are complex, but involve alterations in gonadal steroid hormones. An alternative hypothesis is that the energetics of pregnancy and full immune function are incompatible (Chapter 6).

HPG Axis and Immune Function

The effects of gonadal steroids on lymphoid tissues is well known (Hammar 1929). Even before the importance of the thymus in immune function had been recognized, researchers noted thymic hypertrophy in response to gonadectomy, particularly in female animals (Hammar 1929; Eidinger & Garrett 1972). Consequently, receptors for estrogens and androgens have been found on lymphoid tissue, suggesting that gonadal steroids can act directly on immune organs to influence immune function (Abraham & Buga 1976; Gillette & Gillette 1979; Grossman et al. 1979; Brodie et al. 1980; Barr et al. 1984).

Androgens. Sex differences in immune function are mediated, in part, by the suppressive effects of testosterone on the immune system (Alexander & Stimson 1988; Zuk et al. 1995; Olsen and Kovacs 1996; Hillgarth & Wingfield 1997). Typically, gonadectomized male rodents display immune responses that are similar to those of conspecific females (i.e., higher than gonadally intact males). Gonadectomized male rodents have elevated antibody production, increased immunological intolerance to skin grafts, and heavier lymphoid organs (e.g., thymus, spleen, and lymph nodes) than gonadally intact males (Grossman 1984; Schuurs & Verheul 1990). Gonadectomized male rodents mount higher immune responses and show greater resistance to fungal, protozoan, and helminth infections, compared with gonadally intact males or gonadectomized males injected with testosterone propionate (Eidinger & Garrett 1972; Rifkind & Frey 1972; Kittas & Henry 1979; Kamis et al. 1992; Tiuria et al. 1994). Gonadally intact male reindeer (*Rangifer t. tarandus*) have higher incidences of warble fly (*Hypoderma tarandi*) infestation than both females and castrated male reindeer (Folstad et al. 1989). These data illustrate that immunosuppression by testosterone is one proximate mechanism underlying increased susceptibility of male mammals to infection. Although sex differences in immune function and susceptibility to infection have been reported in nonmammalian species, the precise role of sex steroids has not been adequately addressed (Klein 2000a, b). The few studies that have been conducted suggest that testosterone may induce atrophy of lymphoid organs and reduce immune responsiveness in birds, reptiles, and fish (Zuk et al. 1995; Hillgarth & Wingfield 1997; Hasselquist et al. 1999; Evans et al. 2000). To date, however, whether testosterone mediates susceptibility to infection among nonmammalian vertebrates remains unspecified.

Although testosterone administration can increase the susceptibility to infection, adminstration of androgens (e.g., testosterone, 5α-dihydrotestosterone) in animal models of the autoimmune disease, SLE, ameliorates symptoms, whereas estrogen has potentiating effects on symptoms (Levine & Madin 1962; Roubinian et al. 1977, 1978; Verheul et al. 1981). In human patients with SLE, androgen treatment has been somewhat successful in treating the disease, whereas estrogens have been found to exacerbate symptoms (Amor et al. 1983; Morimoto 1978).

Estrogens. Estrogens modulate immune function in females and may contribute to resistance against infection. Estrogens generally enhance both cell-mediated and humoral immune responses; there are, however, reports of estrogens suppressing some cell-mediated immune responses (Kalland 1980; Luster et al. 1984). Injection of rats or treatment of human lymphocytes with 17β-estradiol elevates immunoglobulin synthesis by B-cells (Paavonen et al. 1981; Alexander & Stimson 1988). Conversely, the same estrogenic treatment suppresses proliferation of human lymphocytes in response to T-cell mitogens among both men and women (Luster et al. 1984; Alexander & Stimson 1988). Similarly, NK cell activity is reduced by estrogen treatment (Ablin et al. 1974; Seaman & Gindhart 1979). Several autoimmmune diseases that are considered primarily T-cell-mediated diseases also are suppressed by estrogen treatment. For example, arthritic symptoms in experimental allergic encephalomyelitis are abated by estrogen treatment (Trooster et al. 1993). Likewise, collagen arthritis and adjuvent arthritis are suppressed by treatment with estrogens (Jansson et al. 1990; Holmdahl et al. 1989). Estrogenic treatment facilitates cutaneous wound healing in female rodents. Ovariectomized female rats have slower wound healing followed by increased scarring, compared with to intact females (Ashcroft et al. 1997). The effect of gonadectomy can be reversed by topical application of 17β-estradiol to the wound (Ashcroft et al. 1997).

Estrogens also alter the course of infection. Female mice infected with *Paracoccidiodes brasiliensis* during proestrous (i.e., when estrogen concentrations are high) are less susceptible to infection than either females infected during other stages of the estrous cycle or males (Sano et al. 1992). Following inoculation with either coxsackievirus or encephalomyocarditis, females are more likely to survive infection with less severe pathology associated with disease than males (Friedman et àl. 1972; Lyden et al. 1987). Female mice also are more resistant to parasite infections, including *Leishmania mexicana*, than conspecific males (Alexander 1988). These data illustrate that estrogens may make females less susceptible to infection than males, possibly by enhancing host immunocompetence. Although females of other vertebrate species have higher immune responses and reduced susceptibility to infection than male conspecifics, the precise role of estrogens has not been identified (Zuk & McKean 1996; Hillgarth & Wingfield 1997).

GnRH. In addition to gonadal steroids influencing immune function, GnRH from hypothalamic sources and produced locally by immune cells may modulate immune function. For example, T-cells synthesize GnRH locally in the thymus (Maier et al. 1992). In addition, GnRH receptors have been characterized on T-cells, mouse lymphocytes, in rat mast cells, thymocytes, and spenocytes (reviewed in Costa et al. 1990; Marchetti et al. 1989, 1998). Apparently, these receptors allow GnRH to influence immune function directly. For example, treatment with a GnRH agonist reverses thymic atrophy and the age-related decline in thymic function (Marchetti et al. 1989). There is some evidence that GnRH agonists may also decrease absolute thymocyte numbers (Rao et al. 1993). These effects are not necessarily mediated by downstream effects on the gonads; GnRH antagonists decrease serum Ig and anti-DNA levels in castrated male and female mice prone to SLE (i.e., SWR $\times$ NZB F_1) mice and improve survival (Jacobson et al. 1994).

GnRH may also be produced locally in immune tissue or cells to modulate immune function; GnRH mRNA and GnRH-like molecules have been localized to rat thymus, thymocytes, splenocytes, and human peripheral T-lymphocytes (reviewed in Marchetti et al. 1998). Thus, GnRH from sources other than hypothalamic neurons (i.e., produced directly by immune cells) may act to regulate immune function. Taken together, these data suggest that gonadal steroids and upstream components of the HPG axis can act both directly and indirectly to influence immune function. Because production and release of HPG hormones change seasonally, we hypothesize that these hormones are likely mediators of seasonal adjustments in immune function and disease. Future studies are required to determine the precise roles of

HPG hormones and the regulation of their receptors in the seasonal patterns of immunity.

Immune System Effects on the HPG Axis

Not only do gonadal steroids influence immune function, but the HPG axis also is modulated by the immune system. This reciprocal cross-talk allows the maintenance of both the immune and endocrine systems to be regulated within narrow limits, and may contribute to seasonal variation in these systems. For example, IL-1β is capable of mediating the release of FSH and LH in cultured pituitary cells (Murata & Ying 1991; Rettori et al. 1991; Cunningham & DeSouza 1993). In addition, IL-1β receptors are present in the testes and ovary (reviewed in Cunningham & De Souza 1993). Apparently, these receptors allow for direct actions of IL-1 on the gonads; IL-1β can inhibit gonadal steroid production by acting directly on the testes (Calkins et al. 1988) and ovaries (Hurwitz et al. 1991).

The effects of cytokines on the gonads has been extensively studied within the past several years. *In vitro*, IL-1 (α and β) stimulates ovarian secretion of progesterone, whereas TNF-α inhibits secretion (Roby & Terranova 1990; Sjogren et al. 1991). Similarly, IL-1 inhibits gonadotropin-induced secretion of estradiol and testosterone; TNF-α also inhibits gonadotrpin-induced estradiol secretion (reviewed in Fauser et al. 1989; Best et al. 1994; Besedovsky & del Ray 1996). Results for testosterone *in vivo* and *in vitro* have been equivocal. For example, TNF stimulates testosterone secretion *in vitro*, yet inhibits testosterone secretion *in vivo* (Mealy et al. 1990; Warren et al. 1990) (Table 7.2).

Because immune factors can influence the HPG axis, one logical assumption is that patients with chronic immune disorders may experience gonadal abnormalities or dysfunction. In studies of human immunodeficiency virus (HIV) patients, hypogonadism is not uncommon. Likewise, HIV patients often report decreased libido and impotency (reviewed in Sellmeyer & Grunfeld 1996). In accord with these symptoms, HIV patients often have low serum/plasma concentrations of testosterone and gonadal steroids. This reduction in testosterone and LH appears to by due to hypothalamic abnormalities; HIV patients respond normally to a GnRH challenge (reviewed in Sellmeyer & Grunfeld 1996). Women have been far less studied, but a high incidence of amenorrhea is reported for female patients with HIV (Widy-Wirski et al. 1988). The extent to which symptoms are exacerbated seasonally has not been examined.

Data from acquired immunodeficiency syndrome (AIDS) patients combined with those from controlled laboratory studies provide convincing evidence that the immune system regulates the HPG axis. In addition, there is reciprocal cross-talk from the HPG axis to the immune system. As mentioned, reciprocal cross-talk exists between the HPA and immune systems. Immune regulation is further complicated by the fact that there is cross-talk between the HPA and HPG axes. These complex interactions help to maintain homeostasis within the immune and endocrine system. As will be seen, these interactions become further complicated with additional endocrine factors that act to regulate immune function.

Seasonal fluctuations in the relationships among hormones and immune function have only recently been examined. In one series of studies, the effects of photoperiod and steroid hormones on immune function were assessed in male and female deer mice (*Peromyscus maniculatus*) (Demas & Nelson 1998a, b). In one experiment, male deer mice were castrated, castrated and given testosterone replacement, or were sham-operated. Some animals from each of the two experimental groups were subsequently housed in either long days (LD 16:8) or short days (LD 8:16) for 10 weeks. Short-day deer mice underwent reproductive regression and displayed elevated lymphocyte proliferation in response to the mitogen Con A, compared with long-day mice, regardless of hormonal condition. In a related experiment, female deer mice were ovariectomized, ovariectomized and given estrogen replacement, or sham-operated. Animals from each of these experimental groups were subsequently housed in either LD 16:8 or LD 8:16 for 10 weeks. Short-day female deer mice also underwent reproductive regression and displayed reduced serum estradiol concentrations and elevated lymphocyte proliferation in response to Con A, compared with long-day mice. Surgical manipulation had no effect on lymphocyte proliferation in either male or female deer mice. Neither photoperiod nor surgical manipulation affected serum corticosterone concentrations. These results confirm that both male and female deer mice housed in short days enhance immune function relative to long-day animals (Demas & Nelson 1998a, b). Additionally, short-day elevation in splenocyte proliferation appears to be independent of the influence of steroid hormones in this species. These findings and the results of other studies (e.g., Nelson & Blom 1994) suggest that other factors in addition to sex steroids (and other HPG hormones) may be involved in mediating seasonal adjustments in immune function.

7.1.4. Prolactin (PRL)

PRL Effects on Immune Function

In common with sex steroid hormones, PRL has well-known effects on reproduction ranging from promoting lactation and corpus luteum function in females to influencing parental behavior in males (Gubernick & Nelson 1989; Neill & Nagy 1994). PRL concentrations display dramatic seasonal fluctuations; virtually all mammals examined show reduced PRL concentrations when exposed to short-day lengths (Goldman & Nelson 1993). Seasonal stressors can affect PRL, as well as glucocorticoids; for example, PRL is suppressed in low temperatures (Smith et al. 1977). Because PRL concentrations change seasonally in response to changes in extrinsic factors, PRL may be considered a likely mediator of seasonal changes in immune function.

As with hormones of the the HPA and HPG axis, PRL influences immune function, and immune factors can influence PRL synthesis/secretion. PRL may act directly on immune cells; receptors for PRL have been identified on lymphocytes and thymic epithelial cells (Russell et al. 1985; Dardenne et al. 1989; Hooghe et al. 1993). In general, PRL enhances immune function. For example, hypophysectomy suppresses hematopoiesis and immune cell proliferation; these effects are reversed after administration of PRL (Berczi et al. 1991). In addition, bromocriptine, a drug that acts to reduce PRL concentrations, leads to immunodeficiency (i.e., suppresses antibody formation and the delayed hypersensitivity reaction), and this decrement in immune function caused by bromocriptine treatment is reversed with PRL (Nagy et al. 1983). Finally, PRL has been shown to stimulate several immune parameters in untreated animals (Nagy et al. 1983).

The studies described herein suggest that pituitary PRL can enhance immune function. However, there is evidence that PRL is produced locally in immune cells to regulate immune function. For example, antibodies directed against PRL inhibit lymphocyte proliferation *in vitro*, suggesting that PRL may enhance lymphocyte proliferation (Hartmann et al. 1989). PRL mRNA is expressed in mitogen-stimulated lymphocytes, providing evidence for this hypothesis (Hiestand et al. 1986). Additional work using B-cell lines revealed that the transcript for immune-cell PRL is 150 nucleotides longer than pituitary PRL, but the resulting proteins are identical (Jurcovicova et al. 1992).

Immune Effects on PRL

In addition to PRL influencing immune function, cytokines are capable of influencing PRL secretion. For example, IL-1β mediates the release of PRL from cultured pituitary cells (Rettori et al. 1987). Likewise, IL-1β can influence PRL *in vivo*; intracerebroventricular injections of IL-1β lead to an increase in plasma PRL concentrations (reviewed in Besedovsky & del Rey 1996). In addition, TNF-α leads to a rapid increase in PRL (Yamaguchi et al. 1991). Rapid increases in PRL are also seen after administration of Freund's complete adjuvant to rats (Neidhart & Larson 1990). Taken together, these findings suggest that immune components can also act to influence PRL production/secretion. In combination with data suggesting that PRL can have pronounced effects on the immune system, these data provide evidence for an additional endocrine component that acts to modulate and is regulated by the immune system.

7.1.5. Thyroid Hormone

Effects of Thyroid Hormones on Immune Parameters

Thyroid hormones, like glucocorticoids, are important for adapting to changes in temperature. Specifically, thyroid activity of nonhibernating mammals is increased during winter, compared with summer. In contrast, thyroid activity is decreased in hibernating mammals during winter (e.g., Johansson 1978; Tomasi et al. 1998). The extent to which differences in thyroid activity between hibernating and nonhibernating animals affect immune function has not been directly examined. Furthermore, the extent to which the metabolic effects of thyroid hormones are critical for shunting energy among competing energetic needs, such as immune function, growth, or reproduction has not been well specified. However, there are well-established effects of thyroid hormones on immune function.

Generally, thyroid hormones enhance immune function (reviewed in Fabris et al. 1995). For example, thyroid hormones promote thymocyte maturation and differentiation (Gala 1991; Johnson et al. 1992). Hypothyroid mice have a reduced proliferative response, along with reduced thymic and splenic mass, and decreased numbers of $CD8^+$ T-cells (Erf 1993). Thyroid hormone replacement restores thymic development and function in these hypothyroid mice (Johnson et al. 1992). Earlier evidence for the involvement of thyroid hormones in immune function came from studies in which mice and rats were thyroidectomized neonatally and exhibited decreased thymus mass and prolonged survival of skin grafts (Pierpaoli et al. 1970; Alqvist 1976). The effects of thyroidectomy were reversed by administration of thyroid hormone (Fabris 1973).

The effects of thyroid hormones on immune function may be modulated through the thymic peptide called thymulin. Thyroidectomy results in reduced thymulin concentrations and thyroid hormone replacement restores thymulin concentrations to control values (Fabris & Mocchegiani 1985). Depressed thymulin concentrations are modulated by triiodothyronine (T_3) rather than thyroxine (T_4); T_3 is positively correlated with thymulin concentrations, whereas T_4 is not (Fabris et al. 1987). Several human disorders characterized by reductions in thyroid hormones also are associated with reduced thymulin concentrations concomitant with immune deficiency (reviewed in Fabris et al. 1995).

Because thyroid hormones increase immune function, humans with hyper-thyroidism may exhibit enhanced immune status and a higher incidence of autoimmune disorders. In humans, there is no consistent correlation between the number of T- and B-lymphocytes and hyperthyroidism associated with Grave's disease (reviewed in Fabris et al. 1995). Equivocal results have been obtained for NK cell activity and number; patients with Grave's disease had higher NK numbers, whereas NK cells were reduced in Grave's disease patients in another study (Calder et al. 1976; Iwatani et al. 1984). In healthy human populations, however, T_3 appears to have a direct stimulative effect on B-cell differentiation (Paavonen 1982).

Upstream components of the hypothalamic-pituitary-thyroid axis may also act to influence immune function. For example, receptors for thyrotropin (TSH) have been found on lymphocytes that possess many similarities with TSH binding sites found on the thyroid (Pekonen & Weintraub 1978). In addition, TSH enhances the antibody response to T-cell-dependent and T-cell-independent antigens *in vitro* (Blalock et al. 1985; Kruger & Blalock 1986). These observations suggest that TSH plays a potentiating role in the immune response to specific antigens. Additionally, TSH alone does not enhance proliferation independent of antigen stimulation (Provinciali et al. 1992). Presumably, because thyroid hormones are responsive to seasonal fluctuations, these hormones may mediate seasonal changes in immune function. We speculate that thyroid effects may act secondarily to the effects of steroid hormones and melatonin. For example, thyroid hormone deficiency often results in a blunted responsiveness of the HPA axis to CRH, as well as reduced corticosteroid concentrations (Spinedi et al. 1991, 1992; Da Silva et al. 1993; Gallucci 1993).

Immune Effects on the Thyroid

Not only do thyroid hormones modulate immune function, but the immune system can affect thyroid hormones (reviewed in Fabris et al. 1995). In addition, thyroid hormones can affect, and be influenced by, other endocrine products that also influence immune function. Thyroid hormone deficiency often results in a blunted responsiveness of the HPA axis to CRH, as well as reduced corticosteroid concentrations (Spinedi et al. 1991, 1992; Da Silva et al. 1993; Gallucci 1993). Thus, thyroid hormones may affect immune function by reducing HPA axis activity which, as mentioned previously, suppresses immune function. Because the HPA axis and thyroid gland modulate the activity of one another, abnormalities in HPA function also can affect thyroid and immune function. For example, corticosteroid depletion after treatment for Cushing syndrome leads to the development of autoimmune thyroiditis (Takasu et al. 1990, 1993). As described previously, high circulating glucocorticoid concentrations suppress immune function. Complex interactions between the HPA axis and thyroid physiology could mediate seasonal changes in immune function.

Thymectomized mice and athymic nude mice exhibit pronounced alterations in thyroid cell morphology and reductions in thyroid hormone production that are reversed by thymus grafts (Pierpaoli & Besedovsky 1975; Pierpaoli & Sorkin 1972). Studies using innocuous antigens such as sheep red blood cells suggest that antigen-stimulated immune responses cause a concomitant decline in thyroid hormone concentrations (Besedovsky et al. 1975, 1985). In humans, daily administration of lymphokines such as IFN-α leads to a progressive decrease in circulating T_3 and T_4 concentrations independent of changes in TSH (Orava et al. 1983). Some studies have reported reductions in thyroid hormones after IL-2 administration; humans treated with IL-2 develop hypothyroidism (Atkins et al. 1988). In rats, treatment with human IL-1β decreases thyroid hormone and TSH concentrations (Dubuis et al. 1988). These effects appear to be due to the direct actions of cytokines on thyroid cells; IL-1 receptors have been identified on cultured rat, human, and porcine thyroid cells (reviewed in Fabris et al. 1995). Thus, complex interactions between the immune system and hypothalamic-pituitary-thyroid axis appear to help maintain homeostasis within the immune and endocrine systems. Thyroid hormones appear to act by both indirect and direct means to influence immune function, whereas immune effects on the thyroid are likely modulated by direct actions of cytokines on high affinity receptors on thyroid cells.

7.1.6. Growth Hormone (GH)

GH Effects on Immune Function

During the winter energy bottleneck, animals often suspend reproduction and growth to conserve energy for cellular maintenance, thermoregulation, and, if possible, immune function. Because growth is a competing energetic cost that must compete with full immune function, we would predict that hormones associated with growth might impede immune function. Data suggest, however, that GH typically stimulates immune function.

Initial evidence for the involvement of GH came from studies in which mice injected with an antibody to GH developed thymic atrophy. Likewise, dwarf mice that are deficient in GH have a reduced ability to synthesize antibodies (reviewed in Weigent et al. 1988; Weigent & Blalock 1995). Most reports to date suggest that GH could have pronounced stimulative effects on immune function. For example, hypophysectomy suppresses hematopoiesis and immune cell proliferation, and these effects are reversed after administration of GH (Kelley 1991; reviewed in Besedovsky & del Ray 1991). Hypophysectomy also leads to reductions in $CD4^+$ lymphocytes in circulation and an increase in the number of $CD8^+$ cells, and both of these effects are reversed with GH treament.

In vivo, GH stimulates the production of IL-1β, IL-2, TNF-α, and thymulin, (Kelley 1989). GH may be produced by immune cells to modulate their own function; the production of GH has been reported for both human and rodent lymphocytes (reviewed in Weigent & Blalock 1995). Additionally, the level of GH produced by lymphocytes is increased in response to mitogen stimulation (Weigent et al. 1988). The effects of endogenous GH are not clear because most studies of GH have used exogenous sources of the hormone to study the role of GH on immune function. However, treatment of rat lymphocytes with an antisense oligodeoxynucleotide directed against GH is capable of inhibiting lymphocyte proliferation *in vitro*, and addition of GH to the culture reverses these effects (Weigent et al. 1991).

The observation that GH receptors have been identified on immune cells (see Table 7.2) lends support for direct actions of GH on the immune system. GH also may act indirectly through other mediators of immune function, such as insulin-like growth factor-1, IL-1β, and thymulin (reviewed in Marsh & Scanes 1994). For example, GH stimulates IL-2 production, and the immunoenhancing effects of IL-2 have been well-established (Kelley et al. 1986; Schimpff & Repellin 1989). Likewise, GH is also capable of increasing thymulin (Goff et al. 1987), suggesting that GH may act through a mechanism similar to that of thyroid hormone. Indeed, there is some evidence that GH and thyroid hormones may interact to modulate immune function (reviewed in Marsh & Scanes 1994). Thus, GH may act both directly as well as indirectly to influence immune function.

Immune Effects on GH

As with other endocrine systems discussed previously, not only does GH act to modulate immune function, but the immune system also affects GH production/secretion. For example, IL-1β modulates pituitary release of GH *in vitro* (Payne et al. 1992). Similarly, IL-6 increases the release of GH from cultured rat hemipituitaries (Lyson & McCann 1991), whereas IL-2 inhibits the release of GH from hemipituitaries *in vitro* (Karanth & McCann 1991). Likewise, in rats, injection of IL-1β into the arcuate nucleus leads to pronounced reductions in GH (Lumpkin & McDonald 1989). *In vivo* injections of IFN-γ results in a delayed inhibition of GH (Gonzalez et al. 1991), whereas TNF α stimulates GH *in vivo* (Bernardini et al. 1990). INF-γ is also capable of inhibiting basal GH release from cultured pituitary cells, as well as GH release stimulated by GH-releasing hormone (Vankelecom et al. 1990).

Taken together, these data suggest that GH, in common with other hormonal systems discussed previously, represents an important regulatory hormone in the neuroendocrine-immune network. Data presented herein suggest that immune factors can act to modulate GH production/secretion. In turn, GH from pituitary sources, as well as local sources in immune cells, acts to regulate the immune response. Thus, reciprocal feedback between these systems, along with those presented here, add to the framework for the complex interactions between the immune and endocrine systems.

7.1.7. Dehydroepiandrosterone (DHEA)

Recently, DHEA has been identified as a significant endocrine mediator of immune function (Danenberg et al. 1995; Hughs et al. 1995; Rasmussen et al. 1995; Loria et al. 1996; Padgett et al. 1997). DHEA is an adrenocortical hormone possessing weak androgenic properties. The major biological function of DHEA is thought to be its role as a precursor to other steroid hormones produced in the adrenal cortex (Rosenfeld et al. 1971). In most cases reported, DHEA appears to enhance immune function (reviewed in Shealy 1995). Specifically, DHEA administration enhances immune function in mice with experimentally induced bacterial, viral, and parasitic infections (Rasmussen et al. 1995; Loria et al. 1996; Padgett et al. 1997). Additionally, DHEA concentrations are deficient in patients with cancer, atherosclerosis, Alzheimer disease, and AIDS (Centurelli & Abate 1997; reviewed in Shealy 1995). These experimental observations combined with correlational evidence linking lowered immune function with reduced DHEA concentrations, suggest that DHEA may be a potent modulator of immune function.

Because DHEA concentrations decline with age, many speculations regarding DHEA and age-related changes in immune status have been offered. Some experimental evidence suggests that DHEA administration may ameliorate some of the effects of aging on immune function. For example, IL-6 is suppressed after DHEA treatment, whereas IL-2 production by T-cells is restored in aged mice and humans (Suzuki et al. 1991; Daynes et al. 1993; Weksler 1993). Likewise, DHEA abates the thymic atrophy caused by glucocorticoids (Blauer et al. 1991). DHEA receptors have been localized to murine T-cells, suggesting that the effects of DHEA on immune function are not necessarily mediated by metabolites of DHEA (Meikle et al. 1992).

The study of the effects of DHEA on immune function is still in its infancy. Numerous popular press reports suggest that DHEA acts to increase immune function and even enhances libido. However, DHEA serves as a prohormone for numerous other steroid hormones synthesized in the adrenal cortex. Thus, from information in the preceding sections, indirect effects of DHEA primarily on other hormonal systems can have far-ranging implications for immune function. Likewise, the long-term effects of DHEA have not been fully investigated. Although DHEA may act to stimulate one component of the immune system, other aspects of the immune system (or other systems) may be adversely affected. Thus, self-medication with DHEA in attempts to "reverse" the effects of aging on immune function should await the outcomes of long-term clinical studies.

DHEA does not seem to be a candidate as a mediator of seasonal changes in immune function. Because DHEA increases immune function in virtually all studies reported to date, we sought to determine if DHEA concentrations might be influenced by photoperiod, thereby suggesting a mechanism whereby short photoperiod may enhance immune function (Kriegsfeld & Nelson 1998). Male deer mice (*P. maniculatus*) were exposed to either short or long days for 10 weeks. Short photoperiods caused significant reduction in all reproductive organs measured relative to animals housed in long days. However, DHEA concentrations did not differ between short- and long-day mice. This study suggests that short-day enhancement of immune function in deer mice is independent of DHEA concentrations (Kriegsfeld & Nelson 1998).

7.1.8. Conclusions

Interactions between the endocrine and immune systems are well defined. Because endocrine function changes seasonally, seasonal changes in immune function may be modulated by the endocrine system. Data presented in this chapter suggest that steroid and peptide hormones may underlie seasonal patterns in host responses to infection and disease. Although steroids have been the primary focus of studies examining seasonal changes in host immune function and infection, several other hormones, such as FSH, LH, adrenocorticotropin hormone, and PRL may influence seasonal patterns in immune function. Because peptide hormones are released in such small concentrations into circulation, they are often difficult to manipulate and study empirically. Because peptide hormones are directly affected by steroid hormone concentrations via feedback mechanisms, it is equally likely that seasonal patterns in immunity and infection may represent a complex interplay between peptide and steroid hormones. Thus, the endocrine system may mediate seasonal changes in immunity and infection via several mechanistic pathways.

The studies presented in this chapter suggest that endocrine factors affect immune parameters and, in turn, the immune system influences endocrine function. In addition, this chapter highlights the potential role of hormones in seasonal alterations in immune function. Evaluation of these studies is complicated by the observation that endocrine factors can also affect other endocrine components which, in turn, influence immune function. These intricate interactions add an additional layer of complexity to endocrine–immune interactions. It is often difficult to determine if the effect of one endocrine factor on immune function is due to direct effects on immune cells, or indirect effects by acting to increase/decrease other endocrine factors that have direct/indirect influences on immune function. Future laboratory studies using *in vitro* techniques to reduce the interactions between hormonal systems will help to clarify these issues.

The effects of stress and stress hormones on other endocrine glands and factors have been the most extensively studied endocrine–endocrine interaction. Thus, this system provides the best-studied example of hormone–immune interactions. Stress can have profound influences on the hypothalamus-pituitary axes. Stress or stress hormones can either directly or indirectly inhibit pituitary secretion of GH, TSH, and PRL (Stephanou et al. 1992; reviewed in Wilder 1995). Likewise, activation of the stress response can inhibit the HPG axis at several levels. For example, CRH can suppress secretion of GnRH from the hypothalamus either directly or indirectly through the secretion of corticosteroids. Additionally, corticosteroids can inhibit pituitary LH secretion and have direct effects on gonadal production of sex steroids (Rivier & Rivest 1991; Ferin 1993; Reichlin 1993; Vermeulen et al. 1993).

In addition to stress and HPA regulation of endocrine function, thyroid and sex steroid hormones can also modulate HPA function. For example, hypothyroidism results in a blunted response to CRH in rats (Spinedi et al. 1991). Ovariectomy suppresses the corticosteroid response to CRH, whereas orchidectomy potentiates this response (Spinedi et al. 1992; Da Silva et al. 1993; Gallucci et al. 1993). This sexual dimorphism in the response to castration on the HPA axis may provide further insight into the mechanisms mediating sex differences in immune function.

Numerous species, including humans, exhibit seasonal alterations in a variety of hormones (Bronson 1995). Thus, seasonal alterations in immune function likely result from seasonal changes in hormonal systems. For example, in long-day breeders, such as rodent species, short-day lengths of winter induce gonadal involution (concomitant with a reduction in gonadal steroids), along with decreased concentrations of PRL (Goldman & Nelson 1993; Bronson & Heideman 1994). As described previously, androgens generally inhibit immune function, whereas PRL typically enhances most immune parameters. In the field, seasonal alterations in immune function are often quite variable; some species enhance immune function during winter, whereas others show a decrement or no change (reviewed in Nelson & Demas 1996; see Chapter 8). Thus, the regulation of seasonal changes in immune function by numerous interacting physiological systems remains a research topic requiring much additional work. This issue is further complicated by the fact that the effects of a hormone on its receptor often involves an interaction between the concentration of the hormone in the bloodstream, the number of available receptors, as well as the availability of carrier proteins to prevent degradation of the hormone. Thus, additional research is necessary to elucidate the dynamic relationship between seasonal alterations in endocrine factors, as well as their receptor and carrier proteins, and the resultant effect on the immune system.

As a final caveat when evaluating studies on endocrine–immune interactions, because hormones are capable of traveling long distances, their effects are widespread and often difficult to evaluate. Controlled laboratory studies where immune cells are exposed to hormones *in vitro* will begin to clarify the direct versus indirect effects of various endocrine components on seasonal modulation of immune function. Recall that some immune cells are capable of producing their own hormones, thereby making *in vitro* results difficult to interpret. It is hoped that this chapter provides a basic overview of endocrine–immune interactions and provides a theoretical framework in which to evaluate studies on endocrine–immune regulation.

8

Clinical Significance of Seasonal Patterns of Immune Function and Disease

For every ailment under the sun, There is a remedy, or there is none,
If there be one, try to find it; If there be none, never mind it.

Mother Goose

8.1. Introduction

The previous chapters have each included some examples of the clinical significance of the seasonal patterns of immune function and disease. The goal of this chapter is to focus on the clinical aspects of the interaction between intrinsic and extrinsic factors that are associated with seasonal patterns of immune function and disease. From our perspective, intrinsic factors include genes, hormones, cytokines, and metabolic or other biochemical signals. Extrinsic factors include environmental conditions, including stressors. This relationship between intrinsic and extrinsic factors serves as the fulcrum between the extensive field literature, emphasizing life history strategies and the dynamic relationship between current and future reproductive efforts, and the clinical literature, emphasizing the effects of malnutrition and other stressors on immune function.

In the previous chapters, we have underscored the importance of various stressors as underlying extrinsic factors in these patterns, and how hormones and other chemical mediators serve as the intrinsic factors that coordinate extrinsic factors with immune function. Importantly, seasonal patterns of immune function and disease represent complex interactions among harsh environmental conditions, variation in coping abilities, and energy balance; and we predict that seasonal patterns of immune function and disease should be most pronounced at the most extreme habitats. Despite individual differences in coping strategies, seasonal stressors are likely to affect all individuals to some extent, and, if sufficiently severe, may impair immune function of an entire population.

Intrinsic factors, such as hormones, often have biphasic actions on immune function. For example, estrogens are generally immunoenhancing (Styrt & Sugarman 1991), but estrogens seem to suppress natural killer cell activity, and perhaps other aspects of innate immune responses at high doses (Seaman et al. 1979; Kalland 1980). Similarly, glucocorticoids are well-known for their immunosuppressive actions during long-term exposure to stressors; but, during short-term or acute stress, glucocorticoids can enhance immune function (Wiegers et al. 1994; Dhabhar & McEwen 1996). The end result of this endocrine calculus is that a host can become resistant or susceptible to infection at different times of the year. By examining the proximate mechanisms that mediate the seasonal patterns of disease, the role of extrinsic factors in resistance and susceptibility may become salient. The interaction between intrinsic and extrinsic factors may also lead to more effective treatments. An obvious example is the emerging appreciation of the relationship among cytokines (e.g., tumor necrosis factor), energy intake, and immune function (Grimble 1994; Johnson 1997), and the strategies for treating body wasting (cachexia) in chronic infections such as human immunodeficiency virus (Mulligan et al. 1999; Yeh & Schuster 1999).

Establishing how extrinsic factors, such as seasonal stressors, affect host physiology and behavior is important in understanding which factors affect the probabilities of contracting specific infections (e.g., Wysocki et al. 1999). We anticipate that these factors might have negligible effects on healthy adults living among technological benefits, such as insulated parkas and central heating. But, for young children, elderly persons, or individuals with challenged immune function, seasonal energetic stressors – such as food shortages or low temperatures – can be sufficiently taxing (even in the presence of technological buffering of the weather) to divert energy from immune function to thermogenesis or other physiological maintenance functions. We expect that seasonal patterns of immune function and disease processes are more salient among individuals who are living without the benefit of technological buffers against the elements. We expect that immune function will be affected by seasonal energetic stressors in healthy adults in some situations, but particularly when they are living under extreme climatic conditions.

8.2. Climatic Stressors and Immune Function

The most hostile environment on the planet that is inhabited by humans is Antarctica. Most individuals in Antarctica only stay during the summer when conditions are relatively mild [average daily temperatures are about $-30°C$

in the interior of the continent and about $-5°C$ on the coast (McMurdo Station)]. But approximately 1,000 hardy folks stay at 1 of the 43 research stations at Antarctica over the entire year. A number of research projects have focused on the physiological responses of these antarctic inhabitants, and several projects have reported changes in immune function. One of the best series of studies of immune function in Antarctica has been conducted on the Australian National Antarctic Research Expedition populations mainly at Casey Station in east Antarctica (66°17'S 110°3'E). One consistent finding is that individuals in Antarctica display suppressed cutaneous immune responses, compared with their responses in Australia before or after they live on the ice (Roberts-Thomson et al. 1985; Williams et al. 1986; Tingate et al. 1997). Personnel who overwinter in the Antarctic stations also show a 50% reduction in T-cell proliferation responses to parathyroid hormone, as well as severe reductions in circulating tumor necrosis factror-α and changes in the production of interleukin (IL)-1β, IL-2, IL-6, and IL-10 (Figure 8.1) (Tingate et al. 1997). Similar results have been reported for temporary personnel of arctic habitats (Dobrodeeva et al. 1998).

Plasma concentrations of testosterone have been reported to be elevated during the winter in male Antarctic inhabitants (Takagi 1986; Sawhney et al. 1998). Androgens such as testosterone are generally immunosuppressive (Grossman 1984), but the elevated testosterone concentrations do not seem to account for the depressed immune responses observed in the Antarctic (Pitson et al. 1996) (Chapter 7). One leading hypothesis that has been put forth to account for these depressed immune responses is "social isolation" stress (Tingate et al. 1997). Several studies have indicated that social isolation or confinement anxiety may suppress immune responses (e.g., Benestad et al. 1990; Meehan et al. 1992, 1993; Sonnenfeld et al. 1992; Konstantaninova et al. 1993; Schmitt & Schaffer 1993a, b). Psychological tests scores for anxiety, however, do not correlate with immune suppression for Antarctic expeditioners (Pitson et al. 1996). It remains possible that the energetic thermoregulatory demands of the extreme climate of Antarctica (despite much assistance from technology) reduces energy expenditure to the immune system to the extent that immune function is compromised.

A positive energy balance is required for survival and subsequent reproductive success. In some cases, a marginal energetic balance can weaken animals to the extent that they become susceptible to diseases (Berczi 1986). Immunological defense against invading pathogens requires cascades of mitotic processes that presumably demand substantial energy (Demas et al. 1997a, b; Spurlock 1997; Nelson & Klein 1999). The energetic costs associated with immunity may be a critical factor in seasonal fluctuations in immune function.

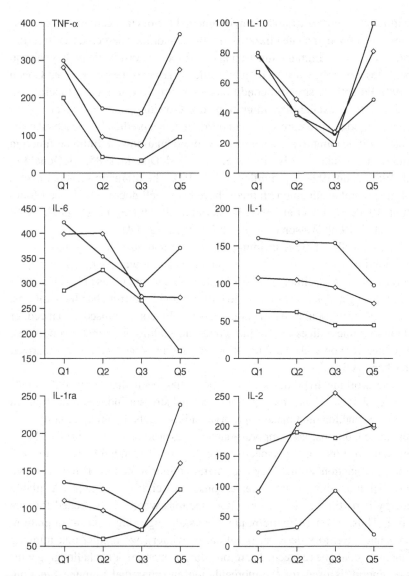

Fig. 8.1. Cytokine production in parathyroid hormone-stimulated lymphocytes. □, 24-hour cultures; ◇, 48-hour cultures; ○,72-hour cultures. The y-axes are pg/ml, except for IL-1ra, which is in pg/ml $\times 10^{-2}$ Q1, beginning of isolation (May 1993); Q2, midpoint of isolation (July 1993); Q3, end of isolation (November 1993); Q5, follow-up study in Australia (September 1994). TNF, tumor necrosis factor. (From Tingate et al. 1997.)

Furthermore, climatic or other environmental factors that can interrupt breeding (e.g., a flood or a late blizzard), or other conditions perceived as stressful, can compromise immune function and promote opportunistic pathogens and parasites possibly causing premature death (Berczi 1986; Ader & Cohen 1993). Potentially stressful conditions, including low ambient temperatures, reduced food availability, migration, overcrowding, lack of cover, or increased predator pressure, can recur on a somewhat predictable, seasonal basis potentially leading to seasonal changes in populationwide immune function and death (Fänge & Silverin 1985; Lee & McDonald 1985; McDonald et al. 1988; John 1994; Lochmiller et al. 1994). The dramatic seasonal cycles of illness and death among humans have been well documented (see Monto et al. 1970; Boulay et al. 1999; Kloner et al. 1999; Rosenbauer et al. 1999; Ross et al. 1999; Watson et al. 1999; Chapters 3 and 4).

Seasonal fluctuations in immune function and survivorship may not be observed every year or in every population. Some winters may not be perceived as stressful, either because of mild ambient conditions or because energetic coping adaptations succeed to buffer individuals from harsh conditions. Consequently, populations in nature may exhibit compromised, enhanced, or static immune indices during the winter, depending on climatic conditions. The literature on seasonal fluctuations in immune function was summarized in Chapters 3 and 4.

The working hypothesis of this book has been that some individuals have evolved mechanisms to predict seasonal stressor-induced reductions in immune function, and make appropriate adjustments in anticipation of challenging conditions, as a temporal adaptation to promote winter survival. From an adaptive functional perspective, we propose that individuals "optimize" immune function so that they can tolerate minor infections if the energetic costs of mounting an immune response outweigh the benefits. Available energy is partitioned into competing functions where most needed for survival. Thus, when energetic requirements are high (e.g., during migration, pregnancy, territory defense, lactation, or winter), we predict that immune function should be reduced. Therefore, during energetically challenging winters, energy is used for thermoregulation and maximal immune function, rather than growth, reproduction, or other nonessential processes.

We hypothesize that exposure to short-day lengths enhances immune function. Many field and clinical studies are consistent with this hypothesis; i.e., lymphatic tissue size and immune function are elevated during the winter. Laboratory studies are also consistent with this hypothesis; that is, immune function is enhanced in animals housed in short days, compared with animals maintained in long-day length conditions (reviewed in Nelson & Demas 1996; but see Yellon et al. 1999). These results suggest that mechanisms have evolved to allow animals to anticipate immunologically challenging conditions by monitoring photoperiod. Presumably, this ability permits individuals to cope with these health-threatening seasonal conditions by bolstering lymphatic tissue development and immune function directly.

Several infectious diseases occur seasonally, and this seasonal pattern in diseases could reflect seasonal patterns in the pathogen (e.g., influenza or rhinovirus), in the vector (e.g., malaria), or seasonal patterns in both. For example, approximately half of all colds are caused by rhinoviruses, whereas the other "colds" are caused by pathogens, including parainfluenza, respiratory syncytial virus, and coronaviruses (Monto, 1994). Rhinoviruses replicate best at low temperatures *in vitro*, and in the nasal passages when temperatures are less than 34°C (Lwoff 1969). Treatment with warm humid air on the first day of symptoms can ameliorate many subsequent cold symptoms (Macknin et al. 1990; Katzman 1994; Monto 1994). Thus, the seasonal patterns of colds may reflect the ease with which rhinoviruses replicate in the low temperature conditions of winter. Perhaps the so-called "summer colds" are due to the nonrhinovirus causative agents. However, some of the seasonal pattern of colds could reflect the energetic effects of low temperatures, as well as the facilitative effects on viral replication (Doyle et al. 1994). We hypothesize that exposure to environmental stressors mediates whether a host will be able to overcome an infection or not; in other words, exposure to stressors mediates seasonal fluctuations in resistance and susceptibility to infection.

8.3. Psychological Stressors and Disease

We hypothesize that in order for stressor-induced disease to develop, at least one of the following conditions must occur: (1) metabolic stores must be low at the onset (e.g., malnutrition or low fat stores in combination with low ambient temperatures) – Darwin wrote about the detrimental effects of extreme cold would be greatest on those that obtained the least amount of food before winter; (2) several environmental stressors occur simultaneously; or (3) a single environmental stressor must be unusually severe. Studies of stress physiology suggest that stress-induced disease ensues based on an individual's perception of a stressor. Perception of stressors reflects several factors, some of which are described herein.

8.3.1. Individual Differences in the Stress Response

Early individual experiences can modulate the responses to stressors in adulthood. For instance, how children in 1934 responded to a cold pressor test (a test of bodily reactions to a hand placed in a tub of iced water for 1 minute) predicted health problems as adults (Wood et al. 1984). Over 70% of children that "overreacted" to the test developed cardiovascular disease as adults, compared with less than 20% of individuals who had displayed little or no reaction to the test 50 years previously (Wood et al. 1984). Approximately 50% of women incarcerated in prison camps during World War II stopped menstruating. Obviously, the prison camps were extremely stressful environments. In such a stressful environment, the interesting question is not why 50% of the women stopped menstruating, but why were only 50% of the women affected? Subsequent stress research has illustrated that individual differences in stress perception exist. One experiment examined two monkeys to study individual differences in response to stress (Sapolsky 1992). In this study, two monkeys were deprived of food. Presumably because both individuals were deprived of nutrients, both should show an equal energy deficit and equivalent rise in stress hormones. One monkey was fed a nonnutritive-flavored placebo and did not display elevated glucocorticoid secretion, but the other unfed individual did. It seems that the critical factor is that the first monkey did not *perceive* things to be as stressful as the second one (Sapolsky 1992).

Most scientists now accept the power of psychological variables to modulate stress physiology (Calabrese et al. 1987; Herbert & Cohen 1993; Sapolsky 1994). What are these variables? Among individuals of vertebrate species, the main psychological factors that modulate stress responses include: (1) lack of control; (2) lack of outlets for dealing with frustration; and (3) lack of predictability (Calabrese et al. 1987; Herbert & Cohen 1993). These factors may be contributory to seasonal stressors.

8.3.2. Psychological Variables: Lack of Control

Individuals suffer fewer stress-related pathologies if they can control the situation causing the stress (Gatchel et al. 1989). Controlling a stressful situation is a type of coping behavior. Rats were subjected to intermittent electric shocks in one early study (Weiss 1968). One rat could control the situation because it was able to depress a lever to decrease the rate of shocks. Another rat received a shock whenever the first did, but could not control the frequency of the shocks. Subsequently, the second rat had elevated glucocorticoid secretion, compared with the rat that could moderate the shocks (Weiss 1968).

8.3.3. Psychological Variables: Outlets for Frustration

The ability to engage in displacement behavior also ameliorates the effects of stress. When rats are shocked, there is relatively low glucocorticoid secretion if they can chew on a piece of wood or attack another rat (Sapolsky 1992). It is common knowledge that the effects of stress in humans can be ameliorated by displacement behaviors such as engaging in a hobby or moderate exercise. Changes in ingestive behavior are commonly reported coping behaviors for stressors. Some seasonal stressors may elicit changes in food intake, which may enhance or decrease immune function.

8.3.4. Psychological Variables: Loss of Predictability

Predictable stress is easier to cope with than random presentation of stressors. In studies comparable to the one just described, predictability reduces the stress response of individuals. For example, if a warning stimulus is given prior to a shock, rats secrete less glucocorticoids, compared with rats that receive no warning stimuli (Sapolsky 1992).

We have argued that the energetic stressors of winter are recurrent and thus predictable. Evidence was presented in earlier chapters that changes in the annual cycle of photoperiod can be used by individuals to anticipate the time of stressful conditions and physiologically prepare for these stressors. The development of various energy-saving adaptations during the onset of short days, for instance, permits energy to be diverted to immune function. Immune function might be critically compromised if the individual encountered low temperatures or restricted food availability without the "programmed" bolstering of immunity by autumnal patterns (i.e., durations) of melatonin secretion (Nelson & Demas 1996). Thus, anticipation of recurrent stressors should permit coping that is not possible with unpredictable stressors and that presumably are more harmful to health. It is difficult to ascertain the genesis of individual differences in coping. For the most part, stressors occur at random times during people's lives and differ in their perceived strength among people. What serves as a stressor for one person (e.g., heights) might be sought out as recreation (e.g., hang gliding, mountain climbing, parachuting) for another. In contrast to these random exposures to individualized stressors, with random timing of immunosuppression, the entire general population is exposed (to some degree) to these recurrent, seasonal stressors. Thus, the effects on health, especially on those living on the energetic edge – including the elderly, small children, and immunocompromised individuals – should show some seasonal components (Haag-Weber et al. 1992; Moustschen et al. 1992; Evans et al. 1998; Touitou et al. 1986).

One obvious example involves the common cold. Although the prevalence of respiratory illnesses is increased during the fall and winter (Van Loghem 1928; Lidwell & Sommerville 1951; Monto et al. 1970; Doyle et al. 1994),

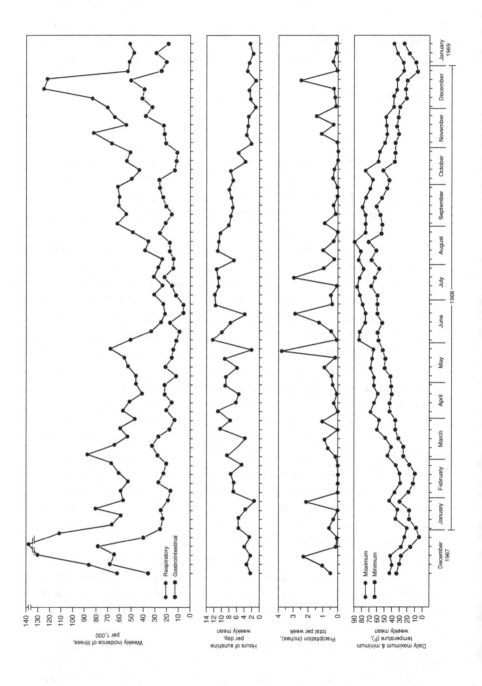

the causal factors underlying this relationship remain unspecified. Positive correlations between the onset of cold symptoms with ambient temperature, humidity, atmospheric pollution, or social conditions (e.g., returning to school) have been reported (Figure 8.2) (reviewed in Lidwell et al. 1965; Fellowes & Proctor 1973). We would argue that low temperatures require additional energy to maintain thermogenesis and that this additional energy is provided by reducing energy expended for immune function, and probably growth. Thus, immune function is somewhat compromised during low temperature conditions. If there has not been sufficient immunoenhancement in anticipation of the low temperatures, then immune function compromise might be severe. Similarly, if temperatures are very low so that thermogenesis requires significant energy "theft" from other systems, then immune function compromise might be severed.

Direct tests of the effects of low temperature on the onset of cold symptoms have been reported. For example, a few studies have placed individuals in low temperature conditions and infected them with rhinoviruses (reviewed in Monto 1994). Most of these studies have been flawed, however, because of a lack of proper controls, double-blinding techniques, and adequate subjects per experimental group (e.g., only three per group). The general consensus of these studies has been that low temperatures are not related to the onset or severity of symptoms of rhinovirus infection (Monto 1994). If the studies were conducted with proper controls, under double-blind conditions and with appropriate numbers of subjects, then we anticipate the results would show that low temperatures would indeed be associated with an increase in infection rates, as well as viral replication. This is even more likely to be the case if the experimental subjects are not overfed American adults, but rather, small children, the elderly, or individuals who might be living on the energetic "edge."

As described previously, rhinoviruses replicate best at low temperatures *in vitro*, and in the nasal passages when temperatures are less than 34°C (Lwoff 1969). Treatment with warm humid air on the first day of symptoms can ameliorate many subsequent cold symptoms (Macknin et al. 1990; Katzman 1994; Monto 1994). Thus, the seasonal patterns of colds may simply reflect the ease with which rhinoviruses replicate in the low temperature

---

Fig. 8.2. Relationship among temperature, precipitation, hours of sunshine to respiratory and gastrointestinal illnesses in Tecumseh, MI, 1968. (From *Archives of Environmental Health*, vol. 21, pp. 408–17, 1970. Reprinted with permission of the Helen Dwight Reid Educational Foundation. Published by Heldref Publications, 1319 Eighteenth St., N. W., Washington, DC 20036-1802. Copyright © 1970.)

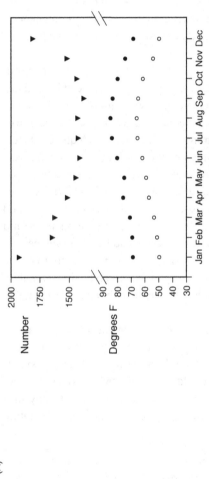

(A)

Number

2000
1750
1500
90
80
70
60
50
40
30

Degrees F

Jan Feb Mar Apr May Jun Jul Aug Sep Oct Nov Dec

▼ Average CAD Deaths
● Average Maximum Temperature
○ Average Minimum Temperature

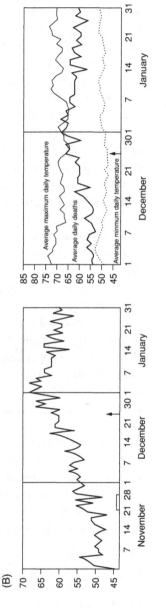

(B)

70
65
60
55
50
45

7 14 21 28 1
November

7 14 21 30 1
December

7 14 21 31
January

85
80
75
70
65
60
55
50
45

Average maximum daily temperature

Average daily deaths

Average minimum daily temperature

1 7 14 21 30 1
December

7 14 21 31
January

228

conditions of winter. Additional studies will be necessary to discriminate among these hypotheses.

Seasonal patterns in other diseases could be due to other nonenergetic causes. For example, a strong seasonal pattern of coronary deaths has been reported in several studies (Bundeson & Falk 1926; Douglas et al.1991; Boulay et al. 1999; Kloner et al. 1999). Human mortality shows a seasonal pattern in the Northern Hemisphere (e.g., United States and Scotland), with the highest number of deaths occurring in January (i.e., winter) and the lowest number of deaths in May and June (i.e., summer) (Figure 1.2). The seasonal pattern is reversed in the Southern Hemisphere (e.g., New Zealand), in which mortality from coronary heart disease and cerebrovascular disease peaks around July (i.e., winter) and declines to its lowest rate in February (i.e., summer) in both men and women (Figure 3.4) (Douglas et al. 1990).

The seasonal pattern of deaths from circulatory disease, occurring both in the Northern and Southern Hemispheres, have been hypothesized to be caused by changes in temperature (Chapter 3). Deaths from coronary disease increase by 40% at low temperatures, compared with mild temperatures; however, the seasonal rise in deaths from coronary diseases during the winter is reportedly independent of temperature, especially among men and women younger than 60 years of age (Enquselassie et al. 1993). An analysis of death records in Los Angeles County for death caused by coronary disease from 1985 to 1996 ($n = 222{,}265$ cases) showed a peak in December and January (Figure 8.3a) (Kloner et al. 1999). Because mean ambient temperatures of the Los Angeles area does not change appreciably across the year (Figure 8.3b), the authors ruled out temperature variables as a contributing cause of the seasonal pattern of coronary death. Rather, the authors speculated that stressors associated with the December holidays and the New Year contributed to the coronary deaths (Kloner et al. 1999). If true, a study of a South American population inhabiting a climate similar to Los Angeles should show the same pattern of coronary deaths (i.e., peaking around Christmas and New Year's Day); otherwise, the peak in cardiac deaths should be phase-shifted 6 months.

Fig. 8.3. (A) Number of coronary artery deaths (CAD) in Los Angeles county per month plotted with average monthly maximum and minimum temperatures for Los Angeles. (B) (Left) Total coronary artery deaths by day for November, December, and January averaged over 12 years to see annual increase of deaths around Thanksgiving and a decrease after 1 January. Arrow indicates Christmas. (Right) Average daily coronary artery deaths in December and January plotted with average maximum and minimum temperatures for Los Angeles. Arrow indicates Christmas. (From Kloner et al. 1999.)

Cardiac disease may be associated with chronic infection (e.g., Mendall et al. 1994; Danesh et al. 1999; Lockhart & Durack, 1999; Smith et al. 1999). Depending on the population under study, the most common infection seems to involve either *Helicobacter pylori* or *Chlamydia pneumonia*; both pathogens have been isolated in atherosclerotic plaques (e.g., Bauriedel et al. 1999; Danesh et al. 1999; Gasbarrini et al. 1999). The extent to which these pathogens vary seasonally and the contribution of these pathogens to the seasonal pattern of cardiac problems require additional studies. Taken together, these data illustrate the clinical relevance for studies of seasonality in infection and the underlying mechanisms involved.

8.4. Future Directions

8.4.1. Where are the Gaps in Our Knowledge?

Although the common cold is the most frequent infection an individual is likely to encounter, prevention, treatment, and natural history of this disease remain elusive. The most obvious feature of the epidemiology of the common cold, as well as influenza and a number of other infectious diseases, is its seasonal pattern of prevalence (Lidwell et al. 1965). Yet, despite many attempts to explain why these infections are more prevalent during the autumn and winter than at other times of the year, the subtitle of a 35-year-old *Science* paper on the epidemiology of infectious diseases remains true: "We do not yet understand how seasonal and other factors affect the incidence of colds and influenza" (Andrews 1964). The same is true for many other diseases. We hope that our testable hypothesis of winter immunoenhancement will generate studies to test this approach, both clinically and in the field.

8.4.2. What will Studies have to Address in the Future?

The effects of stressors on disease must be established on a mechanistic level. Several studies have shown phenomenological associations between stressors and disease susceptibility. For example, exposure to stressors tends to increase the prevalence and aggressiveness of several types of cancer (Selye 1979). Thus, we propose that both the intrinsic and extrinsic factors associated with seasonal cycles in disease must be considered by clinicians and basic biologists alike. By considering the causes of seasonality in infection and immunity, we may begin to understand the impact that the environment has on health and well-being.

References

Abbas, A. K., Lichtman, A. H. & Pober, J. S. 1994. *Cellular and Molecular Immunology.* Philadelphia, PA: W.B. Saunders, Co.

Abioye, A. A. 1973. Fatal amoebic colitis in pregnancy and puerperium: A new clinic-pathological entity. *J. Trop. Med. Hyg.*, **76**, 97–100.

Ablin, R. J., Bartkus, J. M. & Gonder, M. J. 1988. Immunoquantitation of factor VIII-related antigen (von Willebrand factor antigen) in prostate cancer. *Cancer Lett.*, **30**, 283–9.

Ablin, R. J., Bhatti, R. A., Guinan, P. D. & Khin, W. 1979. Modulatory effects of oestrogen on immunological responsiveness. II. Suppression of tumour-associated immunity in patients with prostatic cancer. *Clin. Exp. Immunol.*, **38**, 83–91.

Ablin, R. J., Bruns, G. R., Guinan, P. D. & Bush, I. M. 1974. Afferent blockage, oestrogenic therapy, and malignancy. *Lancet*, **17**, 413.

Abo, T., Kawate, T., Hinuma, S., Itoh, K., Abo, W., Sato, J. & Kumagai, K. 1980. The circadian periodicities of lymphocyte subpopulations and the role of corticosteroid in human beings and mice. In: *Recent Advances in the Chronobiology of Allergy and Immunology* (ed. M. H. Smolensky), pp. 301–6. Oxford: Pergamon Press.

Abo, T. & Kumagai, K. 1978. Studies of surface immunoglobulins on human B lymphocytes. III. Physiological variations of sIg$^+$ cells in peripheral blood. *Clin. Exp. Immunol.*, **33**, 441–52.

Abraham, A. D. & Buga, G. 1976. ^{3}H-testosterone distribution and binding in rat thymus cells *in vivo. Mol. Cell. Biochem.*, **13**, 157–63.

Abu-Zeid, Y. A., Abdulhadi, N. H., Theander, T. G., Hviid, L., Saeed, B. O., Jepsen, S., Jensen, J. B. & Bayoumi, R. A. 1992. Seasonal changes in cell mediated immune responses to soluble *Plasmodium falciparum* antigens in children with haemoglobin AA and haemoglobin AS. *Trans. R. Soc. Trop. Med. Hyg.*, **86**, 20–2.

Acheson, E. D., Bachrach, C. A. & Wright, F. M. 1960. Some comments on the relationship of the distribution of multiple sclerosis to latitude, solar radiation and other variables. *Acta Psychiatr. Neurol. Scand. Suppl.*, **35**, 132–47.

Ader, R. & Cohen, N. 1981. *Psychoneuroimmunology.* New York: Academic Press.

Ader, R. & Cohen, N. 1982. Conditioned suppression of humoral immunity in the rat. *J. Comp. Physiol. Psychol.*, **96**, 517–21.

Ader, R. & Cohen, N. 1993. Psychoneuroimmunology: Conditioning and stress. *Annu. Rev. Psychol.*, **44**, 53–85.

Ader, R., Felten, D. L. & Cohen, N. 2001. *Psychoneuroimmunology*, 3rd ed. New York: Academic Press.

Afoke, A. O., Eeg-Olofsson, O., Hed, J., Hjellman, N.-I. M., Lindblom, B. & Ludvigsson, J. 1993. Seasonal variation and sex differences of circulating macrophages, immunoglobulins and lymphocytes in healthy school children. *Scand. J. Immunol.*, **7,** 209–15.

Aimé, P. 1912. Note sur le thymus chez les cheloniens. *C. R. Soc. Seances Soc. Biol. Fil.*, **72,** 889–90.

Alexander, J. 1988. Sex differences and cross-immunity in DBA/2 mice infected with *L. mexicana* and *L. major. Parasitology*, **96,** 297–302.

Alexander, J. & Stimson, W. H. 1988. Sex-hormones and the course of parasitic infection. *Parasitol. Today*, **4,** 189–93.

Alexander, J. W., MacMillan, B. G., Stinnett, J. D., Ogle, C. K., Bozian, R. C., Fischer, J. E., Oakes, J. B., Morris, M. J. & Krummel, R. 1980. Beneficial effects of aggressive protein feeding in severely burned children. *Ann. Surg.*, **192,** 505–17.

Alfonso, C. P., Cowen, B. S. & Campen, H. V. 1995. Influenza A viruses isolated from waterfowl in two wildlife management areas of Pennsylvania. *J. Wildlife Dis.*, **31,** 179–85.

Alqvist, J. 1976. Endocrine influences on lymphatic organs, immune responses, inflamation autoimmunity. *Acta. Endocrinol.*, **206(Suppl),** 64–70.

Alvarez, F., Razquin, B., Villena, A., Fierro, P. L. & Zapata, A. A. 1988. Alterations in the peripheral lymphoid organs and differential leukocyte counts in saprolegnia-infected brown trout, *Salmo trutta fario. Vet. Immunol. Immunopath.*, **18,** 181–93.

Alvarez, F., Razquin, B. E., Villena, A. J. & Zapata, A. G. 1998. Seasonal changes in the lymphoid organs of wild brown trout, *Salmo trutta* L.: A morphometrical study. *Vet. Immunol. Immunopathol.*, **64,** 267–78.

Alvarez-Mon, M., Kehrl, J. H. & Fauci, A. S. 1985. A potential role for adrenocorticotropin in regulating human B lymphocyte functions. *J. Immunol.*, **135,** 3823–6.

Amin, O. M. 1978. Intestinal helminths of some Nile fishes near Cairo, Egypt with redescriptions of *Camallanus kirandensis* Baylis 1928 (Nematoda) and *Bothriocephalus aegyptiacus Rysavy and* Moravec 1975 (Cestoda). *J. Parasitol.*, **64,** 93–101.

Amino, N., Tanizawa, O., Miyai, K., Tanaka, F., Hayashi, C., Kawashima, M. & Ichihara, K. 1978. Changes of serum immunoglobulins IgG, IgA, IgM and IgE during pregnancy. *Obstet. Gynecol.*, **52,** 415–20.

Amor, B., Dougados, M., Benhamou, L., Kuhn, J. M. & Laudat, M. H. 1983. Failure of androgen therapy in a flare-up of acute disseminated lupus erythematosus. *Presse Med.*, **12,** 1726.

Ancel, A., Visser, H., Handrich, Y., Masman, D. & LeMaho, Y. 1997. Energy saving in huddling penguins. *Nature*, **385,** 304–5.

Anderson, K. D., Nachman, R. J. & Turek, F. W. 1988. Effects of melatonin and 6-methoxybenzoxazolinone on photoperiodic control of testis size in adult male golden hamsters. *J. Pineal Res.*, **5,** 351–65.

Andreis, P. G., Neri, G., Belloni, A. S., Mazzocchi, G., Kasprzak, A. & Nussdorfer, G. G. 1991. Interleukin-1 beta enhances corticosterone secretion by acting directly on the rat adrenal gland. *Endocrinology*, **129,** 53–7.

Andrews, C. H. 1964. The complex epidemiology of respiratory virus infections. *Science*, **146,** 1274–8.

Angeli, A., Gatti, G., Sartori, M. L., Del Ponte, D. & Cerignoa, R. 1988. Effect of exogenous melatonin on human natural killer (NK) cell activity. An approach to the immunomodulatory role of the pineal gland. In: *The Pineal Gland and Cancer* (eds. D. Gupta, A. Attanasio, & R. J. Reiter), pp. 145–57. Tübingen: Müller and Bass.

Angeli, A., Gatti, G., Sartori, M. L. & Masera, R. G. 1992. Chronobiological aspects of the neuroendocrine-immune network. Regulation of natural killer NK cell activity as a model. *Chronobiologia*, **19,** 93–110.

Antoni, F. 1986. Hypothalamic control of adrenocorticotropic hormone secretion: Advances since the discovery of 41-residue corticotropin releasing factor. *Endocr. Rev.*, **7,** 351–73.

Ardawi, M. S. M. & Newsholme, E. A. 1985. Metabolism in lymphocytes and its importance in the immune response. *Essays Biochem.*, **21,** 1–44.

Arderiu, F. A., Coll, F. P. & Bada, A. J. L. 1979. Enfermedad de los legionarios. *Med. Clin.*, **73,** 157–63.

Arendt, J. 1995. *Melatonin: Basic and Clinical Applications.* Boca Raton: CRC Press.

Arendt, J., Wirz-Justice, A. & Bardtke, J. 1977. Annual rhythm of serum melatonin in man. *Neurosci. Lett.*, **7,** 327–30.

Arrenbrecht, S. 1974. Specific binding of growth hormone to thymocytes. *Nature*, **252,** 255–7.

Arzt, E. S., Ferandez-Castelo, S., Finocchario, L. M., Criscuolo, M. E., Dia, A., Finkleman, S. & Nahmod, V. E. 1988. Immunomodulation by indoleamines: Serotonin and melatonin action on DNA and interferon-gamma synthesis by human peripheral blood mononuclear cells. *J. Clin. Immunol.*, **8,** 513–20.

Asanji, M. F. 1988. Haemonchosis in sheep and goats in Sierra Leone. *J. Helminthol.*, **62,** 243–9.

Aschoff, J. 1981. Annual rhythms in man. In J. Aschoff (ed.), *Handbook of Behavioral Neurobiology*, Vol. 4, pp. 475–90. New York: Plenum Press.

Ashcroft, G. S., Dodsworth, J., van Boxtel, E., Tarnuzzer, R. W., Horan, M. A., Schultz, G. S. & Ferguson, M. W. J. 1997. Estrogen accelerates cutaneous wound healing associated with an increase in TGF-β1 levels. *Nature Med.*, **3,** 1209–15.

Atkins, M. B., Mier, J. W., Parkinson, D. R., Gould, J. A., Berkman, E. M. & Kaplan, M. M. 1988. Hypothyroidism after treatment with interleukin-2 and lymphokine-activated killer cells. *N. Engl. J. Med.*, **318,** 1557–63.

Aubert, C., Janiaud, P. & Lecalvez, J. 1980. Effect of pinealectomy and melatonin on mammary tumor growth in Sprague-Dawley rats under different conditions of lighting. *J. Neural Transm.*, **47,** 121–30.

Audhya, T., Jain, R. & Hollander, C. S. 1991. Receptor-mediated immunomodulation by corticotropin-releasing factor. *Cell. Immunol.*, **134,** 77–84.

Avitsur, R. & Yirmiya, R. 1999. The immunobiology of sexual behavior: Gender differences in the suppression of sexual activity during illness. *Pharmacol. Biochem. Behav.*, **64,** 787–96.

Azevedo, E., Ribeiro, J. A., Lopes, F., Martins, R. & Barros, H. 1995. Cold: A risk factor for stroke? *J. Neurol.*, **242,** 217–21.

Baboonian, C. & Griffiths, P. 1983. Is pregnancy immunosuppressive? Humoral immunity against viruses. *Br. J. Obstet. Gynaecol.*, **90,** 1168–75.

Baker, J. R. 1938. The evolution of breeding seasons. In: G. R. deBeer (ed). *Evolution: Essays on Aspects of Evolutionary Biology* (ed. G. R. deBeer), pp. 161–77. London: Oxford University Press.

Baker, J. R. & Ranson, R. M. 1932. Factors affecting the breeding of the field mouse (*Microtus agrestis*). Part I. Light. *Proc. R. Soc. Lond.*, **112**, 313–23.

Baker, J. R. & Ranson, R. M. 1932. Factors affecting the breeding of the field mouse (*Microtus agrestis*). Part I. Temperature and food. *Proc. R. Soc. (London) B*, **112**, 39–46.

Baker, N. F. & Fisk, R. A. 1986. Seasonal occurrence of infective nematode larvae in California Sierra foothill pastures grazed by cattle. *Am. J. Vet. Res.*, **47**, 1680–5.

Balashov, K. E., Olek, M. J., Smith, D. R., Khoury, S. J. & Weiner, H. L. 1998. Seasonal variation of interferon-gamma production in progressive multiple sclerosis. *Ann. Neurol.*, **44**, 824–8.

Bamford, C. R., Sibley, W. A. & Thies, C. 1983. Seasonal variation of MS exacerbations in Arizona. *Neurology*, **33**, 1537–44.

Barnard, C. J., Behnke, J. M. & Sewell, J. 1994. Social behaviour and susceptibility to infection in house mice (*Mus musculus*): Effects of group size, aggressive behaviour and status-related hormonal responses prior to infection on resistance to *Babesia microti*. *Parasitology*, **108**, 487–96.

Barnett, S. A. 1973. Maternal processes in the cold-adaptation of mice. *Biol. Rev. Cambr. Philos. Soc.*, **48**, 477–508.

Barr, I. G., Pyke, K. W., Pearce, P., Toh, B. H. & Funder, J. W. 1984. Thymic sensitivity to sex hormones develops post-natally; an in vivo and an in vitro study. *J. Immunol.*, **132**, 1095–9.

Barrell, G. K., Thrun, L. A., Brown, M. E., Viguie, C. & Karsch, F. J. 2000. Importance of photoperiodic signal quality to entrainment of the circannual reproductive rhythm of the ewe. *Biol. Reprod.*, **63**, 769–74.

Bartness, T. J. 1995. Short day-induced depletion of lipid stores is fat pad- and gender-specific in Siberian hamsters. *Physiol. Behav.*, **58**, 539–50.

Bartness, T. J. & Goldman, B. D. 1989. Mammalian pineal melatonin: A clock for all seasons. *Experientia*, **45**, 939–45.

Bartness, T. J., Powers, J. B., Hastings, M. H., Bittman, E. L. & Goldman, B. D. 1993. The timed infusion paradigm for melatonin delivery: What has it taught us about the melatonin signal, its reception, and the photoperiodic control of seasonal responses? *J. Pineal Res.*, **15**, 161–90.

Bartness, T. J. & Wade, G. N. 1985. Photoperiodic control of seasonal body weight cycles in hamsters. *Neurosci. Biobehav. Rev.*, **9**, 599–612.

Bartsch, H. & Bartsch, C. 1981. Effect of melatonin on experimental tumors under different photoperiods and times of administration. *J. Neural Transm.*, **52**, 269–79.

Baum, M. & Liesen, H. 1997. Exercise and immunology. *Orthopade*, **26**, 976–80.

Baum, M., Liesen, H. & Enneper, J. 1994. Leucocytes, lymphocytes, activation parameters and cell adhesion molecules in middle-distance runners under different training conditions. *Int. J. Sports Med.*, **15**, S122–6.

Bauriedel, G., Andrie, R., Likungu, J. A., Welz, A, Braun, P., Welsch, U., & Luderitz, B. 1999. Persistence of Chlamydia pneumoniae in coronary plaque tissue. [A contribution to infection and immune hypothesis in unstable angina pectoris] [English summary]. *Dtsch. Med. Wochenschr.*, **124**, 1408–13.

Beaudoin, R. L., Applegate, J. E., Davis, D. E. & McLean, R. G. 1971. A model for the ecology of avian malaria. *J. Wildlife Dis.*, **7**, 5–13.

Beck, J., Rondot, P., Catinot, L., Falcoff, E., Kirchner, H. & Wietzerbin, J. 1988. Increased production of interferon gamma and tumor necrosis factor precedes clinical manifestation in multiple sclerosis: Do cytokines trigger off exacerbations? *Acta Neurol. Scand.*, **78**, 318–23.

Beck-Friis, J., Von Rosen, D., Kjellman, B. F., Ljunggren, J. G. & Wetterberg, L. 1984. Melatonin in relation to body measures, sex, age, season and use of drugs in patients with major affective disorders and healthy subjects. *Psychoneuroendocrinology*, **9**, 261–77.

Behnke, J. M., Barnard, C. J. & Wakelin, D. 1992. Understanding chronic nematode infections: Evolutionary considerations, current hypotheses and the way forward. *Int. J. Parasitol.*, **22**, 861–907.

Benestad, H. B., Hersleth, I. B., Hardersen, H. & Molvaer, O. I. 1990. Functional capacity of neutrophil granulocytes in deep-sea divers. *Scand. J. Clin. Lab. Invest.*, **50**, 9–18.

Ben-Nathan, D., Maestroni, G. J., Lustig, S. & Conti, A. 1995. Protective effects of melatonin in mice infected with encephalitis viruses. *Arch. Virol.*, **140**, 223–30.

Berczi, I. 1986. The influence of the pituitary-adrenal axis on the immune system. In: *Pituitary Function and Immunity* (ed. I. Berczi). Boca Raton: CRC Press.

Berczi, I., Nagy, E., de Toledo, S. M., Matusik, R. J. & Friesen, H. G. 1991. Pituitary hormones regulate c-myc and DNA synthesis in lymphoid tissue. *J. Immunol.*, **146**, 2201–6.

Berger, P. J., Negus, N. C., Sanders, E. H. & Gardner, P. D. 1981. Chemical triggering of reproduction in *Microtus montanus*. *Science*, **214**, 69–70.

Berginer, V. M., Goldsmith, J., Batz, U., Vardi, H. & Shapiro, Y. 1989. Clustering of strokes in association with meteorologic factors in the Negev Desert of Israel: 1981–1983. *Stroke*, **20**, 65–9.

Berkenbosch, F., van Oer, J., del Rey, A., Tilders, F. & Besedovsky, H. 1987. Cortico-tropin-releasing factor-producing neurons in the rat activated by interleukin-1. *Science*, **238**, 524–6.

Bernardini, R., Kammilaris, T. C., Calogero, A. E., Johnson, E. O., Gomez, M. T., Gold, P. W. & Chrousos, G. P. 1990. Interactions between tumor necrosis factor-alpha, hypothalamic corticotropin releasing hormone, and adrenocorticotropin secretion in the rat. *Endocrinology*, **126**, 2876–81.

Bertini, R., Bianci, M. & Ghezzi, P. 1988. Adrenalectomy sensitizes mice to the lethal effects of interleukin-1 and tumor necrosis factor. *J. Exp. Med.*, **167**, 1708–12.

Besedovsky, H. O. & del Rey, A. 1991. Feed-back interactions between immunolog-ical cells and the hypothalamus-pituitary-adrenal axis. *Netherlands J. Med.*, **39**, 274–80.

Besedovsky, H. O. & del Rey, A. 1996. Immune-neuro-endocrine interactions: Facts and hypotheses. *Endocr. Rev.*, **17**, 64–102.

Besedovsky, H. O., del Rey, A., Klusman, I., Furukawa, H., Monge Arditi, G. & Kabiersch, A. 1991. Cytokines as modulators of the hypothalamus-pituitary-adrenal axis. *J. Steroid Biochem. Mol. Biol.*, **40**, 613–8.

Besedovsky, H. O., del Rey, A. & Sorkin, E. 1981. Lyphokine-containing supernatants from Con-A stimulated cells increase corticosterone blood levels. *J. Immunol.*, **126**, 385–7.

Besedovsky, H. O., del Rey, A., Sorkin, E. & Dinarello, C. A. 1986. Immunoregula-tory feedback between interleukin-1 and glucocorticoid hormones. *Science*, **233**, 652–4.

Besedovsky, H. O., del Rey, A., Sorkin, E., Lotz, W. & Schwulera, U. 1985. Lymphoid cells produce an immunoregulatory glucocorticoid increasing factor (GIF) acting through the pituitary gland. *Clin. Exp. Immunol.*, **59**, 622–8.

Besedovsky, H. O., Sorkin, E. & Mueller, J. 1975. Hormonal changes during the immune response. *Proc. Soc. Exp. Biol.*, **150**, 466–79.

Best, C. L., Pudney, J., Anderson, D. J. & Hill, J. A. 1994. Modulation of human granulosa cell steroid production in vitro by tumor necrosis factor alpha: Implications of white blood cells in culture. *Obstet. Gynecol.*, **84**, 121–7.

Beutler, B. & Cerami, A. 1988a. Cachectin (tumor necrosis factor): A macrophage hormone governing cellular metabolism and inflammatory response. *Endocr. Rev.*, **9**, 57–66.

Beutler, B. & Cerami, A. 1988b. The common mediator of shock, cachexia, and tumor necrosis. *Adv. Immunol.*, **42**, 213–31.

Bigaj, J. & Plytycz, B. 1984. Endogenous rhythm in the thymus gland of *Rana temporaria* (morphological study). *Thymus*, **6**, 369–73.

Bijlsma, J. W. J., Cutolo, M., Masi, A. T. & Chikanza, I. C. 1999. The neuroendocrine immune basis of rheumatic diseases. *Trends Immunol.*, **20**, 298–301.

Bilbo, S. D., Drazen, D. L. & Nelson, R. J. 2001. Short days alleviate symptoms of infection in *Siberian hamsters*. (In review).

Bilbo, S. D. & Nelson, R. J. 2001. Sex steroid hormones enhance immune function in male and female Siberian hamsters. *Am. J. Physiol.*, **280**, R207–13.

Biller, J., Jones, M. P., Bruno, A., Adams, H. P. & Banwart, K. 1988. Seasonal variation of stroke – Does it exist? *Neuroepidemiology*, **7**, 89–98.

Billingham, R. E. 1986. Immunologic advantages and disadvantages of being a female. In: *Reproductive Immunology* (eds. D. A. Clarke & B. A. Croy), pp. 1–9. New York: Elsevier Science Publishers.

Billiteri, H. & Bindoni, M. 1969. Accrescimento e moltiplicazione cellulare del tumore di Erlich nel toppo, dopop azportazione della thiandola pincale. *Boll. Soc. Ital. Biol. Sper.*, **45**, 1647–50.

Bindoni, M. 1971. Relationship between the pineal gland and the mitotic activity of some tissues. *Arch. Sci. Biol.*, **55**, 3–21.

Bindoni, M., Jutisz, M. & Ribot, G. 1976. Characterization and partial purification of a substance in the pineal gland which inhibits cell multiplication in vitro. *Biochem. BioPhys. Acta.*, **437**, 577–88.

Biondi, M. 2001. Effects of stress on immune function: An overview. In: *Psychoneuroimmunology*, 3rd ed., Vol. 2 (eds. R. Ader, D. L. Felten & N. Cohen), pp. 189–226. New York: Academic Press.

Birkeland, S. A. & Kristoffersen, K. 1980. Lymphocyte transformation with mitogens and antigens during normal human pregnancy: A longitudinal study. *Scand. J. Immunol.*, **11**, 321–5.

Bittman, E. L. & Goldman, B. D. 1979. Serum levels of gonadotrophins in hamsters exposed to short photoperiods: Effects of adrenalectomy and ovariectomy. *J. Endocrinol.*, **83**, 113–8.

Bittman, E. L. & Karsch, F. J. 1984. Nightly duration of pineal melatonin secretion determines the reproductive response to inhibitory day length in ewe. *Biol. Reprod.*, **30**, 588–93.

Black, P. H. 1994. Central nervous system-immune system interactions: Psychoneuroendocrinology of stress and its immune consequences. *Antimicrob. Agents Chemother.*, **38**, 1–6.

Blalock, J. E., Bost, K. L. & Smith, E. M. 1985. Neuroendocrine peptide hormones and their receptors in the immune system. Production, processing and action. *J. Neuroimmunol.*, **10**, 31–40.

Blalock, J. E., Harbour-McMenamin, D. & Smith, E. M. 1985. Peptide hormones shared by the neuroendocrine and immunologic systems. *J. Immunol.*, **135**(2 Suppl.), 858s–61s.

Blank, J. L. & Desjardins, C. 1984. Spermatogenesis is modified by food intake in mice. *Biol. Reprod.*, **30**, 410–5.

Blank, J. L. & Desjardins, C. 1985. Differential effects of food restriction on pituitary-testicular function in mice. *Am. J. Physiol.*, **248**, R181–9.

Blank, J. L. & Ruf, T. 1992. Effect of reproductive function on cold tolerance in deer mice. *Am. J. Physiol.*, **263**, R820–6.

Blask, D. E. 1984. The pineal: An oncostatic gland? In: *The Pineal Gland* (ed. R. J. Reiter), pp. 253–84. New York: Raven Press.

Blauer, K. L., Poth, M., Rogers, W. M. & Bernton, E. W. 1991. Dehydroepiandrosterone antagonizes the suppressive effects of dexamethasone on lymphocyte proliferation. *Endocrinology*, **129**, 3174–9.

Blom, J. M. C., Gerber, J. & Nelson, R. J. 1994. Immune function in deer mice: Developmental and photoperiodic effects. *Am. J. Physiol.*, **267**, R596–601.

Blom, J. M. C., Tamarkin, L., Shiber, J. R. & Nelson, R. J. 1995. Learned immunosuppression is associated with an increased risk of chemically-induced tumors. *Neuroimmunomodulation*, **2**, 92–9.

Blom, L. & Dahlquist, G. 1985. Epidemiological aspects of the natural history of childhood diabetes. *Acta Pædiatr. Scand. Suppl.*, **320**, 20–5.

Boccia, M. L., Scanlan, J. M., Laudenslager, M. L., Berger, C. L., Hijazi, A. S. & Reite, M. L. 1997. Juvenile friends, behavior, and immune responses to separation in bonnet macaque infants. *Physiol. & Behav.*, **61**, 191–8.

Boctor, F. N., Charmy, R. A. & Cooper, E. L. 1989. Seasonal differences in the rhythmicity of human male and female lymphocyte blastogenic responses. *Immunol. Invest.*, **18**, 775–84.

Boivin, D. B., Duffy, J. F., Kronauer, R. E. & Czeisler, C. A. 1996. Dose-response relationships for resetting of human circadian clock by light. *Nature*, **379**, 540–2.

Bolinger, M., Olson, S. L., Delagrange, P. & Turek, F. W. 1996. Melatonin agonist attenuates stress response and permits growth hormone release in male golden hamsters. Society for the Study of Biological Rhythms, 5th Meeting, Amelia Island.

Boonstra, R., Krebs, C. J. & Beacham, T. D. 1980. Impact of botfly parasitism on *Microtus townsendii* populations. *Can. J. Zool.*, **58**, 1683–92.

Borhanmanesh, F., Haghighi, P., Hekmat, K., Rezaizadeh, K., & Ghavami, A. G. 1973. Viral hepatitis during pregnancy. Severity and effect on gestation. *Gastroenterology*, **64**, 304–12.

Borysenko, M. 1987. Area review: Psychoneuroimmunology. *Ann. Behav. Med.*, **9**, 3–10.

Boulay, F., Berthier, F., Sisteron, O., Gendreike, Y. & Gibelin, P. 1999. Seasonal variation in chronic heart failure hospitalizations and mortality in France. *Circulation*, **100**, 280–6.

Box, E. D. 1966. Blood and tissue protozoa of the English sparrow (*Passer domesticus domesticus*) in Galveston, Texas. *J. Protogy*, **13**, 204–8.

Brabin, L. & Brabin, B. J. 1992. Parasitic infections in women and their consequences. *Adv. Parasitol.*, **31**, 1–81.

Brackowski, R., Zubelewicz, B., Romanowski, W., Lissoni, Q., Barni, S., Tancini, G. & Maestroni, G. J. 1994. Preliminary study on modulation of the biological effects of tumor necrosis factor-alpha in advanced cancer patients by the pineal hormone melatonin. *J. Biol. Regul. Homeost. Agents*, **8**, 77–80.

Brainard, G. C., Knobler, R. L., Podoin, P. I., Lavasa, M. & Lubin, F. D. 1985. Neuroimmunology: modulation of the hamster immune system by photoperiod. *Life Sci.*, **40**, 1319–26.

Bratescu, A. & Teodorescu, M. 1981. Circannual variations in the B cell/T cell ratio in normal human peripheral blood. *J. Allergy Clin. Immunol.*, **68**, 273–80.

Bray, R. S. & Anderson, M. J. 1979. Falciparum malaria and pregnancy. *Trans. R. Soc. Trop. Med. Hyg.*, **73**, 427–31.

Brenner, G. J. & Moynihan, J. A. 1997. Stressor-induced alternations in immune response and viral clearance following infection with herpes simplex virus-type 1 in BALB/c and C57B1/6 mice. *Brain Behav. Immunity*, **11**, 9–23.

Brenner, I. K., Shek, P. N. & Shephard, R. J. 1994. Infection in athletes. *Sports Med.*, **17**, 86–107.

Brent, L. & Medawar, P. 1966. Quantitative studies on tissue transplantation immunity. 8. The effects of irradiation. *Proc. R. Soc. Lond. B Biol. Sci.*, **165**, 413–23.

Brick, J. E., Wilson, D. A. & Walker, S. E. 1985. Hormonal modulation of responses to thymus-independent and thymus-dependent antigens in autoimmune NZB/W mice. *J. Immunol.*, **134**, 3693–8.

Bridges, R. S. & Goldman, B. D. 1975. Diurnal rhythms in gonadotropins and progesterone in lactating and photoperiod induced acyclic hamsters. *Biol. Reprod.*, **13**, 617–22.

Brodie, J. Y., Hunter, I. C., Stimson, W. H. & Green, B. 1980. Specified oestradiol binding in cytosols from the thymus glands from normal and hormone-treated male rats. *Thymus*, **1**, 337–45.

Bronson, F. H. 1979. The reproductive ecology of the house mouse. *Q. Rev. Biol.*, **54**, 265–99.

Bronson, F. H. 1989. *Mammalian Reproductive Biology*. Chicago: University of Chicago Press.

Bronson, F. H. 1995. Seasonal variation in human reproduction: Environmental factors. *Q. Rev. Biol.*, **70**, 141–64.

Bronson, F. H. & Heideman, P. N. 1994. Seasonal regulation of reproduction in mammals. In: *The Physiology of Reproduction* (eds. E. Knobil & J. D. Neil), pp. 541–84. New York: Raven Press.

Brosche, P. & Sünderman, J. 1990. *Earth's Rotation from Eons to Days*. Berlin: Springer-Verlag.

Bubenik, G. A., Blask, D. E., Brown, G. M., Maestroni, G. J., Pang, S. F., Reiter, R. J., Viswanathan, M. & Zisapel, N. 1998. Prospects of the clinical utilization of melatonin. *Biol. Signals Recep.*, **7**, 195–219.

Bucknell, D. G., Gasser., R. B. & Beveridge, I. 1995. The prevalence and epidemiology of gastrointestinal parasites of horses of Victoria, Australia. *Int. J. Parasitol.*, **25**, 711–24.

Bundeson, H. N. & Falk, S. I. 1926. Low temperature, high barometer and sudden death. *J. Am. Med. Assoc.*, **87**, 1987–90.

Bünning, E. 1973. *The Physiological Clock, Circadian Rhythms and Biological Chronometry.* Springer-Verlag, New York.

Burrells, C., Nettleton, P. F., Reid, H. W., Miller, H. R. P., Hopkins, J., McConnell, I., Gorrell, M. D. & Brandon, M. R. 1989. Lymphocyte subpopulations in the blood of sheep persistently infected with border disease virus. *Clin. Exp. Immunol.*, **76,** 446–51.

Burrells, C., Wells, P. W. & Sutherland, A. D. 1978. Reactivity of ovine lymphocytes to phytohaemagglutinin and pokeweed mitogen during pregnancy and in the immediate post-parturient period. *Clin. Exp. Immunol.*, **33,** 410–5.

Buswell, R. S. 1975. The pineal and neoplasia. *Lancet*, **1,** 34–5.

Butterworth, M. B., McClellan, B. & Alansmith, M. 1967. Influence of sex on immunoglobulin levels. *Nature*, **214,** 1224–5.

Calabrese, J. R., Kling, M. A. & Gold, P. W. 1987. Alterations in immunocompetence during stress, bereavement, and depression: Focus on neuroendocrine regulation. *Am. J. Psychiatry*, **144,** 1123–34.

Calder, E. A., Irvine, W. J., Davidson, N. M. & Wu, F. 1976. T, B and K cells in autoimmune thyroid disease. *Clin. Exp. Immunol.*, **25,** 17–22.

Calkins, J. H., Sigel, M. M., Nankin, H. R. & Lin, T. 1988. Interleukin-1 inhibits Leydig cell steroidogenesis in primary culture. *Endocrinology*, **123,** 1605–10.

Calvo, J. R., Rafii-el-Idrissi, M. & Guerrero, J. M. 1995. Immunomodulatory role of melatonin: Specific binding sites in human and rodent lymphoid cells. *J. Pineal Res.*, **18,** 119–26.

Campbell, C. S., Schwartz, N. B. & Firlit, M. G. 1977. The role of adrenal and ovarian steroids in the control of serum LH and FSH. *Endocrinology*, **101,** 162–72.

Campbell, R. M. & Scanes, C. G. 1995. Endocrine peptides 'moonlighting' as immune modulators: Roles for somatostatin and GH-releasing factor. *J. Endocrinol.*, **187,** 383–96.

Cannon, W. B. 1939. *Wisdom of the Body.* Philadelphia: Routledge and Kegan.

Cardinali, D. P., Brusco, L. I., Cutrera, R. A., Castrillon, P. & Esquifino, A. I. 1999. Melatonin as a time-meaningful signal in circadian organization of immune response. *Biol. Signals Recep.*, **8,** 41–8.

Cardinali, D. P., Cutera, R. A. & Esquifino, A. I. 2000. Psychoimmune neuroendocrine integrative mechanisms revisited. *Biol. Signals Recep.*, **9,** 215–30.

Caroleo, M. C., Frasca, D., Nistico, G. & Doria, G. 1992. Melatonin as immunomodulator in immunodeficient mice. *Immunopharmacology*, **23,** 81–9.

Carter, C. S., Getz, L. L., Gavish, L., McDermott, J. L. & Arnold, P. 1980. Male-related pheromones and the activation of female reproduction in the prairie vole (*Microtus ochrogaster*). *Biol. Reprod.*, **23,** 1038–45.

Carter, D. S. & Goldman, B. D. 1983. Antigonadal effects of timed melatonin infusion in pinealectomized male Djungarian hamsters (*Phodopus sungorus sungorus*): Duration is the critical parameter. *Endocrinology*, **113,** 1261–7.

Castle, K. T. & Wunder, B. A. 1995. Limits to food-intake and fiber utilization in the prairie vole, *Microtus ochrogaster*– Effects of food quality and energy need. *J. Comp. Physiol. B*, **164,** 609–17.

Cavallaro, J. J. & Monto, A. S. 1970. Community-wide outbreak of infection with a 229E-like coronavirus in Tecumseh, Michigan. *J. Infect. Dis.*, **122,** 272–9.

Cello, J. & Svennerholm, B. 1994. Detection of enterovirus-specific total and polymeric

IgA antibodies in serum using a synthetic peptide or heated virion antigen in ELISA. *J. Med. Virol.*, **44**, 422–7.

Centers for Disease Control. 1991. Influenza activity–Worldwide, 1990–91. *J. Am. Med. Assoc.*, **266**, 2534.

Centurelli, M. A. & Abate, M. A. 1997. The role of dehydroepiandrosterone in AIDS. *Ann. Pharmacother.*, **31**, 639–42.

Champney, T. H. & McMurray, D. N. 1991. Spleen morphology and lymphoproliferative activity in short photoperiod exposed hamsters. In: *Role of Melatonin and Pineal Peptides in Neuroimmunomodulation* (eds. F. Franschini & R. J. Reiter), pp. 19–225. New York: Plenum Publishing Company.

Chandra, R. K. 1975. Immunocompetence in undernutrition. *Brit. Med. J.*, **81**, 1194–200.

Chandra, R. K. 1979. Interactions of nutrition, infection and immune response. Immunocompetence in nutritional deficiency, methodological considerations and intervention strategies. *Acta Paediatr. Scand.*, **68**, 137–44.

Chandra, R. K. 1991. Immunocompetence is a sensitive and functional barometer of nutritional status. *Acta Paediatr. Scand. Suppl.*, **374**, 129–32.

Chandra, R. K. 1992. Protein-energy malnutrition and immunological responses. *J. Nutr.*, **122**, 597–600.

Chandra, R. K. 1993. Nutrition and the immune system. *Proc. Nutr. Soc.*, **52**, 77–84.

Chandra, R. K. 1997. Nutrition and the immune system: An introduction. *Am. J. Clin. Nutr.*, **66**, 460–3.

Chandra, S., Agarwal, G. P., Singh, S. P. & Saxena, A. K. 1990. Seasonal changes in a population of *Menacanthus eurysternus* (Mallophaga, Amblycera) on the common myna *Acridotheres tristis*. *Int. J. Parasitol.*, **20**, 1063–5.

Chariyalertsak, S., Sirisanthana, T., Supparatpinyo, K. & Nelson, K. E. 1996. Seasonal variation of disseminated *Penicillium marneffei* infections in Northern Thailand: A clue to the reservoir? *J. Infect. Dis.*, **173**, 1490–3.

Chernin, E. 1952. The relapse phenomenon in the *Leucocytozoon simondi* infection of the domestic duck. *Am. J. Hyg.*, **56**, 101–18.

Cheung, K. & Morris, B. 1984. The respiration and energy metabolism of sheep lymphocytes. *Austr. J. Exp. Biol. Med. Sci.*, **62**, 671–85.

Chevalier, P., Sevilla, R., Sejas, E., Zalls, L., Belmonte, G. & Parent, G. 1998. Immune recovery of malnourished children takes longer than nutritional recovery: Implications for treatment and discharge. *J. Trop. Pediatr.*, **44**, 304–7.

Chleboun, J. O. & Gray, B. N. 1987. The profile of breast cancer in Western Australia. *Med. J. Austr.*, **147**, 331–4.

Chougnet, C., Deloron, P., Lepers, J. P., Tallet, S., Rason, M. D., Astagneau, P., Savel, J. & Coulanges, P. 1990. Humoral and cell-mediated immune responses to the *Plasmodium falciparum* antigens PF155/RESA and CS proteins: Seasonal variations in a population recently reexposed to endemic malaria. *Am. J. Trop. Med. Hyg.*, **43**, 234–42.

Christau, B., Kromann, H., Christy, M., Andersen, O. O. & Nerup, J. 1979. Incidence of insulin-dependent diabetes mellitus (0–29 years at onset) in Denmark. *Acta Med. Scand. Suppl.*, **624**, 54–60.

Chrousos, G. P. 1992. Regulation and dysregulation of the hypothalamic-pituitary-adrenal axis. The corticotropin-releasing hormone perspective. *Endocrinol. Metab. Clin. North Am.*, **21**, 833–58.

Chrousos, G. P. & Gold, P. W. 1992. The concepts of stress and stress system disorders. *J. Am. Med. Assoc.*, **267**, 1244–52.

Claman, H. N. 1972. Corticosteroids and lymphoid cells. *N. Engl. J. Med.*, **287**, 388–97.

Clarke, B. L. & Bost, K. L. 1989. Differential expression of functional adrenocorticotropic hormone receptors by subpopulations of lymphocytes. *J. Immunol.*, **143**, 464–9.

Cohen, P., Wax, Y. & Modan, B. 1983. Seasonality in the occurrence of breast cancer. *Cancer Res.*, **43**, 892–6.

Coleman, E. S., Elsasser, T. H., Kemppainen, R. J., Coleman, D. A. & Sartin, J. L. 1993. Effect of endotoxin on pituitary hormone secretion in sheep. *Neuroendocrinology*, **58**, 111–22.

Collazos, M. E., Ortega, E. & Barriga, C. 1995. Influence of the temperature upon the proliferative response of lymphocytes of tench (*Tinca tinca*) during winter and summer. *Comp. Immunol. Microbiol. Infect. Dis.*, **18**, 209–14.

Compton, M. M., Caron, L. A. & Cidlowski, J. A. 1987. Glucocorticoid action on the immune system. *J. Steroid Biochem.*, **27**, 201–8.

Compton, M. M., Gibbs, P. S. & Johnson, L. R. 1990. Glucocorticoid activation of deoxyribonucleic acid degradation in bursal lymphocytes. *Poult. Sci.*, **69**, 1292–8.

Connors, V. A. & Nickol, B. B. 1991. Effects of *Plagiorhynchus cylindraceus* (*Acanthocephala*) on the energy metabolism of adult starlings, *Sturnus vulgaris*. *Parasitology*, **103**, 395–402.

Costa, O., Mulchahey, J. J., and Blalock, J. E. 1990. Structure and function of luteinizing hormone releasing hormone (LHRH) receptors on lymphocytes. *Prog. Neuroendocrinol. Immunol.*, **3**, 55–60.

Cowart, R. P., Boessen, C. R. & Kliebenstein, J. B. 1992. Patterns associated with season and facilities for atrophic rhinitis and pneumonia in slaughter swine. *J. Am. Vet. Med. Assoc.*, **200**, 190–3.

Crevel, R. W. R., Friend, J. V., Goodwin, B. F. J. & Parish, W. E. 1992. High-fat diets and the immune response of C57 BL mice. *Br. J. Nutr.*, **67**, 17–26.

Csaba, G., Sudar, F. & Dobozy, O. 1977. Triiodothyronine receptors in lymphocytes of newborn and adult rats. *Horm. Metab. Res.*, **9**, 499–501.

Culpepper, J. A. & Lee, F. 1985. Regulation of IL 3 expression by glucocorticoids in cloned murine T lymphocytes. *J. Immunol.*, **135**, 3191–7.

Cunningham, E. T., Jr. & De Souza, E. B. 1993. Interleukin 1 receptors in the brain and endocrine tissues. *Immunol. Today*, **14**, 171–6.

Cunningham, E. T., Jr., Wada, E., Carter, D. B., Tracey, D. E., Battey, J. F. & De Souza, E. B. 1992. In situ histochemical localization of type I interleukin-1 receptor messenger RNA in the central nervous system, pituitary, and adrenal gland of the mouse. *J. Neurosci.*, **12**, 1101–4.

Czaba, G. & Barath, P. 1975. Morphological changes of thymus and thyroid gland after postnatal extirpation of the pineal body. *Endocrinol. Exp.*, **9**, 59–67.

Czaba, G., Dunay, C., Fischer, J. & Bodoky, M. 1968. Hormonal relationships of mastocytogenesis in lymphatic organs. 3. Effect of the pineal body-thyroid-thymus system on mast cell production. *Acta Anat.*, **71**, 565–8.

Czeisler, C. A. 1995. The effect of light on the human circadian pacemaker. *Ciba Found. Symp.*, **183**, 254–90.

Dagan, R., Englehard, D., Piccard, E. & the Israeli Pediatric Bacteremia and Meningitis Group. 1992. Epidemiology of invasive childhood pneumococcal infections in Israel. *J. Am. Med. Assoc.*, **268**, 3328–32.

Dajani, Y. F. & Halabi, M. K. 1986. 19 culture positive cases of brucellosis in Jordan. A pilot study. *Bull. Consult. Med. Lab.*, **4**, 1.

Dajani, Y. F., Masoud, A. A. & Barakat, H. F. 1989. Epidemiology and diagnosis of human brucellosis in Jordan. *J. Trop. Med. Hyg.*, **92**, 209–14.

Danenberg, H. D., Ben-Yehuda, A., Zakay-Rones, Z. & Friedman, G. 1995. Dehydroepiandrosterone (DHEA) treatment reverses the impaired immune response of old mice to influenza vaccination and protects from influenza infection. *Vaccine*, **13**, 1445–8.

Danesh, J., Youngman, L., Clark, S., Parish, S., Peto, R. & Collins, R. 1999. *Helicobacter pylori* infection and early onset myocardial infarction: Case-control and sibling pairs study. *Br. Med. J.*, **319**, 1157–62.

Daniels, C. W. & Belosevic, M. 1994. Serum antibody responses by male and female C57Bl/6 mice infected with *Giardia muris*. *Clin. Exp. Immunol.*, **97**, 424–9.

Dardenne, M., Itoh, T. & Homo-Delarche, F. 1986. Presence of glucocorticoid receptors in cultured thymic epithelial cells. *Cell Immunol.*, **100**, 112–8.

Dardenne, M., Savino, W., Gagnerault, M. C., Itoh, T. & Bach, J. F. 1989. Neuroendocrine control of thymic hormonal production. I. Prolactin stimulates in vivo and in vitro the production of thymulin by human and murine thymic epithelial cells. *Endocrinology*, **125**, 3–12.

Dark, J., Miller, D. R. & Zucker, I. 1994. Reduced glucose availability induces torpor in Siberian hamsters. *Am. J. Physiol.*, **267**, R496–501.

Dark, J. & Zucker, I. 1983. Short photoperiods reduce winter energy requirements of the meadow vole, *Microtus pennsylvanicus*. *Physiol. Behav.*, **31**, 699–702.

Da Silva, J. A., Peers, S. H., Perretti, M. & Willoughby, D. A. 1993. Sex steroids affect glucocorticoid response to chronic inflammation and to interleukin-1. *J. Endocrinol.*, **136**, 389–97.

Davenport, C. B. 1922. Multiple sclerosis: From the standpoint of geographic distribution and race. *Arch. Neurol. Psychiatr.*, **8**, 51–8.

Davidson, L. M. & Baum, A. 1986. Chronic stress and post traumatic stress disorder. *J. Consult. Clin. Psychol.*, **54**, 303–8.

Davis, D. E. 1976. Hibernation and circannual rhythms of food consumption in marmots and ground squirrels. *Q. Rev. Biol.*, **51**, 477–514.

Davis, R. L. & Lochmiller, R. L. 1995. Quantitative and qualitative numerical alterations in splenocyte subpopulations of the cotton rat (*Sigmodon hispidus*) across seasons. *Biol. Rhythm Res.*, **26**, 20–31.

Davis, S. L. 1998. Environmental modulation of the immune system via the endocrine system. *Domest. Anim. Endocrinol.*, **15**, 283–9.

Daynes, R. A. & Araneo, B. A. 1989. Contrasting effects of glucocorticoids on the capacity of T cells to produce the growth factors interleukin-2 and interleukin-4. *Eur. J. Immunol.*, **19**, 2319–25.

Daynes, R. A., Araneo, B. A., Ershler, W. B., Maloney, C., Li, G. Z. & Ryu, S. Y. 1993. Altered regulation of IL-6 production with normal aging. Possible linkage to the age-associated decline in dehydroepiandrosterone and its sulfated derivative. *J. Immunol.*, **150**, 5219–30.

Daynes, R. A., Araneo, B. A., Hennebold, J., Enioutina, E. & Mu, H. H. 1995. Steroids as regulators of the mammalian immune response. *J. Invest. Dermatol.*, **105**, 14S–9S.

D'Cruz, I. A., Balani, S. G. & Iyer, L. S. 1968. Infectious hepatitis and pregnancy. *Obstet. Gynecol.*, **31**, 449–55.

DeBont, J., Vercruysse, J., Sabbe, F., Southgate, V. R. & Rollins, D. 1995. *Schistosoma mattheei* infections in cattle: Change associated with season and age. *Vet. Parasitol.*, **57**, 299–307.

Deerenberg, C., Apanius, V., Daan, S. & Bos, N. 1997. Reproductive effort decreases antibody responsiveness. *Proc. R. Soc. Lond. B*, **264**, 1021–9.

Delafuente, J. C. 1991. Nutrients and immune responses. *Rheum. Dis. Clin. North Am.*, **17**, 203–12.

Delage, G., Montplaisir, S., Remy-Prince, S. & Pierri, E. 1986. Prevalence of hepatitis B virus infection in pregnant women in the Montreal area. *CMAJ.*, **134**, 897–901.

Del Gobbo, V., Bibri, V., Villani, N., Calio, R. & Nistico, G. 1989. Pinealectomy inhibits interleukin-2 production and natural killer activity in mice. *Int. J. Immunopharmacol.*, **11**, 567–77.

del Rey, A., Besedovsky, H., Sorkin, E. & Dinarello, C. A. 1987. Interleukin-1 and glucocorticoid hormones integrate an immunoregulatory feedback circuit. *Ann. N. Y. Acad. Sci.*, **496**, 85–90.

Demas, G. E., Chefer, V., Talan, M. C. & Nelson, R. J. 1997a. Metabolic costs of an antigen-stimulated immune response in adult and aged C57BL/6J mice. *Am. J. Physiol.*, **273**, R1631–7.

Demas, G. E., DeVries, A. C. & Nelson, R. J. 1997b. Effects of photoperiod and 2-deoxy-D-glucose-induced metabolic stress on immune function in female deer mice (*Peromyscus maniculatus*). *Am. J. Physiol.*, **272**, R1762–7.

Demas, G. E., Drazen, D. L., Jasnow, A. M., Bartness, T. J. & Nelson, R. J. 2001. Sympathoadrenal system differentially affects photoperiodic changes in humoral immunity of Siberian hamsters (*Phodopus sungorus*). *J. Neuroendocrinol.* (In press).

Demas, G. E., Klein, S. L. & Nelson, R. J. 1996. Reproductive and immune responses to photoperiod and melatonin are linked in *Peromyscus* subspecies. *J. Comp. Physiol. A*, **179**, 819–25.

Demas, G. E. & Nelson, R. J. 1996. The effects of photoperiod and temperature on immune function of adult male deer mice (*Peromyscus maniculatus*). *J. Biol. Rhythms*, **11**, 94–102.

Demas, G. E. & Nelson, R. J. 1998a. Exogenous melatonin enhances cellular, but humoral immune function in adult male deer mice (*Peromyscus maniculatus*). *J. Biol. Rhythms*, **13**, 245–52.

Demas, G. E. & Nelson, R. J. 1998b. Short-day enhancement of immune function is not mediated by gonadal steroid hormones in adult deer mice (*Peromyscus maniculatus*). *J. Comp. Physiol. B*, **168**, 419–26.

Descôteaux J.-P. & Mihok, S. 1986. Serologic study on the prevalence of murine viruses in a population of wild meadow voles (*Microtus pennsylvanicus*). *J. Wildlife Dis.*, **22**, 314–9.

Desjardins, C. & Lopez, M. J. 1983. Environmental cues evoke differential responses in pituitary-testicular function in deer mice. *Endocrinology*, **112**, 1398–406.

Dhabhar, F. S. & McEwen, B. S. 1996. Stress-induced enhancement of antigen-specific cell-mediated immunity. *J. Immunol.*, **156**, 2608–15.

Dhabhar, F. S., McEwen, B. S. & Spencer, R. L. 1993. Stress response, adrenal steroid receptor levels and corticosteroid-binding globulin levels—A comparison between Sprague-Dawley, Fischer 344 and Lewis rats. *Brain Res.*, **9,** 89–98.

Dhom, G. 1991. Epidemiology of hormone-depending tumors. In: *Endocrine Dependent Tumors* (eds. K. Voigt & C. Knabbe), pp. 1–42. New York: Raven Press.

DiStefano, A. & Paulescu, L. 1994. Inhibitory effect of melatonin production on INF gamma or TNF alpha in peripheral blood mononuclear cells of some blood donors. *J. Pineal Res.*, **17,** 164–9.

Doblhammer, G. & Vaupel, J. W. 2001. Lifespan depends on month of birth. *Proc. Natl. Acad. Sci. U.S.A.*, **98,** 2934–9.

Dobrodeeva, L. K., Tkachev, A. V., Tipisova, E. V. & Kashutin, S. L. 1998. Immunologic characteristics of individuals working on the Spitzbergen archipelago. *Ross Fiziologi Zh Im I. M. Sechenova*, **84,** 119–24.

Dobrowolska, A. & Adamczewska-Andrzejewska, K. A. 1991. Seasonal and long-term changes in serum gamma-globulin levels in comparing the physiology and population density of the common vole, *Microtus arvalis* Pall. *J. Interdiscipl. Cycle Res.*,

Dobrowolska, A., Rewkiwwicz-Dziarska, A., Szarska, I. & Gill, J. 1974. Seasonal changes in haematological parameters, level of serum proteins and glycoproteins, activity of the thyroid gland, suprarenals and kidney in the common vole (*Microtus arvalis* Pall.). *J. Interdiscipl. Cycle Res.*, **5,** 347–54.

d'Offay, J. M. & Rosenquist, B. D. 1988. Combined effects of fasting and diet on interferon production and virus replication in calves infected with a vaccine strain of infectious bovine rhinotracheitis virus. *Am. J. Vet. Res.*, **49,** 1311–5.

Doherty, M. L., Monaghan, M. L., Bassett, H. F., Quinn, P. J. & Davis, W. C. 1996. Effect of dietary restriction on cell-mediated immune responses in cattle infected with *Mycobacterium bovis. Vet. Immunol. Immunopathol.*, **49,** 307–20.

Douglas, A. S., Allan, T. M. & Rawles, J. M. 1991. Composition of seasonality of disease. *Scottish Med. J.*, **36,** 76–82.

Douglas, A. S., Russell, D. & Allan, T. M. 1990. Seasonal, regional and secular variations of cardiovascular and cerebrovascular mortality in New Zealand. *Austral. N. Z. J. Med.*, **20,** 669–76.

Doyle, W. J., Skoner, D. P., Seroky, M. A., Fireman, P. & Gwaltney, J. M. 1994. Effect of experimental rhinovirus 39 infection on the nasal response to histamine and cold air challenges in allergic and nonallergic subjects. *J. Allergy Clin. Immunol.*, **93,** 534–42.

Dozier, M. M., Ratajczak, H. V., Sothern, R. B. & Thomas, P. T. 1997. The influence of vehicle gavage on seasonality of immune system parameters in the B6C3F1 mouse. *Fundam. Appl. Toxicol.*, **38,** 116–22.

Drača, S. R. 1995. Endocrine-immunological homeostasis: The interrelationship between the immune system and sex steroids involves the hypothalamo-pituitary-gonadal axis. *Panminerva Med.*, **37,** 71–6.

Drazen, D. L. 2001. *Neuroendocrine mechanisms underlying seasonal changes in immune function and energy balance.* PhD. Dissertation. Baltimore, MD: Johns Hopkins University.

Drazen, D. L., Bilu, D., Bilbo, S. D. & Nelson, R. J. 2001. Melatonin enhancement of splenocyte proliferation is attenuated by luzindole, a melatonin receptor antagonist. *Am. J. Physiol.*, **280,** R1476–82.

Drazen, D. L., Demas, G. E., & Nelson, R. J. 2001. Leptin effects on immune function and energy balance are photoperiod dependent in Siberian hamsters (*Phodopus sungorus*). *Endocrinology*, **142**, 2768–75.

Drazen, D. L., Klein, S. L., Yellon, S. M. & Nelson, R. J. 2000. In vitro melatonin enhances splenocyte proliferation in female prairie voles (*Microtus ochrogaster*). *J. Pineal Res.*, **28**, 34–40.

Drazen, D. L. & Nelson, R. J. 2001. Melatonin receptor subtype 1a is unnecessary for melatonin-induced enhancement of cell-mediated and humoral immune function. *Neuroendocrinology* (In press).

Dubocovich, M. L., Yun, K., Al-Ghoul, W. M., Benloucif, S. & Masana, M. I. 1998. Selective MT2 melatonin receptor antagonists block melatonin-mediated phase advances of circadian rhythms. *FASEB J.*, **12**, 1211–20.

Dubuis, J. M., Dayer, J. M., Siegrist-Kaiser, C. A. & Burger, A. G. 1988. Human recombinant interleukin-1 beta decreases plasma thyroid hormone and thyroid stimulating hormone levels in rats. *Endocrinology*, **123**, 2175–81.

Duncan, M. E., Pearson, J. M., Ridley, D. S., Melsom, R. & Bjune, G. 1982. Pregnancy and leprosy: the consequences of alterations of cell-mediated and humoral immunity during pregnancy and lactation. *Int. J. Lepr. Other Mycobact. Dis.*, **50**, 425–35.

Dunn, A. J. 1989. Psychoneuroimmunology for psychoneuroendocrinologist: A review of animal studies of nervous system-immune system interactions. *Psychoneuroendocrinology*, **14**, 251–74.

Dunnigan, M. G., Harland, W. A. & Fyfe, T. 1970. Seasonal incidence and mortality of ischæmic heart-disease. *Lancet*, **295**, 793–7.

Dunsmore, J. D. & Jue Sue, L. P. 1985. Prevalence and epidemiology of the major gastrointestinal parasites of horses in Perth, western Australia. *Equine Vet. J.*, **17**, 208–13.

Durruty, P., Ruiz, F. & García de los Ríos, M. 1979. Age at diagnosis and seasonal variation in the onset of insulin-dependent diabetes in Chile (Southern Hemisphere). *Diabetologia*, **17**, 357–60.

Ebihara, S., Marks, T., Hudson, D. J. & Menaker, M. 1986. Genetic control of melatonin synthesis in the pineal gland of the mouse. *Science*, **231**, 491–3.

Edson, D. C., Stiefel, H. E., Wentworth, B. B. & Wilson, D. L. 1979. Prevalence of antibodies to Legionnaires' disease. *Ann. Int. Med.*, **90**, 691–3.

Eidinger, D. & Garrett, T. J. 1972. Studies of the regulatory effects of the sex hormones on antibody formation and stem cell differentiation. *J. Exp. Med.*, **136**, 1098–116.

Elliott, J. A. 1976. Circadian rhythms and photoperiodic time measurement in mammals. *Fed. Proc.*, **35**, 2339–46.

Elliott, J. A. & Goldman, B. D. 1981. Seasonal reproduction: Photoperiodism and biological clocks. In: *Neuroendocrinology of Reproduction* (ed. N. T. Adler), pp. 377–426. New York: Plenum Press.

Elliott, J. A., Stetson, M. H. & Menaker, M. 1972. Regulation of testis function in golden hamsters: A circadian clock measures photoperiodic time. *Science*, **178**, 771–3.

Ellis, G. B. & Turek, F. W. 1980. Photoperiodic regulation of serum luteinizing hormone and follicle-stimulating hormone in castrated and castrated-adrenalectomized male hamsters. *Endocrinology*, **106**, 1338–44.

El Ridi, R., Badir, N. & El Rouby, S. 1981. Effect of seasonal variations on the immune system of the snake, *Psammophis schokari*. *J. Exp. Zool.*, **216**, 357–65.

El Ridi, R., Zada, S., Afifi, A., el Deeb, S., el Rouby, S., Farag, M. & Saad, A. H. 1988. Cyclic changes in the differentiation of lymphoid cells in reptiles. *Cell Differ.*, **24**, 1–8.

Elsasser, T. H., Rumsey, T. S., Hammond, A. C. & Fayer, R. 1988. Influence of parasitism on plasma concentrations of growth hormone, somatomedin-C and somatomedin-binding proteins in calves. *J. Endocrinol.*, **116**, 191–200.

Enquselassie, F., Dobson, A. J., Alexander, H. M. & Steele, P. L. 1993. Seasons, temperature and coronary disease. *Int. J. Epidemiol.*, **22**, 632–6.

Epstein, W. W., Rowesmitt, C. N., Berger, P. J. & Negus, N. C. 1986. Dynamics of 6-methoxybenzoxazolinone in winter wheat: Effects of photoperiod and temperature. *J. Chem. Ecol.*, **12**, 2011–20.

Erf, G. F. 1993. Immune development in young-adult C.RF-hyt mice is affected by congenital and maternal hypothyroidism. *Proc. Soc. Exp. Biol. Med.*, **204**, 40–8.

Eshet, R., Peleg, S. & Laron, Z. 1984. Direct visualization of binding, aggregation and internalization of human growth hormone in cultured human lymphocytes. *Acta. Endocrinol. (Copenh).*, **107**, 9–15.

Eskola, J., Frey, H., Molnar, G. & Sppi, E. 1976. Biological rhythm of cell mediated immunity in man. *Clin. Exp. Immunol.*, **26**, 253–7.

Eskola, J., Takala, A. K., Kela, E., Pekkanen, E., Kalliokoski, R. & Leinonen, M. 1992. Epidemiology of invasive pneumococcal infections in children in Finland. *J. Am. Med. Assoc.*, **268**, 3323–7.

Eure, H. 1976. Seasonal abundance of *Proteocephalus ambloplitis* (Cestoidea: Proteocephalidea) from largemouth bass living in a heated reservoir. *Parasitology*, **73**, 205–12.

Evans, M. R., Goldsmith, A. R. & Norris, S. R. A. 2000. The effects of testosterone on antibody production and plumage coloration in male house sparrows (*Passer domesticus*). *Behav. Ecol. Sociobiol.*, **47**, 156–63.

Evans, W. J., Roubenoff, R. & Shevitz, A. 1998. Exercise and the treatment of wasting: Aging and human immunodeficiency virus infection. *Semin. Oncol.*, **25**, 112–22.

Ewald, P. W. 1994. *Evolution of Infectious Disease*. Oxford University Press: Oxford.

Fabris, N. 1973. Immunodepression in thyroid-deprived animals. *Clin. Exp. Immunol.*, **15**, 601–11.

Fabris, N. & Mocchegiani, E. 1985. Endocrine control of thymic serum factor production in young-adult and old mice. *Cell. Immunol.*, **91**, 325–35.

Fabris, N., Mocchegiani, E., Mariotti, S., Caramia, G., Braccili, T., Pacini, F. & Pinchera, A. 1987. Thymulin deficiency and low 3,5,3'triiodothyronine syndrome in infants with low birth weight syndromes. *J. Clin. Endocrinol. Metab.*, **65**, 247–52.

Fabris, N., Mocchegiani, E. & Provinciali, M. 1995. Pituitary-thyroid axis and immune system: A reciprocal neuroendocrine-immune interaction. *Horm. Res.*, **43**, 29–38.

Fairbairn, D. J. 1977. Why breed early? A study of reproductive tactics in *Peromyscus*. *Can. J. Zool.*, **55**, 862–71.

Fakhir, S., Ahmad, P., Faridi, M. A. & Rattan, A. 1989. Cell-mediated immune responses in malnourished host. *J. Trop. Pediatr.*, **35**, 175–8.

Falkoff, R. 1987. Maternal immunologic changes during pregnancy: A critical appraisal. *Clin. Rev. Allergy*, **5**, 287–300.

Fan, J., Char, D., Bagby, G. J., Gelato, M. C. & Lang, C. H. 1995. Regulation of insulin-like growth factor-I (IGF-I) and IGF-binding proteins by tumor necrosis factor. *Am. J. Physiol.*, **38**, R1204–12.

Fan, J., Molina, P. E., Gelato, M. C. & Lang, C. H. 1994. Differential tissue regulation of insulin-like growth factor-I content and binding-proteins after endotoxin. *Endocrinology*, **134**, 1685–92.

Fan, J., Wojnar, M. M., Theodorakis, M. & Lang, C. H. 1996. Regulation of insulin-like growth factor (IGF)-I mRNA and peptide and IGF-binding proteins by inter-leukin-1. *Am. J. Physiol.*, **270**, R621–R9.

Fänge, M. A. & Silverin, B. 1985. Variation of lymphoid activity in the spleen of a migratory bird, the pied flycatcher (*Ficdula hypoleuca*; Aves, Paseriformes). *J. Morphol.*, **181**, 33–40.

Farag, M. A. & El Ridi, R. 1985. Mixed leukocyte reaction (MLR) in the snake *Psammophis sibilans*. *Immunology*, **55**, 173–81.

Faulkner, M., Halton, D. W. & Montgomery, W. I. 1989. Sexual, seasonal and tissue variation in the encystment of *Cotylurus variegatus* metacercariae in perch, *Perca fluviatilis*. *Int. J. Parasitol.*, **19**, 285–90.

Fauser, B. C., Soto, D., Czekala, N. M. & Hsueh, A. J. 1989. Granulosa cell aromatase bioassay: Changes of bioactive FSH levels in the female. *J. Steroid Biochem.*, **33**, 721–6.

Fellowes, D. S. B. & Proctor, I. R. D. 1973. The incidence of the common cold in relation to certain meteorological parameters. *Int. J. Biometerol.*, **17**, 193–203.

Felsner, P., Hofer, D., Rinner, I., Mangge, H., Gruber, M., Korsatko, W. & Schauenstein, K. 1992. Continuous in vivo treatment with catecholamines suppresses in vitro reactivity of rat peripheral blood T-lymphocytes via α-mediated mechanisms. *J. Neuroimmunol.*, **37**, 47–57.

Felsner, P., Hofer, D., Rinner, I., Porta, S., Korsatko, W. & Schauenstein, K. 1995. Adrenergic suppression of peripheral blood T cell reactivity in the rat is due to activation of peripheral α_2-receptors. *J. Neuroimmunol.*, **57**, 27–34.

Fenlon, D. R. 1985. Wild birds and silage as reservoirs of *Listeria* in the agricultural environment. *J. Appl. Bacteriol.*, **59**, 537–43.

Ferin, M. 1993. Stress and the gonadal axis in the female rhesus monkey: Interface between the immune and neuroendocrine systems. *Hum. Reprod.*, **Suppl 2**, 147–50.

Fernandes, G., Halberg, F., Yunis, E. J. & Good, R. A. 1976. Circadian rhythmic plaque-forming cell response of spleens from mice immunized with SRBC. *J. Immunol.*, **117**, 962–66.

Festa-Bianchet, M. 1989. Individual differences, parasites, and the costs of reproduction for bighorn ewes (*Ovis canadensis*). *J. Anim. Ecol.*, **58**, 785–95.

Finck, B. N., Kelley, K. W., Dantzer, R. & Johnson, R. W. 1998. In vivo and in vitro evidence for the involvement of tumor necrosis factor-alpha in the induction of leptin by lipopolysaccharide. *Endocrinology*, **139**, 2278–83.

Finn, R., St. Hill, C. A., Govan, A. J., Ralfs, I. G., Gurney, F. J. & Denye, V. 1972. Immunological responses in pregnancy and survival of fetal homograft. *Br. Med. J.*, **3**, 150–2.

Finocchiaro, L. M. E., Nahmod, V. E. & Aunay, J. M. 1991. Melatonin biosynthesis and metabolism in peripheral blood mononuclear leucocytes. *Biochem. J.*, **280**, 727–31.

Fischer, T., Schobel, H. & Barenbrock, M. 1996. Specific immune tolerance during pregnancy after renal transplantation. *Eur. J. Obstet. Gynecol. Reprod. Biol.*, **70**, 217–9.

Fiske, R. A. & Adams, L. G. 1985. Immune responsiveness and lymphoreticular morphology in cattle fed hypo- and hyperalimentative diets. *Vet. Immunol. Immunopathol.*, **8**, 225–44.

Fitzgerald, T. J. & Veal, A. 1976. Melatonin antagonizes colchicine-induced mitotic arrest. *Experientia*, **32**, 372–3.

Fleegler, F. M., Rogers, K. D., Drash, A., Rosenbloom, A. L., Travis, L. B. & Court, J. M. 1979. Age, sex, and season of onset of juvenile diabetes in different geographic areas. *Pediatrics*, **63**, 374–9.

Flower, R. J., Parente, L., Persico, P. & Salmon, J. A. 1986. A comparison of the acute inflammatory response in adrenalectomised and sham-operated rats. *Br. J. Pharmacol.*, **87**, 57–62.

Folstad, I., Nilssen, A. C., Halvorsen, O. & Andersen, J. 1989. Why do male reindeer (*Rangifer t. tarandus*) have higher abundance of second and third instar larvae of *Hypoderma tarandi* than females? *Oikos*, **55**, 87–92.

Forbes, M. R. L. 1993. Parasitism and host reproductive effort. *Oikos*, **67**, 444–50.

Forbes, M. R. L. & Baker, R. L. 1991. Condition and fecundity of the damselfly, *Enallagma ebrium* (Hagen): The importance of ectoparasites. *Oecologia*, **86**, 335–41.

Forrester, D. J., Portor, J. H., Belden, R. C. & Frankenberger, W. B. 1982. Lungworms of feral swine in Florida. *J. Am. Vet. Med. Assoc.*, **181**, 1278–80.

Franklin, A., Hinsull, S. M. & Bellamy, D. 1983. The effect of dietary restriction on thymus autonomy. *Thymus*, **5**, 345–54.

Fraschini, F., Scaglione, F., Demartini, G., Lucini, V. & Sacerdote, P. 1990. Melatonin action on immune responses. In: *Advances in Pineal Research*, Vol. 4. (eds. R. J. Reiter and A. Lukaszyk), pp. 225–33, London: John Libbey.

Freeman, D. A. & Zucker, I. 2000. Temperature-independence of circannual variations in circadian rhythms of golden-mantled ground squirrels. *J. Biol. Rhythms*, **15**, 336–43.

Friedman, S. B., Grota, L. J. & Glasgow, L. A. 1972. Differential susceptibility of male and female mice to encephalomyocarditis virus: Effects of castration, adrenalectomy, and the administration of sex hormones. *Infect. Immun.*, **5**, 637–44.

Gadient, R. A., Lachmund, A., Unsicker, K. & Otten, U. 1995. Expression of interleukin-6 (IL-6) and IL-6 receptor mRNAs in rat adrenal medulla. *Neurosci. Lett.*, **194**, 17–20.

Gadient, R. A. & Otten, U. H. 1997. Interleukin-6 (IL-6)–A molecule with both beneficial and destructive potentials. *Progr. Neurobiol.*, **52**, 379–90.

Gala, R. R. 1991. Prolactin and growth hormone in the regulation of the immune system. *Proc. Soc. Exp. Biol. Med.*, **198**, 513–27.

Galea, M. H. & Blamey, R. W. 1991. Season of initial detection in breast cancer. *Br. J. Cancer*, **63**, 157.

Gallucci, W. T., Baum, A., Laue, L., Rabin, D. S., Chrousos, G. P., Gold, P. W. & Kling, M. A. 1993. Sex differences in sensitivity of the hypothalamic-pituitary-adrenal axis. *Health Psychol.*, **12**, 420–5.

Garcia-Maurino, S., Gonzales-Haba, M. G., Calvo, J. R., Rafii-El-Idrissi, M., Sanchez-Margalet, V., Goberna, R. & Guerreo, J. M. 1997. Melatonin enhances IL-2, IL-6, and IFN-gamma production by human circulating CD4+ cells: A possible nuclear receptor-mediated mechanism involving T-helper type 1 lymphocytes and monocytes. *J. Immunol.*, **159**, 574–81.

Gardiner, E. E., Hunt, J. R. & Newsberry, R. C. 1988. Relationships between age, body weight, and season of the year and the incidence of sudden death syndrome in male broiler chickens. *Poult. Sci.*, **67**, 1243–9.

Garre, M. A., Boles, J. M. & Youinou, P. Y. 1987. Current concepts in immune derangement due to undernutrition. *J. Parenter. Enter. Nutr.*, **11**, 309–13.

Gasbarrini, A., Cremonini, F., Armuzzi, A., Ojetti, V., Candelli, M., Di Campli, C., Sanz-Torre, E., Pola, R., Gasbarrini, G. & Pola, P. 1999. The role of *Helicobacter pylori* in cardiovascular and cerebrovascular diseases. *J. Physiol. Pharmacol.*, **50**, 735–42.

Gatchel, R. J., Baum, A. & Krantz, D. S. 1989. *An Introduction to Health Psychology*, 2nd ed. Newberry: New York.

Gatti, S. & Bartfai, T. 1993. Induction of tumor necrosis factor-alpha mRNA in the brain after peripheral endotoxin treatment: Comparison with interleukin-1 family and interleukin-6. *Brain Res.*, **624**, 291–4.

Gilkerson, J., Jorm, L. R., Love, D. N., Lawrence, G. L. & Whalley, J. M. 1994. Epidemiological investigation of equid herpes virus-4 (EHV-4) excretion assessed by nasal swabs taken from thoroughbred foals. *Vet. Microbiol.*, **39**, 275–83.

Gillette, M. U. 1986. The suprachiasmatic nuclei: Circadian phase-shifts induced at the time of hypothalamic slice preparation are preserved in vitro. *Brain Res.*, **379**, 176–81.

Gillette, S. & Gillette, R. 1979. Changes in thymic estrogen receptor expression following orchidectomy. *Cell. Immunol.*, **42**, 194–6.

Gillis, S., Crabtree, G. R. & Smith, K. 1979. Glucocorticoid induced inhibition of T-cell growth factor production. I. The effect on mitogen-induced lymphocyte proliferation. *J. Immunol.*, **123**, 1624–31.

Giordano, M. & Palermo, M. S. 1991. Melatonin-induced enhancement of antibody dependent cellular cytotoxicity. *J. Pineal Res.*, **10**, 117–21.

Giordano, M., Vermeulen, M. & Palermo, M. S. 1993. Seasonal variations in antibody-dependent cellular cytotoxicity regulation by melatonin. *FASEB J.*, **7**, 1052–4.

Glaser, R., Rice, J., Sheridan, J., Fertel, R., Stout, J., Speicher, C., Pinsky, D., Kotur, M., Post, A., Beck, M. & Kiecolt-Glaser, J. 1987. Stress-related immune suppression: Health implications. *Brain Behav. Immun.*, **1**, 7–20.

Glaser, R., Rice, T., Speicher, C. E., Stout, J. C. & Kiecolt-Glaser, J. K. 1986. Stress depresses interferon production by leucocytes concomitant with a decrease in natural killer cell activity. *Behav. Neurosci.*, **100**, 675–8.

Glezen, W. P., Payne, A. A., Snyder, D. N. & Downs, T. D. 1982. Mortality and influenza. *J. Infect. Dis.*, **146**, 313–21.

Glimåker, M., Samuelson, A., Magnius, L., Ehrnst, A., Olcén, P. & Forsgren, M. 1992. Early diagnosis of enteroviral meningitis by detection of specific IgM antibodies with a solid-phase reverse immunosorbent test (SPRIST) and μ-capture EIA. *J. Med. Virol.*, **36**, 193–201.

Goble, F. C. & Konopka, E. A. 1973. Sex as a factor in infectious disease. *Trans. N.Y. Acad. Sci.*, **35**, 325–46.

References

Goff, B. L., Roth, J. A., Arp, L. H. & Incefy, G. S. 1987. Growth hormone treatment stimulates thymulin production in aged dogs. *Clin. Exp. Immunol.*, **68**, 580–7.

Goldman, B. D. 2001. Mammalian photoperiodic system: formal properties and neuroendocrine mechanisms of photoperiodic time measurement. *J. Biol. Rhythms*, **16**, 283–301.

Goldman, B. D. & Nelson, R. J. 1993. Melatonin and seasonality in mammals. In: *Melatonin: Biosynthesis, Physiological Effects, and Clinical Applications*, (eds. H. S. Yu and R. J. Reiter), pp. 225–52, Boca Raton: CRC Press.

Golla, J. A., Larson, L. A., Anderson, C. F., Lucas, A. R., Wilson, W. R. & Tomasi, T. B. 1981. An immunological assessment of patients with anorexia nervosa. *Am. J. Clin. Nutr.*, **34**, 2756–62.

Gomez, E. A. & Hashiguchi, Y. 1991. Monthly variation in natural infection of the sandfly *Lutzomyia ayacuchensis* with *Leishmania mexicana* in an endemic focus in the Ecuadorian Andes. *Ann. Trop. Med. Parasitol.*, **85**, 407–11.

Gonzalez, M. C., Aguila, M. C. & McCann, S. M. 1991. In vitro effects of recombinant human gamma-interferon on growth hormone release. *Prog. Neuroendocrinol. Immunol.*, **4**, 222–7.

Good, R. A. & Lorenz, E. 1992. Nutrition and cellular immunity. *Int. J. Immunopharmacol.*, **14**, 361–6.

Gorman, M. R., Freeman, D. A. & Zucker, I. 1997. Photoperiodism in hamsters: Abrupt versus gradual changes in day length differentially entrain morning and evening circadian oscillators. *J. Biol. Rhythms*, **12**, 122–35.

Gorski, R. A. 1979. The neuroendocrinology of reproduction: An overview. *Biol. Reprod.*, **20**, 111–27.

Goto, M., Oshima, I., Tomita, T. & Ebihara, S. 1989. Melatonin synthesis of the pineal gland in different mouse strains. *J. Pineal Res.*, **7**, 195–204.

Gould, K. G., Akinbami, M. A. & Mann, D. R. 1998. Effect of neonatal treatment with a gonadotropin releasing hormone antagonist on developmental changes in circulating lymphocyte subsets: A longitudinal study in male rhesus monkeys. *Dev. Comp. Immunol.*, **22**, 457–67.

Graff, R. J., Lappe, M. A. & Snell, G. D. 1969. The influence of the gonads and adrenal glands on the immune response to skin grafts. *Transplantation*, **7**, 105–11.

Graunt, J. 1662. *Natural and political observations made upon the bills of mortality*, edited with an introduction by W. F. Willcox. Baltimore: The Johns Hopkins University Press, 1939.

Gregoire, C. 1945. Sur le mechanisme de l'hypertrophie thymique declanchee par la castration. *Arch. Int. Pharmacodynam. Ther.*, **67**, 45–77.

Griebel, P. J., Schoonderwoerd, M. & Babiuk, L. A. 1987. Ontogeny of the immune response: Effect of protein energy malnutrition in neonatal calves. *Can. J. Vet. Res.*, **51**, 428–35.

Grimble, R. F. 1994. Malnutrition and the immune response. 2. Impact of nutrients on cytokine biology in infection. *Trans. R. Soc. Trop. Med. Hyg.*, **88**, 615–9.

Grossman, C. J. 1984. Regulation of the immune system by sex steroids. *Endocr. Rev.*, **5**, 435–54.

Grossman, C. J., Sholiton, L. J., Blaha, G. C. & Nathan, P. 1979. Rat thymic estrogen receptor. II. Physiological properties. *J. Steroid Biochem.*, **11**, 1241–6.

Guarcello, V., Weigent, D. A. & Blalock, J. E. 1991. Growth hormone releasing hormone receptors on thymocytes and splenocytes from rats. *Cell. Immunol.*, **136**, 291–302.

Gubernick, D. J. & Nelson, R. J. 1989. Prolactin and paternal behavior in the biparental California mouse, *Peromyscus californicus. Horm. Behav.*, **23**, 203–10.

Gustafsson, L., Nordling, D., Andersson, M. S., Sheldon, B. C. & Qvarnstrom, A. 1994. Infectious diseases, reproductive effort and the cost of reproduction in birds. *Philos. Trans. R. Soc. Lond. B. Biol. Sci.*, **346**, 323–31.

Gwinner, E. 1986. *Circannual Rhythms.* Berlin: Springer-Verlag.

Haag-Weber, M., Dumann, H. & Horl., W. H. 1992. Effect of malnutrition and uremia on impaired cellular host defense. *Miner. Electrolyte Metab.*, **18**, 174–85.

Hafez, E. S. E. 1993. *Reproduction in Farm Animals*, 6th ed. Philadelphia: Lea & Febiger.

Halberg, F. 1961. Physiologic 24-hour periodicity in human beings and mice, the lighting regimen and daily routine. In: *Photoperiodism and Related Phenomena in Plants and Animals* (ed. R. B. Withrow), pp. 803–78. Washington, DC: American Association for the Advancement of Science.

Hall, C. B. 1991. Respiratory syncytial virus. In: *Textbook of Pediatric Infectious Diseases*, Vol. 2, 3rd ed. (eds. R. D. Feigin & J. D. Cherry), Philadelphia: W. B. Saunders.

Halvorson, D. A., Kelleher, C. J. & Senne, D. A. 1985. Epizootiology of avian influenza: Effect of season on incidence in sentinel ducks and domestic turkeys in Minnesota. *Appl. Environ. Biol.*, **49**, 914–9.

Hambre, D. & Beem, M. 1972. Virologic studies of acute respiratory disease in young adults. V. Coronavirus 229E infections during six years of surveillance. *Am. J. Epidemiol.*, **96**, 94–106.

Hamilton, T. 1969. Influence of environmental light and melatonin upon mammary tumour induction, *Brit. J. Surg.*, **56**, 764–6.

Hamilton, W. D., Axelrod, R. & Tanese, R. 1990. Sexual reproduction as an adaptation to resist parasites. *Proc. Natl. Acad. Sci. U.S.A.*, **87**, 3566–73.

Hamilton, W. D. & Zuk, M. 1982. Heritable true fitness and bright birds: A role for parasites? *Science*, **218**, 384–7.

Hammar, J. A. 1929. Die Menschenthymus in Gesundheit und Krankheit. Teil II. Das Organ unter anormalen Körperverhältnissen. *Zeitung Mikroskopanatomie Forschung*, **16**(Suppl)., pp. 1–49.

Hansson, I., Holmdahl, R. & Mattsson, R. 1992. The pineal hormone melatonin exaggerates development of collagen-induced arthritis in mice. *J. Neuroimmunol.*, **39**, 23–31.

Haranaka, K. & Satomi, N. 1981. Cytotoxic activity of tumor necrosis factor (TNF) on human cancer cells in vitro. *Jpn. J. Exp. Med.*, **51**, 191–4.

Hart, B. L. 1988. Biological basis of the behavior of sick animals. *Neurosci. Biobehav. Rev.*, **12**, 123–37.

Hartmann, D. P., Holaday, J. W. & Bernton, E. W. 1989. Inhibition of lymphocyte proliferation by antibodies to prolactin. *FASEB J.*, **3**, 2194–202.

Hartveit, F. 1984. Do infiltrative and anaplastic growth patterns alternate with season in human breast cancer? *Invasion Metastasis*, **4**, 156–9.

Hasselquist, D., Marsh, J. A., Sherman, P. W. & Wingfield, J. C. 1999. Is avian humoral immunocompetence suppressed by testosterone? *Behav. Ecol. Sociobiol.*, **45**, 167–75.

Hastings, M. 1996. Melatonin and seasonality: Filling the gap. *J. Neuroendocrinol.*, **8**, 482–3.

Hathaway, M. R., Dayton, W. R., White, M. E., Henderson, T. L. & Henningson, T. B. 1996. Serum insulin-like growth factor I (IGF-I) concentrations are increased in pigs fed antimicrobials. *J. Anim. Sci.*, **74**, 1541–7.

Hauger, R. L., Millan, M. A., Lorang, M., Harwood, J. P. & Aguilera, G. 1988. Corticotropin releasing factor receptors and pituitary adrenal responses during immobilization stress. *Endocrinology*, **123**, 396–405.

Haus, E. & Smolensky, M. H. 1999. Biologic rhythms in the immune system. *Chronobiol. Int.*, **16**, 581–622.

Hawley, D. J. & Wolfe, F. 1994. Effect of light and season on pain and depression in subjects with rheumatic disorders. *Pain*, **59**, 227–34.

Heideman, P. D. & Sylvester, C. J. 1997. Reproductive photoresponsiveness in unmanipulated male Fischer 344 laboratory rats. *Biol. Reprod.*, **57**, 134–8.

Heldmaier, G., Steinlechner, S., Ruf, T., Wiesinger, H. & Klingenspor, M. 1989. Photoperiod and thermoregulation in vertebrates: Body temperature rhythms and thermogenic acclimation. *J. Biol. Rhythms*, **4**, 251–65.

Hendley, J. O., Fishburne, H. B. & Gwaltney, J. M. 1972. Coronavirus infections in working adults. Eight-year study with 229E and OC43. *Am. Rev. Respir. Dis.*, **105**, 805–11.

Henken, A. M. & Brandsma, H. A. 1982. The effect of environmental temperature on immune response and metabolism of the young chicken. 2. Effect of the immune response to sheep red blood cells on energy metabolism. *Poult. Sci.*, **61**, 1667–73.

Henken, A. M., Groote-Schaarsberg, A. M. & van der Hel, W. 1983. The effect of environmental temperature on metabolism of the young chicken. 4. Effect of environmental temperature on some aspects of energy and protein metabolism. *Poult. Sci.*, **62**, 59–67.

Herbert, T. B. & Cohen, S. C. 1993. Stress and immunity in humans: A meta-analytic review. *Psychosom. Med.*, **55**, 364–79.

Hiebert, S. M., Thomas, E. M., Lee, T. M., Pelz, K. M., Yellon, S. M. & Zucker, I. 2000. Photic entrainment of circannual rhythms in golden-mantled ground squirrels: Role of the pineal gland. *J. Biol. Rhythms*, **15**, 126–34.

Hiestand, P. C., Mekler, P., Nordmann, R., Grieder, A. & Permmongkol, C. 1986. Prolactin as a modulator of lymphocyte responsiveness provides a possible mechanism of action for cyclosporine. *Proc. Natl. Acad. Sci. U.S.A.*, **83**, 2599–603.

Hillgarth, N. & Wingfield, J. C. 1997. Testosterone and immunosuppression in vertebrates: Implications for parasite-mediated sexual selection. In: *Parasites and Pathogens: Effects on Host Hormones and Behavior* (ed. N. E. Beckage), pp. 143–55. New York: Chapman & Hall.

Hoffstater, R. 1952. Versuche der postoperativen Krebsbehandlung mit Zirbelstoffen. *Krebsarzt.*, **14**, 307–16.

Höhn, E. O. 1947. Seasonal cyclical changes in the thymus of the mallard. *J. Exp. Biol.*, **24**, 184–91.

Höhn, E. O. 1956. Seasonal recrudescence of the thymus in adults birds. *Can. J. Biochem. Physiol.*, **34**, 90–101.

Holmdahl, R., Carlsten, H., Jansson, L. & Larsson, P. 1989. Oestrogen is a potent immunomodulator of murine experimental rheumatoid disease. *Br. J. Rheumatol.*, **28**(Suppl. 1), 54–8 discussion 69–71.

Homo-Delarche, F., Fitzpatrick, F., Christeff, N., Nunez, E. A., Bach, J. F. & Dardenne, M. 1991. Sex steroids, glucocorticoids, stress and autoimmunity. *J. Steroid Biochem. Mol. Biol.*, **40**, 619–37.

Honma, Y. & Tamura, E. 1984. Anatomical and behavioral differences among three-spine sticklebacks–The marine form, the landlocked form and their hybrids. *Acta Zool.*, **65**, 79–87.

Hooghe, R., Delhase, M., Vergani, P., Malur, A. & Hooghe-Peters, E. L. 1993. Growth hormone and prolactin are paracrine growth and differentiation factors in the haemopoietic system. *Immunol. Today*, **14**, 212–4.

Hoover, R. C., Lincoln, S. D., Hall, R. F. & Wescott, R. 1984. Seasonal transmission of *Fasciola hepatica* to cattle in northwestern United States. *J. Am. Vet. Med. Assoc.*, **184**, 695–8.

Horak, P., Ots, I., Tegelmann, L. & Moller, A. 2000. Health impact of phytohae-magglutinin-induced immune challenge on great tit (*Parus major*) nestlings. *Can. J. Zool.*, **78**, 905–10.

Horton, T. H. 1984. Growth and reproductive development of male *Microtus montanus* is affected by the prenatal photoperiod. *Biol. Reprod.*, **31**, 499–504.

Høstmark, J. G., Laerum, O. D. & Farsund, T. 1984. Seasonal variations of symptoms and occurrence of human bladder carcinomas. *Scand. J. Urol. Nephrol.*, **18**, 107–11.

Hrushesky, W. J. 1991. The multifrequency (circadian, fertility cycle, and season) balance between host and cancer. *Ann. N.Y. Acad. Sci.*, **618**, 228–56.

Hrushesky, W. J. M. 1983. The clinical application of chronobiology to oncology. *Am. J. Anat.*, **168**, 519–42.

Hughes, C. S., Gaskell, R. M., Jones, R. C., Bradbury, J. M. & Jordon, F. T. W. 1989. Effects of certain stress factors on the re-excretion of infectious laryngotracheitis virus from latently infected carrier birds. *Res. Vet. Sci.*, **46**, 274–6.

Hughes, T. K., Fulep, E., Juelich, T., Smith, E. M. & Stanton, G. J. 1995. Modulation of immune responses by anabolic androgenic steroids. *Int. J. Immunopharmacol.*, **17**, 857–63.

Hurwitz, A., Payne, D. W., Packman, J. N., Andreani, C. L., Resnick, C. E., Hernandez, E. R. & Adashi, E. Y. 1991. Cytokine-mediated regulation of ovarian function: Interleukin-1 inhibits gonadotropin-induced androgen biosynthesis. *Endocrinology*, **129**, 1250–6.

Husband, A. 1995. The immune system and integrated homeostasis. *Immunol. Cell Biol.*, **73**, 377–82.

Hussein, M. F., Badir, N., El Ridi, R. & El Deeb, S. 1979. Effect of seasonal variation on immune system of the lizard, *Scincus scincus*. *J. Exp. Zool.*, **209**, 91–2.

Hyde, L. L. & Underwood, H. 1993. Effects of night-break, T-cycle, and resonance lighting schedules on the pineal melatonin rhythm of the lizard *Anolis carolinensis*: Correlations with the reproductive response. *J. Pineal Res.*, **15**, 70–80.

Immelmann, K. 1973. Role of the environment as source of 'predictive' information. In: *Breeding Biology of Birds* (ed. D. S. Farner), pp. 121–47. Washington, DC: National Academy of Science.

Inman, R. D. 1978. Immunologic sex differences and the female predominance in systemic lupus erythematosus. *Arthritis Rheum.*, **21**, 849–52.

Inouye, S. T. & Kawamura, H. 1979. Persistence of circadian rhythmicity in a mammalian hypothalamic "island" containing the suprachiasmatic nucleus. *Proc. Natl. Acad. Sci. U.S.A.*, **76**, 5962–6.

Irwin, M. & Hauger, R. L. 1988. Adaptation to chronic stress: Temporal pattern of immune and neuroendocrine correlates. *Neuropsychopharmacology*, **1**, 239–42.

Iverson, S. L. & Turner, B. N. 1974. Winter weight dynamics in *Microtus pennsylvanicus*. *Ecology*, **55**, 1030–1036.

Iwatani, Y., Amino, N., Kabutomori, O., Mori, H., Tamaki, H., Motoi, S., Izumiguchi, Y. & Miyai, K. 1984. Decrease of peripheral large granular lymphocytes in Graves' disease. *Clin. Exp. Immunol.*, **55**, 239–44.

Jacobsen, H. K. & Janerich, D. T. 1977. Seasonal variation in the diagnosis of breast cancer. *Proc. Assoc. Cancer Res.*, **18**, 93.

Jacobson, J. D., Nisula, B. C. & Steinberg, A. D. 1994. Modulation of the expression of murine lupus by gonadotropin-releasing hormone analogs. *Endocrinology*, **134**, 2516–23.

Janeway, C. A., Travers, P., Walport, M. & Capra, J. D. 1999. *Immunobiology: The Immune System in Health and Disease*, 4th ed. London: Garland Science Publishing.

Jankovic, B. D., Isakovic, K. & Petrovic, S. 1970. Effect of pinealectomy on immune reaction in the rat. *Immunology*, **8**, 1–6.

Jansson, L., Mattsson, A., Mattsson, R. & Holmdahl, R. 1990. Estrogen induced suppression of collagen arthritis. V. Physiological level of estrogen in DBA/1 mice is therapeutic on established arthritis, suppresses anti-type II collagen T-cell dependent immunity and stimulates polyclonal B-cell activity. *J. Autoimmun.*, **3**, 257–70.

Jefferies, W. M. 1991. Cortisol and immunity. *Med. Hypotheses*, **34**, 198–208.

Jennings, G., Cruickshank, A. M., Shenkin, A., Wight, D. G. & Elia, M. 1992. Effect of aseptic abscesses in protein-deficient rats on the relationship between interleukin-6 and the acute-phase protein, alpha-2-macroglobulin. *Clin. Sci.*, **83**, 731–5.

Johansson, B. W. 1978. Seasonal variations in the endocrine system of hibernators. *Monograph*, 103–10.

John, J. L. 1994. The avian spleen: A neglected organ. *Q. Rev. Biol.*, **69**, 327–51.

John, T. M. 1966. A histochemical study of adrenal corticoids in the pre- and post-migratory phases in the migratory wagtails *Motacilla alba* and *Motacilla flava*. *Pavo*, **4**, 9–14.

Johnson, B. E., Marsh, J. A., King, D. B., Lillehoj, H. S. & Scanes, C. G. 1992. Effect of triiodothyronine on the expression of T cell markers and immune function in thyroidectomized White Leghorn chickens. *Proc. Soc. Exp. Biol. Med.*, **199**, 104–13.

Johnson, R. W. 1997. Inhibition of growth by pro-inflammatory cytokines: An integrated view. *J. Anim. Sci.*, **75**, 1244–55.

Jonsson, K. I., Korpimaki, E., Pen, I. & Tolonen, P. 1996. Daily energy expenditure and shortterm reproductive costs in free-ranging Eurasian Kestrels (*Falco tinnunculus*). *Funct. Ecol.*, **10**, 475–82.

Jose, D. G. & Good, R. A. 1973. Quantitative effects of nutritional protein and calorie deficiency upon immune responses to tumors in mice. *Cancer Res.*, **33**, 807–12.

Jurcovicova, J., Day, R. N., MacLeod, R. M. 1992. Expression of prolactin in rat lymphocytes. *Prog. Neuroendocrinol. Immunol.*, **5**, 256–63.

Kalland, D. 1980. Decreased and disproportionate T-cell population in adult mice after neonatal exposure to diethylstilbestrol. *Cell. Immunol.*, **51**, 55–63.

Kalland, T. 1980. Reduced natural killer activity in female mice after neonatal exposure to diethylstilbestrol. *J. Immunol.*, **124**, 1297–300.

Kamilaris, T. C., DeBold, C. R., Johnson, E. O., Mamalaki, E., Listwak, S. J., Calogero, A. E., Kalogeras, K. T., Gold, P. W. & Orth, D. N. 1991. Effects of short and long duration hypothyroidism and hyperthyroidism on the plasma adrenocorticotropin and corticosterone responses to ovine corticotropin-releasing hormone in rats. *Endocrinology*, **128**, 2567–76.

Kamis, A. B., Ahmad, R. A. & Badrul-Munir, M. Z. 1992. Worm burden and leukocyte response in *Angiostrongylus malaysiensis*-infected rats: Influence of testosterone. *Parasitol. Res.*, **78**, 388–91.

Kapikian, A. Z., Kim, H. W., Wyatt, R. G., Cline, W. L., Arrobio, J. O., Brandt, C. D., Rodriguez, W. J., Sack, D. A., Chanock, R. M. & Parrott, R. H. 1976. Human reovirus-like agent as the major pathogen associated with "winter" gastroeneritis in hospitalized infants and young children. *N. Engl. J. Med.*, **294**, 965–72.

Karanth, S. & McCann, S. M. 1991. Anterior pituitary hormone control by interleukin 2. *Proc. Natl. Acad. Sci. U.S.A.*, **88**, 2961–5.

Karp, J. D., Moynihan, J. A. & Ader, R. 1993. Effects of differential housing on the primary and secondary antibody responses of male C57BL/6 and BALB/c mice. *Brain Behav. Immun.*, **7**, 326–33.

Karsch, F. J. 1984. The hypothalamus and anterior pituitary gland. In: *Reproduction in Mammals*, 2nd ed. (ed. C. R. Austin & R. V. Short), pp. 1–20. Cambridge: Cambridge University Press.

Katzman, L. 1994. The common cold. The effect of hot humid air on the nasal mucosa. *J. Am. Med. Assoc.*, **272**, 1103–4.

Kaufmann, S. H. E. 1993. Immunity to intracellular bacteria. *Ann. Rev. Immunol.*, **11**, 29–63.

Keenan, R. A., Moldawer, L. L., Yang, R. D., Kawamura, I., Blackburn, G. L. & Bistrian, B. R. 1982. An altered response by peripheral leukocytes to synthesize or release leukocyte endogenous mediator in critically ill, protein-malnourished patients. *J. Lab. Clin. Med.*, **100**, 844–57.

Keller, S. E., Weiss, J. M., Schleifer, S. L., Miller, N. E. & Stern, M. 1983. Stress-induced suppression of immunity in adrenalectomized rats. *Science*, **221**, 1301–4.

Kelley, K. W. 1989. Growth hormone, lymphocytes and macrophages. *Biochem. Pharmacol.*, **38**, 705–13.

Kelley, K. W. 1990. The role of growth-hormone in modulation of the immune-response. *Ann. N.Y. Acad. Sci.*, **594**, 95–103.

Kelley, K. W. 1991. Growth hormone in immunology. In *Psychoneuroimmunology*, 2nd Edition, Ader, R., Felton, D. L., Cohen, N. (ed), pp. 377–402. New York: Academic Press.

Kelley, K. W., Arkins, S., Minshall, C., Liu, Q. & Dantzer, R. 1996. Growth hormone, growth factors and hematopoiesis. *Horm. Res.*, **45**, 38–45.

Kelley, K. W., Brief, S., Westly, H. J., Novakofski, J., Bechtel, P. J., Simon, J. & Walker, E. B. 1986. GH3 pituitary adenoma cells can reverse thymic aging in rats. *Proc. Natl. Acad. Sci. U.S.A.*, **83**, 5663–7.

Kelley, K. W., Johnson, R. W. & Dantzer, R. 1994. Immunology discovers physiology. *Vet. Immunol. Immunopathol.*, **43**, 157–65.

Kelso, A. & Munck, A. 1984. Glucocorticoid inhibition of lymphokine secretion by alloreactive T lymphocyte clones. *J. Immunol.*, **133**, 784–91.

Kenison, D. C., Elsasser, T. H. & Fayer, R. 1991. Tumor necrosis factor as a potential mediator of acute metabolic and hormonal responses to endotoxemia in calves. *Am. J. Vet. Res.*, **52**, 1320–6.

Kern, J. A., Lamb, R. J., Reed, J. C., Daniele, R. P. & Nowell, P. C. 1988. Dexamethasone inhibition of interleukin 1 beta production by human monocytes. Posttranscriptional mechanisms. *J. Clin. Invest.*, **81**, 237–44.

Keusch, G. T. 1982. Immune function in the malnourished host. *Pediatr. Ann.*, **11**, 1004–14.

Khuroo, M. S., Teli, M. R., Skidmore, S., Sofi, M. A. & Khuroo, M. I. 1981. Incidence and severity of viral hepatitis in pregnancy. *Am. J. Med.*, **70**, 252–5.

Kiecolt-Glaser, J. K. 1999. Norman Cousins Memorial Lecture 1998. Stress, personal relationships, and immune function: health implications. *Brain Behav. Immun.*, **13**, 61–72.

Kiecolt-Glaser, J. K., Dura, J. R., Speicher, C. E., Trask, O. J. & Glaser, R. 1991. Spousal caregivers of dementia victims: Longitudinal changes in immunity and health. *Psychosom. Med.*, **53**, 345–62.

Kiecolt-Glaser, J. K., Garner, W., Speicher, C. E., Penn, G. & Glaser, R. 1984. Psychosocial modifiers of immune competence in medical students. *Psychosom Med.*, **46**, 7–14.

Kiecolt-Glaser, J. T., Fisher, L., Ogrocki, P., Stont, J. C., Speichler, C. E. & Glaser, R. 1987. Marital quality, marital disruption and immune function. *Psychosom. Med.*, **49**, 13–34.

Kiecolt-Glaser, J. T. & Glaser, R. 1988. Methodological issues in behavioral immunology research in humans. *Brain Behav. Immunol.*, **2**, 67–78.

Kiess, W. & Butenandt, O. 1985. Specific growth hormone receptors on human peripheral mononuclear cells: Reexpression, identification, and characterization. *J. Clin. Endocrinol. Metab.*, **60**, 740–6.

Kirk, S. J., Hurson, M., Regan, M. C., Holt, D. R., Wasserkrug, H. L., Barbul, A., Daly, J. M. & Flye, M. W. 1993. Arginine stimulates wound-healing and immune function in elderly human beings. *Surgery*, **114**, 155–60.

Kirkham, N., Machin, D., Cotton, D. W. K. & Pike, J. M. 1985. Seasonality and breast cancer. *Eur. J. Surg. Oncol.*, **11**, 143–6.

Kittas, C. & Henry, L. 1979. Effect of sex hormones on the immune system of guinea-pigs and on the development of toxoplasmic lesions in non-lymphoid organs. *Clin. Exp. Immunol.*, **36**, 16–23.

Klasing, K. C. 1998. Nutritional modulation of resistance to infectious diseases. *Poult. Sci.*, **77**, 1119–25.

Klein, D. C., Smoot, R., Weller, J. L., Higa, S., Markey, S. P., Creed, G. H. & Jacobowitz, D. M. 1983. Lesions of the paraventricular nucleus area of the hypothalamus disrupt the suprachiasmatic-spinal cord circuit in the melatonin rhythm generating system. *Brain Res. Bull.*, **10**, 647–52.

Klein, S. L. 2000a. Hormones and mating system affect sex and species differences in immune function among vertebrates. *Behav. Proc.*, **51**, 149–66.

Klein, S. L. 2000b. The effects of hormones on sex differences in infection: From genes to behavior. *Neurosci. Biobehav. Rev.*, **24**, 627–38.

Klein, S. L., Bird, B. H. & Glass, G. E. 2000. Sex differences in Seoul virus infection are not related to adult sex steroid concentrations in Norway rats. *J. Virol.*, **74**, 8213–7.

Klein, S. L., Gamble, H. R. & Nelson, R. J. 1999. Sex differences in *Trichinella spiralis* infection are not mediated by circulating steroid hormones in voles. *Am. J. Physiol.*, **277**, R1362–7.

Klein, S. L. & Nelson, R. J. 1999. Social interactions unmask sex and species differences in humoral immunity in voles. *Anim. Behav.*, **57**, 603–10.

Kliger, C. A., Gehad, A. E., Hulet, R. M., Roush, W. B., Lillehoj, H. S. & Mashaly, M. M. 2000. Effects of photoperiod and melatonin on lymphocyte activities in male broiler chickens. *Poult. Sci.*, **79**, 18–25.

Kloner, R. A., Poole, W. K. & Perritt, R. L. 1999. When throughout the year is coronary death most likely to occur? A 12-year population-based analysis of more than 220,000 cases. *Circulation*, **100**, 1630–4.

Kluger, M. J., Kozak, W., Conn, C. A., Leon, L. R. & Soszynski, D. 1998. Role of fever in disease. *Ann. N.Y. Acad. Sci.*, **856**, 224–33.

Koenig, J. I., Snow, K., Clark, B. D., Toni, R., Cannon, J. G., Shaw, A. R., Dinarello, C. A., Reichlin, S., Lee, S.L. & Lechan, R. M. 1990. Intrinsic pituitary interleukin-1 beta is induced by bacterial lipopolysaccharide. *Endocrinology*, **126**, 3053–8.

Komada, H., Nakabayashi, N., Yoshida, T., Takanari, H., Hara, M., Hara, M., Takahashi, T. & Izutsu, K. 1989. Seasonal variation in mitogenic response and subsets of human tonsillar lymphocytes. *J. Interdiscipl. Cycle Res.*, **20**, 107–22.

Komukai, Y., Amao, H., Goto, N., Kusajima, Y., Sawada, T., Saito, M. & Takahashi, K. W. 1999. Sex differences in susceptibility of IRC mice to oral infection with *Corynebacterium kutsheri*. *Exp. Anim.*, **48**, 37–42.

Konstadoulakis, M. M., Syrigos, K. N., Baxevanis, C. N., Syrigou, E. I., Papamichail, M., Peveretos, P., Anapliotou, M. & Golematis, B. C. 1995. Effect of testosterone administration, pre- and postnatally, on the immune system of rats. *Horm. Metab. Res.*, **27**, 275–8.

Konstantaninova, I. V., Rykova, M. P., Lesnyak, A. T. & Antropova, E. Q. 1993. Immune changes during long-duration missions. *J. Leukoc. Biol.*, **54**, 189–201.

Korn, H. & Taitt, M. J. 1987. Initiation of early breeding in a population of *Microtus townsendii* with secondary plant compound 6-methoxybenzoxazolinone. *Oecologia*, **71**, 593–6.

Kothari, L. S., Shah, P. N. & Mhatre, M. C. 1982. Effect of continuous light on the incidence of 9,10-dimethyl-1,2-benzathracene induced mammary tumors in female Holtzman rats. *Cancer Lett.*, **16**, 313–7.

Kramer, T. R., Moore, R. J., Shippee, R. L., Friedl, K. E., Martinez-Lopez, L., Chan, M. M. & Askew, E. W. 1997. Effects of food restriction in military training on T-lymphocyte responses. *Int. J. Sports Med.*, **18** Suppl. 1, S84–90.

Krause, R. 1922. Die Milz. *Mikroskopische Anatomie der Wirbeltier, Bd. 2: Vogel und Reptilien.* Berlin: Verlag W. De Gruyter.

Kriegsfeld, L. J., Drazen, D. L. & Nelson, R. J. 2001. In vitro melatonin treatment enhances cell-mediated immune function in male prairie voles (*Microtus ochrogaster*). *J. Pineal Res.*, **30**, 193–8.

Kriegsfeld, L. J. & Nelson, R. J. 1998. Short photoperiod affects reproductive function but not dehydroepiandrosterone concentrations in male deer mice (*Peromyscus maniculatus*). *J. Pineal Res.*, **25**, 101–15.

Kriegsfeld, L. J. & Nelson, R. J. 1999. Photoperiod affects the gonadotropin-releasing

258 *References*

hormone neuronal system of male prairie voles (*Microtus ochrogaster*). *Neuroendocrinology*, **69**, 238–44.

Krishnan, L., Guilbert, L. J., Russell, A. S., Wegmann, T. G., Mosmann, T. R. & Belosevic, M. 1996a. Pregnancy impairs resistance of C57BL/6 mice to *Leishmania major* infection and causes decreased antigen-specific IFNγ responses and increased production of T helper 2 cytokines. *J. Immunol.*, **156**, 644–52.

Krishnan, L., Guilbert, L. J., Wegmann, T. G., Belosevic, M. & Mosmann, T. R. 1996b. T helper 1 response against *Leishmania major* in pregnant C57BL/6 mice increases implantation failure and fetal resorptions. *J. Immunol.*, **156**, 653–62.

Kruger, T. E. & Blalock, J. E. 1986. Cellular requirements for thyrotropin enhancement of in vitro antibody production. *J. Immunol.*, **137**, 197–200.

Kruger, T. E., Smith, L. R., Harbour, D. V. & Blalock, J. E. 1989. Thyrotropin: An endogenous regulator of the in vitro immune response. *J. Immunol.*, **142**, 744–7.

Krzych, U., Strausser, H. R., Bressler, J. P. & Goldstein, A. L. 1981. Effects of sex hormones on some T and B cell functions, evidenced by differential immune expression between male and female mice and cyclic pattern of immune responsiveness during the estrous cycle in female mice. *Am. J. Reprod. Immunol.*, **1**, 73–77.

Kubo, C., Gajar, A., Johnson, B. C. & Good, R. A. 1992b. The effects of dietary restriction on immune function and development of autoimmune disease in BXSB mice. *Proc. Natl. Acad. Sci. (U.S.A.)*. **89**, 3145–9.

Kubo, C., Johnson, B. C., Day, N. K. & Good, R. A. 1992a. Effects of calorie restriction on immunologic functions and development of autoimmune disease in NZB mice. *Proc. Soc. Exp. Biol. Med.*, **201**, 192–9.

Kuci, S., Becker, J. & Verr, G. 1983. Circadian variations of the immunomodulatory role of the pineal gland. *Neuroendocrinol. Lett.*, **10**, 65–79.

Kuhl, H., Gross, M., Schneider, M., Weber, W., Mehlis, W., Stegmuller, M. & Taubert, H. D. 1983. The effect of sex steroids and hormonal contraceptives upon thymus and spleen on intact female rats. *Contraception*, **28**, 587–601.

Kurtzke, J. F. 1975. A reassessment of the distribution of multiple sclerosis. *Acta Neurol. Scand.*, **51**, 110–36.

Kurtzke, J. F. 1980. Geographic distribution of multiple sclerosis: An update with special reference to Europe and the Mediterranean region. *Acta Neurol. Scand.*, **62**, 65–80.

Küstner, H. G. V. & Du Plessis, G. 1991. The cholera epidemic in South Africa, 1980–1987. Epidemiological features. *South African Med. J.*, **79**, 539–44.

Lapin, V. 1974. Influence of simultaneous pinealectomy and thymectomy on the growth and formation of metastases of the Yoshida sarcoma in rats. *Exp. Pathol.*, **9**, 108–12.

Lapin, V. 1976. Pineal gland and malignancy. *Oster Z-Onkol*, **3**, 51–60.

Lapin, V. 1978. Effects of reserpine on the incidence of 9,10-dimethyl-1,2-benzanthracene-induced tumors in pinealectomized and thymectomized rats. *Oncology*, **35**, 132–5.

Lapin, V. & Frowein, A. 1981. Effects of growing tumors on pineal melatonin levels in male rats. *J. Neural Transm.*, **50**, 123–36.

Larkin, J. E., Freeman, D. A. & Zucker, I. 2001. Low ambient temperature accelerates short-day responses in Siberian hamsters by altering responsiveness to melatonin. *J. Biol. Rhythms*, **16**, 76–86.

Latshaw, J. D. 1991. Nutrition–Mechanisms of immunosuppression. *Vet. Immunol. Immunopathol.*, **30**, 111–20.

Laudenslager, M. L., Fleshner, M., Hofstadter, P., Held, P. E., Simons, L. & Maier, S. F.

1988. Suppression of specific antibody production by inescapable shock: Stability under varying conditions. *Brain Behav. Immun.*, **2,** 92–101.

Laudenslager, M. L., Ryan, S. M., Drugan, R. C., Hyson, R. L. & Maier, S. F. 1983. Coping and immunosuppression: Inescapable but not escapable shock suppresses lymphocyte proliferation. *Science*, **221,** 568–70.

Laue, L., Kawai, S., Brandon, D. D., Brightwell, D., Barnes, K., Knazek, R. A., Loriaux, D. L. & Chrousos, G. P. 1988. Receptor-mediated effects of glucocorticoids on inflammation: Enhancement of the inflammatory response with a glucocorticoid antagonist. *J. Steroid Biochem.*, **29,** 591–8.

Laugero, K. D. & Moberg, G. P. 2000. Effects of acute behavioral stress and LPS-induced cytokine release on growth and energetics in mice. *Physiol. Behav.*, **68,** 415–22.

Leceta, J., Garrido, E., Torroba, M. & Zapata, A. G. 1989. Ultrastructural changes in the thymus of the turtle *Mauremys caspica* in relation to the seasonal cycle. *Cell Tissue Res.*, **256,** 213–9.

Leceta, J. & Zapata, A. 1985. Seasonal changes in the thymus and spleen of the turtle, *Mauremys caspica*. A morphometrical, light microscopical study. *Dev. Comp. Immunol.*, **9,** 653–68.

Leceta, J. & Zapata, A. 1986. Seasonal variations in the immune response of the tortoise *Mauremys caspica*. *Immunology*, **57,** 483–7.

Lee, A. K. & McDonald, I. R. 1985. Stress and population regulation in small mammals. *Oxford Rev. Reprod. Biol.*, **7,** 261–304.

Lee, T. M. & Zucker, I. 1995. Seasonal variations in circadian rhythms persist in gonadectomized golden-mantled ground squirrels. *J. Biol. Rhythms*, **10,** 188–95.

Leigh, W. H. 1960. The Florida spotted gar, as the intermediate host for *Odhneriotrema incommodom* (Leidy, 1856) from *Alligator missippiensis*. *J. Parasitol.*, **46,** 16.

Lemarchand-Beraud, T. H., Holm, A. C. & Scazziga, B. R. 1977. Triodothyronine and thyroxine nuclear receptors in lyphocytes from normal, hyper-and hypothyroid subjects. *Acta. Endocrinol. (Copenh).*, **85,** 44–54.

Leonard, B. E. & Song, C. 1996. Stress and the immune system in the etiology of anxiety and depression. *Pharmacol. Biochem. Behav.*, **54,** 299–303.

LeSauter, J., Lehman, M. N. & Silver, R. 1996. Restoration of circadian rhythmicity by transplants of SCN "micropunches." *J. Biol. Rhythms*, **11,** 163–71.

Lévi, F., Canon, C., Depres-Brummer, P., Adam, R., Bourin, P., Pati, A., Florentin, I., Misset, J. L. & Bismuth, H. 1992. The rhythmic organization of the immune network: Implications for the chronopharmacologic delivery of interferons, interleukins and cyclosporin. *Adv. Drug Delivery Rev.*, **9,** 85–112.

Levine, H. B. & Madin, S. H. 1962. Enhancement of experimental coccidiomycosis in mice with testosterone and estradiol. *Sabouraudia*, **2,** 47–52.

Lewis, J. W. 1966. *Maritrema apodemicum* sp. nov. Digenea: Microphallidae from the long-tailed field mouse, *Apodemus sylvaticus sylvaticus* (L.) on Skomer Island. *J. Helminthol.*, **40,** 363–74.

Licht, P., McCreery, B. R., Barnes, B. R. & Pang, R. 1983. Seasonal and stress related changes in plasma gonadotropins, sex steroids, and corticosterone in the bullfrog, *Rana catesbeiana*. *Gen. Comp. Endocrinol.*, **50,** 124–45.

Lidicker, W. Z. 1973. Experimental manipulation of the timing of reproduction in the California vole. *Res. Pop. Ecol.*, **18,** 14–27.

Lidwell, O. M., Morgan, R. W. & Williams, R. E. O. 1965. The epidemiology of the common cold. IV. The effect of weather. *J. Hyg. (Cambridge)*, **63,** 427–39.

Lidwell, O. M. & Sommerville, T. 1951. Observations on the incidence and distribution of the common cold in a rural community during 1948 and 1949. *J. Hyg. (Lond.)*, **49**, 365–400.

Liebmann, P. M., Hofer, D., Felsner, P., Wolfler, A. & Schauenstein, K. 1996. Beta-blockade enhances adrenergic immunosuppression in rats via inhibition of melatonin release. *J. Neuroimmunol.*, **67**, 137–42.

Liebmann, P. M., Wolfer, A., Felsner, P., Hofer, D. & Schauenstein, K. 1997 Melatonin and the immune system. *Int. Arch. Allergy Immunol.*, **112**, 203–11.

Limburg, C. D. 1950. The geographic distribution of multiple sclerosis and its estimated prevalence in the United States. *Proc. Assoc. Res. Nerv. Mental Dis.*, **28**, 15–24.

Lin, E., Kotani, J. G. & Lowry, S. F. 1998. Nutritional modulation of immunity and the inflammatory response. *Nutrition*, **14**, 545–50.

Linenger, J. M., Flinn, S., Thomas, B. 1993. Musculoskeletal and medical morbidity associated with rigorous physical training. *Clin. J. Sport Med.*, **3**, 229–34.

Lippman, M. & Barr, R. 1977. Glucocorticoid receptors in purified subpopulations of human peripheral blood lymphocytes. *J. Immunol.*, **118**, 1977–81.

Lisse, I. M., Aabe, P., Whittle, H., Jensen, H., Engelmann, M. & Christensen, L. B. 1997. T-l lymphocyte subsets in West African children: Impact of age, sex, and season. *J. Pediatr.*, **130**, 77–85.

Lissoni, P., Barni, S., Meregalli, S., Fossati, V., Cazzaniga, M., Esposti, D. & Tancini, G. 1995. Modulation of cancer endocrine therapy by melatonin: A phase II study of tamoxifen plus melatonin in metastatic breast cancer patients progressing under tamoxifen alone. *Brit. J. Cancer*, **71**, 854–6

Lissoni, P., Barni, S., Tancini, G., Crispino, S., Paolorossi, F., Lucini, V., Mariani, M., Cattaneo, G., Esposti, D., Espoti, G. & Fraschini, F. 1987. Clinical study of melatonin in untreatable advanced cancer patients. *Tumori*, **73**, 475–80.

Lissoni, P., Morabito, F., Esposti, G., Esposti, D., Ripamonti, G. & Fraschini, F. 1988. A study on the thymus-pineal axis: Interactions between thymic hormones and melatonin secretion. *Acta. Med. Auxol.*, **20**, 35–9.

Liu, Z. M. & Pang, S. F. 1993. [^{125}I]Iodomelatonin binding sites in the bursa of Fabricius of birds: Binding characteristic, subcellular distribution, diurnal variations and age studies. *J. Endocrinol.*, **138**, 51–7.

Lloyd, S. 1983. Effect of pregnancy and lactation upon infection. *Vet. Immunol. Immunopathol.*, **4**, 153–76.

Lochmiller, R. L. & Deerenberg, C. 2000. Trade-offs in evolutionary immunology: Just what is the cost of immunity? *Oikois*, **88**, 87–98.

Lochmiller, R. L., Vestey, M. R. & Boren, J. C. 1993. Relationship between protein nutritional-status and immunocompetence in northern bobwhite chicks. *Auk*, **110**, 503–10.

Lochmiller, R. L., Vesty, M. R. & McMurry, S. T. 1986. Primary immune responses of selected small mammal species to heterologous erythrocytes. *Comp. Biochem. Physiol.*, **100**, 139–43.

Lochmiller, R. L., Vesty, M. R. & McMurry, S. T. 1994. Temporal variation in humoral and cell-mediated immune response in a *Sigmodon hispidus* population. *Ecology*, **75**, 236–45.

Lockhart, P. B. & Durack, D. T. 1999. Oral microflora as a cause of endocarditis and other distant site infections. *Infect. Dis. Clin. North Am.*, **13**, 833–50.

Loffreda, S., Yang, S. Q., Lin, H. Z., Karp, C. L., Brengman, M. L., Wang, D. J., Klein, A. S., Bulkley, G. B., Bao, C., Noble, P. W., Lane, M. D. & Diehl, A. M. 1998. Leptin regulates proinflammatory immune responses. *FASEB J.*, **12**, 57–65.

Lopez-Gonzales, M. A., Calvo, J. R., Osuna, C. & Guerrero, J. M. 1992. Interaction of melatonin with human lymphocytes: Evidence for binding sites coupled to potentiation of cyclic AMP stimulated vasoactive intestinal peptide and activation of cyclic GMP. *J. Pineal Res.*, **12**, 97–104.

Loria, R. M., Padgett, D. A. & Huynh, P. N. 1996. Regulation of the immune response by dehydroepiandrosterone and its metabolites. *J. Endocrinol.*, **150**, S209–20.

Loscher, W., Mevissen, M. & Haussler, B. 1997. Seasonal influence on 7,12-dimethylbenz[a]anthracene-induced mammary carcinogenesis in Sprague-Dawley rats under controlled laboratory conditions. *Pharmacol. Toxicol.*, **81**, 265–70.

Lotfy, O. A., Saleh, W. A. & el-Barbari, M. 1998. A study of some changes in cell-mediated immunity in protein energy malnutrition. *J. Egypt. Soc. Parasitol.*, **28**, 413–28.

Low, T. L. K. & Goldstein, A. L. 1984. Thymosin, peptidic moieties, and related agents. In: *Immune Modulation Agents and Their Mechanisms* (eds. R. L. Fenichel & M. A. Chirigos), pp. 135–62. New York: Marcel Dekker.

Lumpkin, M. D. & McDonald, J. K. 1989. Blockade of growth hormone-releasing factor (GRF) activity in the pituitary and hypothalamus of the conscious rat with a peptidic GRF antagonist. *Endocrinology*, **124**, 1522–31.

Luster, M. I., Hayes, H. T., Korach, K., Tucker, A. N., Dean, J. H., Greenlee, W. F. & Boorman, G. A. 1984. Estrogen immunosuppression is regulated through estrogenic responses in the thymus. *J. Immunol.*, **133**, 110–6.

Lwoff, A. 1969. Death and transfiguration of a problem. *Bacteriol. Rev.*, **33**, 390–400.

Lyden, D. C., Olszewski, J., Feran, M., Job, L. P. & Huber, S. A. 1987. Coxsackievirus B-3-induced myocarditis. *Am. J. Pathol.*, **126**, 432–8.

Lynch, G. R., Heath, H. W. & Johnston, C. M. 1981. Effect of geographical origin on the photoperiodic control of reproduction in the white-footed mouse, *Peromyscus leucopus*. *Biol. Reprod.*, **25**, 475–80.

Lyngbye, J. & Krøll, J. 1971. Quantitative immunoelectrophoresis of proteins in serum from a normal population: Season-, age-, and sex-related variation. *Clin. Chem.*, **17**, 495–500.

Lysle, D. T., Cunnick, J. E., Wu, R., Caggiula, A. R., Wood, P. G. & Rabin. B. S. 1988. 2-Deoxy-D-glucose modulation of T-lymphocyte reactivity: Differential effects on lymphoid compartments. *Brain Behav. Immun.*, **2**, 212–21.

Lyson, K. & McCann, S. M. 1991. The effect of interleukin-6 on pituitary hormone release in vivo and in vitro. *Neuroendocrinology*, **54**, 262–6.

Mackinnon, L. T., Ginn, E. & Seymour, G. J. 1993. Decreased salivary immunoglobulin A secretion rate after intense interval exercise in elite kayakers. *Eur. J. Appl. Physiol.*, **67**, 180–4.

Macknin, M. L., Mathews, S. & Medendorp, S. V. 1990. Effect of inhaling heated vapor on symptoms of the common cold. *J. Am. Med. Assoc.*, **264**, 989–91.

MacMurray, J. P., Barker, J. P., Armstrong, J. D., Bozzetti, L. P. & Kuhn, I. N. 1983. Circannual changes in immune function. *Life Sci.*, **32**, 2363–70.

Madden, K. S. 2001. *Psychoneuroimmunology*, Vol. 2, 3rd ed., R. Ader, D. L., Felten & N. Cohen (eds.), pp. 189–226. New York: Academic Press.

Madison, D. M. 1984. Group nesting & its ecological & evolutionary significance in overwintering microtine rodents. *Bull. Carnegie Museum Nat. Hist.*, **10**, 267–74.

Maes, M., Stevens, W., Scharpé S., Bosmans, E., De Meyer, F., D'Hondt, P., Peeters, D., Thompson, P., Cosyns, P., De Clerck, L., Bridts, C., Neels, H., Wauters, A. & Cooreman, W. 1994. Seasonal variation in peripheral blood leukocyte subsets and in serum interleukin-6, and soluble interleukin-2 and -6 receptor concentrations in normal volunteers. *Experientia*, **50**, 821–9.

Maestroni, G. J. & Conti, A. 1989. Beta-endorphin and dynorphin mimic the circadian immunoenhancing and anti-stress effects of melatonin. *Int. J. Immunopharmacol.*, **11**, 333–40.

Maestroni, G. J., Conti, A. & Pierpaoli, W. 1988. Role of the pineal gland in immunity. III. Melatonin antagonizes the immunosuppressive effect of acute stress via an opiatergic mechanism. *Immunology*, **63**, 465–9.

Maestroni, G. J. M. 1993. The immunoneuroendocrine role of melatonin. *J. Pineal Res.*, **14**, 1–10.

Maestroni, G. J. M. & Conti, A. 1991. Anti-stress role of the melatonin-immuno-opioid network. Evidence for a physiological mechanism involving T cell-derived immunoreactive μ-endorphin and met-enkaphalin binding to thymic opioid receptors. *Int. J. Neurosci.*, **61**, 289–98.

Maestroni, G. J. M. & Conti, A. 1993. Meatonin in relation with the immune system. In: Melatonin, Biosynthesis, Physiological Effects and Clinical Applications. Y. Hing-Su, R.J. Reiter (eds.), pp. 289–311. Boca Raton: CRC Press.

Maestroni, G. J. M., Conti, A. & Pierpaoli, W. 1986. Role of the pineal gland in immunity. Circadian synthesis & release of melatonin modulates the antibody response and antagonizes the immunosuppressive effect of corticosterone. *J. Neuroimmunol.*, **13**, 19–30.

Maestroni, G. J. M., Conti, A. & Pierpaoli, W. 1987. Role of the pineal gland in immunity. II. Melatonin enhances the antibody response via an opiatergic mechanism. *Clin. Exp. Immunol.*, **68**, 384–91.

Maestroni, G. J. M., Conti, A. & Pierpaoli, W. 1988. Pineal melatonin: Its fundamental immunoregulatory role in aging and cancer. *Ann. N.Y. Acad. Sci.*, **521**, 140–8.

Maestroni, G. J. M. & Pierpaoli, W. 1981. Pharmacological control of the hormonally mediated immune-response. In: *Psychoneuroimmunology* (ed. R. Ader), pp. 405–25. New York: Academic Press.

Mahmoud, I., Salman, S. S. & Al-Khateeb, A. 1994. Continuous darkness and continuous light induced structural changes in the rat thymus. *J. Anat.*, **185**, 143–9.

Mahmoud, I. Y. & Licht, P. 1997. Seasonal changes in gonadal activity & the effects of stress on reproductive hormones in the common snapping turtle, *Chelydra serpentina*. *Gen. Comp. Endocrinol.*, **107**, 359–72.

Maier, C. C., Marchetti, B., LeBoeuf, R. D. & Blalock, J. E. 1992. Thymocytes express a mRNA that is identical to hypothalamic luteinizing hormone-releasing hormone mRNA. *Cell. Mol. Neurobiol.*, **12**, 447–54.

Maier, S. F. & Laudensiager, M. L. 1988. Inescapable shock, shock controllability, and mitogen stimulated lymphocyte proliferation. *Brain Beh. Immun.*, **2**, 87–91.

Maier, S. F., Watkins, L. R. & Fleshner, M. 1994. Psychoneuroimmunology. The interface between behavior, brain, and immunity. *Am. Psychol.*, **49**, 1004–17.

Malakhova, R. P. 1961. Seasonal changes in the parasite fauna of certain freshwater fishes of Karelian Lakes (Lake Konche). *Trudy Karelskogo Filiala Akademiya Nauk SSR*, **30**, 55–78.

Malhotra, S. K. 1986. Bioecology of the parasites of high altitude homeothermic host-parasite systems. I. Influence of season & temperature on infections by strobilocerci of three species of *Hydatigera* in Indian house rat. *J. Helminthol.*, **60**, 15–20.

Mann, D. R., Akinbami, M. A., Gould, K. G., Ansari, A. A. 2000a. Seasonal variations in cytokine expression & cell-mediated immunity in male rhesus monkeys. *Cell. Immunol.*, **200**, 105–15.

Mann, D. R., Akinbami, M. A., Lunn, S. F., Fraser, H. M., Gould, K. G. & Ansari, A. A. 2000b. Endocrine-immune interaction: Alterations in immune function resulting from neonatal treatment with a GnRH antagonist & seasonality in male primates. *Am. J. Reprod. Immunol.*, **44**, 30–40.

Mann, P. L. 1978. The effect of various dietary restricted regimes on some immunological parameters in mice. *Growth*, **42**, 87–103.

Marchetti, B., Gallo, F., Farinella, Z., Tirolo, C., Testa, N., Romeo, C. & Morale, M. C. 1998. Luteinizing hormone-releasing hormone is a primary signaling molecule in the neuroimmune network. *Ann. N. Y. Acad. Sci.*, **840**, 205–48.

Marchetti, B., Guarcello, V., Morale, M. C., Bartoloni, G., Raiti, F., Palumbo, G., Jr., Farinella, Z., Cordaro, S. & Scapagnini, U. 1989. Luteinizing hormone-releasing hormone LHRH agonist restoration of age-associated decline of thymus weight, thymic LHRH receptors, and thymocyte proliferative capacity. *Endocrinology*, **125**, 1037–45.

Marsh, J. A. & Scanes, C. G. 1994. Neuroendocrine-immune interactions. *Poult. Sci.*, **73**, 1049–61.

Martin, C. P. 1976. *Textbook of Endocrine Physiology*. Baltimore, MD: Williams & Wilkins.

Martin-Cacao, A., Lopez-Gonzalez, M. A., Reiter, R. J., Calvo, J. R. & Guerrero, J. M. 1993. Binding of 2[^{125}I]melatonin by rat thymus membranes during postnatal development. *Immunol. Lett.*, **36**, 59–64.

Martinet, L. & Meunier. L. 1969. Influence des variations saisonnières de la luzerne sur la croissance, la mortalité, et l'éstablissement de la maturité sexuelle chez le campagnol des champs (*Microtus arvalis*). *Ann. Biol. d'Anim. Biochem. Biophys.*, **9**, 451–62.

Martinez-Lopez, L. E., Friedl, K. E., Moore, R. J. & Kramer, T. R. 1993. A longitudinal study of infections & injuries of Ranger students. *Mil. Med.*, **158**, 433–7.

Martino, E., Buratti, L., Bartalena, L., Mariotti, S., Cupini, C., Aghini-Lombardi, F. & Pinchera, A. 1987. High prevalence of subacute thyroiditis during summer season in Italy. *J. Endocr. Invest.*, **10**, 321–3.

Mason, B. H., Holdaway, I. M., Mullins, P. R., Kay, R. G. & Skinner, S. J. 1985. Seasonal variation in breast cancer detection: Correlation with tumour progesterone receptor status. *Breast Cancer Res. Treat.*, **5**, 171–6.

Mason, B. H., Holdaway, I. M. & Skinner, S. J. 1987. Association between season of first detection of breast cancer and disease progression. *Breast Cancer Res. Treat.*, **9**, 227–32.

Mason, B. H., Holdaway, I. M., Stewart, A. W., Neave, L. M. & Kay, R. G. 1990. Season of initial discovery of tumour as an independent variable predicting survival in breast cancer. *Br. J. Cancer*, **61**, 137–41.

McCruden, A. B. & Stimson, W. H. 1991. Sex hormones and immune function. In R. Ader, J. Cohen (eds), *Psychoneuroimmunology*, New York: Academic Press.

McCruden, A. B. & Stimson, W. H. 1991. Sex hormones and immune function in *Psychoneuroimmunology,* 2nd ed, R. Ader, D. L. Felton & N. Cohen (eds), pp. 475–493. New York: Academic Press.

McDonald, I. R., Lee, A. K., Than, K. A. & Martin, R. W. 1988. Concentrations of free glucocorticoids in plasma and mortality in the Australian bush rat (*Rattus fuscipes* Waterhouse). *J. Mammal.*, **69,** 740–8.

McEwen, B. S., Biron, C. A., Brunson, K. W., Bulloch, K., Chambers, W. H., Dhabhar, F. S., Goldfarb, R. H., Kitson, R. P., Miller, A. H., Spencer, R. L. & Weiss, J. M. 1997. The role of adrenalcorticoids as modulators of immune function in health and disease: Neural, endocrine, and immune interactions. *Brain Res. Rev.*, **23,** 79–133.

McShea, W. J. 1990. Social tolerance and proximate mechanisms of dispersal among winter groups of meadow voles (*Microtus pennsylvanicus*). *Anim. Behav.*, **39,** 346–51.

McWhirter, W. R. & Dobson, C. 1995. Childhood melanoma in Australia. *World J. Surg.*, **19,** 334–6.

Mealy, K., Robinson, B., Millette, C. F., Majzoub, J. & Wilmore, D. W. 1990. The testicular effects of tumor necrosis factor. *Ann. Surg.*, **211,** 470–5.

Meehan, R., Whitson, P. & Sams, C. 1993. The role of psychoneuroendocrine factors on space-flight-induced immunological alterations. *J. Leukoc. Biol.*, **54,** 236–44.

Meehan, R. T., Neale, L. S., Krause, E. T., Stuart, C. A., Smith, M. L., Cintron, N. M. & Sams, C. F. 1992. Alteration in human mononuclear leucocytes following space flight. *Immunology*, **76,** 491–7.

Meikle, A. W., Dorchuck, R. W., Araneo, B. A., Stringham, J. D., Evans, T. G., Spruance, S. L. & Daynes, R. A. 1992. The presence of a dehydroepiandrosterone-specific receptor binding complex in murine T cells. *J. Steroid Biochem. Mol. Biol.*, **42,** 293–304.

Melnikov, O. F., Nikolsky, I. S., Dugovskaya, L. A., Balitskaya, N.A. & Krabvchuk, G. P. 1987. Seasonal aspects of immunological reactivity of the human and animal organism. *J. Hyg. Microbiol. Immunol.*, **31,** 225–30.

Menaker, M., Moreira, L. F. & Tosini, G. 1997. Evolution of circadian organization in vertebrates. *Braz. J. Med. Biol. Res.*, **30,** 305–13.

Mendonca, M. T. & Licht, P. 1986. Seasonal cycles in gonadal activity and plasma gonadotropin in the musk turtle, *Sternotherus odoratus. Gen. Comp. Endocrinol.*, **62,** 459–69.

Merino, S., Moller, A. P. & de Lope, F. 2000. Seasonal changes in cell-mediated immunocompetence and mass gain in nestling barn swallows: A parasite mediated effect? *Oikos*, **90,** 327–32.

Miller, E. S., Klinger, J. C., Akin, C., Koebel, D. A. & Sonnenfeld, G. 1994. Inhibition of murine splenic T lymphocyte proliferation by 2-deoxy-D-glucose-induced metabolic stress. *J. Neuroimmunol.*, **52,** 165–73.

Miller, H. 1992. Respiratory syncytial virus and the use of ribavirin. *MCN Am. J. Matern. Child Nurs.*, **17,** 238–41.

Miura, T., Kawana, H., Karita, K., Ikegami, T. & Katayama, T. 1992. Effect of birth season on case fatality rate of pulmonary tuberculosis: Coincidence or causality? *Tubercle Lung Dis.*, **73,** 291–4.

Mizutani, H., Engelman, R. W., Kurata, Y., Ikehara, S. & Good, R. A. 1994. Energy restriction prevents and reverses immune thrombocytopenic purpura (ITP) and increases life span of ITP-prone (NZW x BXSB) F1 mice. *J. Nutr.*, **124**, 2016–23.

Moffatt, C. A., DeVries, A. C. & Nelson, R. J. 1993. Winter adaptations of male deer mice (*Peromyscus maniculatus*) and prairie voles (*Microtus ochrogaster*) that vary in reproductive responsiveness to photoperiod. *J. Biol. Rhythms*, **8**, 221–32.

Møller, A. P., Christe, P. & Lux, E. 1999. Parasitism, host immune function and sexual selection: A meta-analysis of parasite-mediated sexual selection. *Q. Rev. Biol.*, **74**, 3–20.

Møller, A. P., Dufva, R. & Erritzoc, J. 1998. Host immune function and sexual selection in birds. *J. Evol. Biol.*, **11**, 703–19.

Møller, P., Knudsen, L. E., Frentz, G., Dybdahl, M., Wallin, H. & Nexo, B. A. 1998. Seasonal variation of DNA damage and repair in patients with non-melanoma skin cancer and referents with and without psoriasis. *Mutat. Res.*, **407**, 25–34.

Monjan, A. A. 1981. Stress & immunologic competence: Studies in animals. In: *Psychoneuroimmunology* (ed. R. Ader), pp. 185–228. New York: Academic Press.

Montgomery, W. P., Young, R. C., Allen, M. P. & Harden, K. A. 1968. The tuberculin test in pregnancy. *Am. J. Obstet. Gynecol.*, **100**, 829–31.

Monto, A. S. 1994. The common cold: Cold water on hot news. *J. Am. Med. Assoc.*, **271**, 1122–4.

Monto, A. S., Cavallaro, J. J. & Keller, J. B. 1970. Seasonal patterns of acute infection. *Arch. Environ. Health*, **21**, 408–17.

Moore, R. Y. & Eichler, V. B. 1972. Loss of a circadian adrenal corticosterone rhythm following suprachiasmatic lesions in the rat. *Brain Res.*, **42**, 201–6.

Moore-Ede, M. C., Sulzman, F. M. & Fuller, C. A. 1982. *The Clocks That Time Us*. Cambridge, MA: Harvard University Press.

Morgan, A. H. & Grierson, M. C. 1930. The effects of thymectomy on young fowls. *Anat. Rec.*, **47**, 101–17.

Morimoto, C. 1978. Loss of suppressor T-lymphocyte function in patients with systemic lupus erythematosus SLE. *Clin. Exp. Immunol.*, **32**, 125–33.

Mosmann, T. R. & Sad, S. 1996. The expanding universe of T-cell subsets: Th1, Th2 and more. *Immunol. Today*, **17**, 138–46.

Moustschen, M. P., Scheen, A. J. & Lefebre, P. J. 1992. Impaired immune responses in diabetes mellitus: Analysis of the factors and mechanisms involved. Relevance to the increased susceptibility of diabetic patients to specific infections. *Diabetes Metabo.*, **18**, 187–201.

Mucha, S., Zylinska, K., Zerek-Meten, G., Swietoslawski, J. & Stepien, H. 1994. Effect of interleukin-1 on in vivo melatonin secretion by the pineal gland in rats. *Adv. Pineal Res.*, **7**, 177–81.

Muller, H. K., Lugg, D. J. & Quinn, D. 1995a. Cell mediated immunity in Antarctic wintering personnel; 1984–1992. *Immunol. Cell Biol.*, **73**, 316–20.

Muller, H. K., Lugg, D. J., Ursin, H., Quinn, D. & Donovan, K. 1995b. Immune responses during an Antarctic summer. *Pathology*, **27**, 186–90.

Mulligan, K., Tai, V. W. & Schambelan, M. 1999. Use of growth hormone and other anabolic agents in AIDS wasting. *J. Parenter. Enter. Nutr.*, **23**, S202–9.

Munck, A. & Guyre, P. M. 1986. Glucocorticoid physiology, pharmacology and stress. *Adv. Exp. Med. Biol.*, **196**, 81–96.

Murata, T. & Ying, S. Y. 1991. Effects of interleukin-1 beta on secretion of follicle–stimulating hormone FSH and luteinizing hormone LH by cultured rat anterior pituitary cells. *Life Sci.*, **49**, 447–53.

Muscettola, M., Grasso, G., Fanetti, G., Tanganelli, C. & Borghesi-Nicoletti, C. 1992. Melatonin and modulation of interferon-gamma production. *Neuropeptides*, **22**, 46.

Muscettola, M., Tanganelli, C. & Grasso, G. 1994. Melatonin modulation of interferon-production by human peripheral blood mononuclear cells. *Adv. Pineal Res.*, **7**, 119–23.

Nagy, E., Berczi, I., Wren, G. E., Asa, S. L. & Kovacs, K. 1983. Immunomodulation by bromocriptine. *Immunopharmacology*, **6**, 231–43.

Naitoh, Y., Fukata, J., Tominaga, T., Nakai, Y., Tamai, S., Mori, K. and Imura, H. 1988. Interleukin-6 stimulates the secretion of adrenocorticotropic hormone in conscious, freely-moving rats. *Biochem. Biophys. Res. Commun.*, **155**, 1459–63.

Nakanishi, T. 1986. Seasonal changes in the humoral response and the lymphoid tissues of the marine teleost, *Sebasticus marmoratu. Vet. Immunol. Immunopathol.*, **12**, 213–21.

Nakano, K., Suzuki, S. & Oh, C. 1987. Significance of increased secretion of glucocorticoids in mice and rats injected with bacterial endotoxin. *Brain Behav. Immun.*, **1**, 159–72.

Nanda, K. K. & Hamner, K. C. 1958. Studies on the nature of the endogenous rhythm affecting photoperiodic response of Biloxi soybean. *Botan. Gazette*, **120**, 14–28.

Nazian, S. & Piacsek, B. E. 1977. Maturation of the reproductive system in male rats raised at low ambient temperature. *Biol. Reprod.*, **17**, 668–75.

Negus, N. C. & Berger, P. J. 1987. Mammalian reproductive physiology. In: *Current Mammalogy* (ed. H. H. Genoways), pp. 149–73. New York: Plenum Press.

Neidhart, M. & Larson, D. F. 1990. Freund's complete adjuvant induces ornithine decarboxylase activity in the central nervous system of male rats and triggers the release of pituitary hormones. *J. Neuroimmunol.*, **26**, 97–105.

Neill, J. D., Nagy, G. M. 1994. Prolactin secretion and its control. In: Knobil, E., Neill, J. D. *The Physiology of Reproduction.* 2nd ed. New York: Raven Press, 1833–60.

Nejad, A. G. & Mohagheghpour, N. 1975. Cellular immunity in malnourished mice. *Nutr. Metab.*, **19**, 158–60.

Nelson, R. J. 1987. Photoperiod-nonresponsive morphs: A possible variable in microtine population density fluctuations. *Am. Naturalist*, **130**, 350–69.

Nelson, R. J. 1990. Photoperiodic responsiveness in laboratory house mice. *Physiol. Behav.*, **48**, 403–8.

Nelson, R. J. 1993. The effects of simulated drought on reproductive function of deer mice. *Physiol. Zool.*, **66**, 99–114.

Nelson, R. J. & Blom, J. M. C. 1993. 6-Methoxy-2-benzoxazolinone and photoperiod: Prenatal and postnatal influences on reproductive development in prairie voles (*Microtus ochrogaster*). *Can. J. Zool.*, **71**, 771–89.

Nelson, R. J. & Blom, J. M. 1994. Photoperiodic effects on tumor development and immune function. *J. Biol. Rhythms*, **9**, 233–49.

Nelson, R. J., Dark, J. & Zucker, I. 1983. Influence of photoperiod, nutrition and water availability on reproduction of male California voles (*Microtus californicus*). *J. Reprod. Fertil.*, **69**, 673–77.

Nelson, R. J. & Demas, G. E. 1996. Seasonal changes in immune function. *Q. Rev. Biol.*, **71,** 511–48.

Nelson, R. J., Demas, G. E., Klein, S. L. & Kriegsfeld, L. J. 1995a. The influence of season, photoperiod, and pineal melatonin on immune function. *J. Pineal Res.*, **19,** 149–65.

Nelson, R. J. & Desjardins, C. 1987. Water availability affects reproduction in deer mice. *Biol. Reprod.*, **37,** 257–60.

Nelson, R. J. & Drazen, D. L. 1999. Melatonin mediates seasonal adjustments in immune function. *Reprod. Nutr. Dev.*, **39,** 383–98.

Nelson, R. J., Fine, J. B., Demas, G. E. & Moffatt, C. A. 1996. Photoperiod and population density interact to affect reproductive and immune function in male prairie voles. *Am. J. Physiol.*, **270,** R571–7.

Nelson, R. J., Frank, D., Smale, L. & Willoughby, S. B. 1989. Photoperiod and temperature affect reproductive and nonreproductive functions in male prairie voles (*Microtus ochrogaster*). *Biol. Reprod.*, **40,** 481–5.

Nelson, R. J., Gubernick, D. J. & Blom, J. M. C. 1995b. Influences of photoperiod, green food, and water intake on reproduction in male California mice (*Peromyscus californicus*). *Physiol. Behav.*, **57,** 1175–80.

Nelson, R. J., Kita, M., Blom, J. M. C. & Rhyne-Grey, J. 1992. Photoperiod influences the critical caloric intake necessary to maintain reproduction among male deer mice (*Peromyscus maniculatus*). *Biol. Reprod.*, **46,** 226–32.

Nelson, R. J. & Klein, S. L. 1999. Immune function, mating systems, and seasonal breeding. In: *Reproduction in Context.* Edited by K. Wallen & J. E. Schneider, Cambridge, MA: MIT Press. pp. 219–56.

Nelson, R. J., Moffatt, C. A. & Goldman, B. D. 1994. Reproductive and nonreproductive responsiveness to photoperiod in laboratory rats. *J. Pineal Res.*, **17,** 123–31.

Neri, B., Fiorelli, C., Moroni, F., Nicita, G., Paoletti, M. C., Ponchietti, R., Raugei, A., Santoni, G., Trippitelli, A. & Grechi, G. 1994. Modulation of human lymphoblastoid interferon activity by melatonin in metastatic renal cell carcinoma. A phase II study. *Cancer*, **73,** 3015–9.

Nesse, R. & Williams, G. 1994. *Why We Get Sick: The New Science of Darwinian Medicine.* New York: First Vintage Books.

Newsholme, E. A., Calder, P. & Yaqoob, P. 1993. The regulatory, informational, and immunoimmodulatory roles of fat fuels. *Am. J. Clin. Nutr.*, **57**(Suppl.):738S–50S.

Newson, J. 1962. Seasonal differences in reticulocyte count, hemoglobin levels and spleen weight in wild voles. *Br. J. Haematol.*, **8,** 296–302.

Nieman, D. C., Johanssen, L. M., Lee, J. W. & Arabatzis, K. 1990. Infectious episodes in runners before and after the Los Angeles Marathon. *J. Sports Med. Phys. Fitness*, **30,** 316–28.

Nieman, D. C., Nehlsen-Cannarella, S. I., Henson, D. A., Butterworth, D. E., Fagoaga, O. R., Warren, B. J. & Rainwater, M. K. 1996. Immune response to obesity and moderate weight loss. *Int. J. Obesity Relat. Metab. Disord.*, **20,** 353–60.

Nieman, R. E. & Lorber, B. 1980. Listeriosis in adults: a changing pattern. Report of eight cases and review of the literature, 1968–1978. *Rev. Infect. Dis.*, **2,** 207–27.

Nordby, G. L. & Cassidy, J. T. 1983. Seasonal effect on the variability of summer immunoglobulin levels. *Hum. Biol.*, **55,** 797–809.

Nordling, D., Andersson, M., Zohari, S. & Gustafsson, L. 1998. Reproductive effort reduces specific immune response and parasite resistance. *Proc. R. Soc. Lond. B*, **265**, 1291–8.

Norris, K., Anwar, M. & Read, A. F. 1994. Reproductive effort influences the prevalence of parasites in great tits. *J. Anim. Ecol.*, **63**, 601–10.

Nussdorfer, G. G. & Mazzocchi, G. 1998. Immune-endocrine interactions in the mammalian adrenal gland: Facts and hypotheses. *Int. Rev. Cytol.*, **183**, 143–84.

Oakson, B. B. 1953. Cyclic changes in liver and spleen weights in migratory white-crowned sparrows. *Condor*, **55**, 3–16.

Oakson, B. B. 1956. Liver and spleen weights in migratory white-crowned sparrows. *Condor*, **58**, 3–16.

Odum, E. P. 1955. An eleven year history of a *Sigmodon* population. *J. Mammal.*, **36**, 368–78.

Officer, C. & Page, J. 1993. *Tales of the Earth*. New York: Oxford University Press.

Ogura, M., Ogura, H., Lorenz, E., Ikehara, S. & Good, R. A. 1990. Undernutrition without malnutrition restricts the numbers and proportions of Ly-1 B lymphocytes in autoimmune (MRL/I and BXSB) mice. *Proc. Soc. Exp. Biol. Med.*, **193**, 6–12.

O'Leary, A. 1990. Stress, emotion, and human immune function. *Psychol. Bull.*, **108**, 363–82.

Olsen, N. J., Nicholson, W. E., DeBold, C. R. & Orth, D. N. 1992. Lymphocyte-derived adrenocorticotropin is insufficient to stimulate adrenal steroidogenesis in hypophysectomized rats, *Endocrinology*, **130**, 2113–9.

Olsen, N. J. & W. J. Kovacs. 1996. Gonadal steroids and immunity. *Endocr. Rev.*, **17**, 369–84.

Olsen, Y. A., Reitan, I. J. & Roed, K. H. 1993. Gill Na+, K+ ATPase activity, plasma cortisol level, and non-specific immune response in Atlantic salmon (*Salmo salar*) during parrsmolt transformation. *J. Fish. Biol.*, **43**, 559–73.

Olson, D. P. & Bull, R. C. 1986. Antibody responses in protein-energy restricted beef cows and their cold stressed progeny. *Can. J. Vet. Res.*, **50**, 410–7.

Omenaas, E., Bakke, P., Haukenes, G., Hanoa, R. & Gulsvik, A. 1995. Respiratory virus antibodies in adults of a Norwegian community: Prevalence and risk factors. *Int. J. Epidemiol.*, **24**, 223–31.

Orava, M., Cantell, K., Kauppila, A. & Vihko, R. 1983. Interferon and serum thyroid hormones. *Int. J. Cancer*, **31**, 671–2.

Orchinik, M., Licht, P. & Crews, D. 1988. Plasma steroid concentrations change in response to sexual behavior in *Bufo marinus*. *Hormones Behav.*, **22**, 338–50.

Ortaldo, J. R. & Heberman, R. B. 1984. Heterogeneity of natural killer cells. *Ann. Rev. Immunol.*, **2**, 359–77.

Ortega, C., Muzquiz, J. L., Docando, J., Planas, E., Alonso, J. L. & Simon, M. C. 1995. Ecopathology in aquaculture: Risk factors in infectious disease outbreak. *Vet. Res.*, **26**, 57–62.

Ostensen, M., Lundgren, R., Husby, G. & Rekvig, O. P. 1983. Studies on humoral immunity in pregnancy: immunoglobulins, alloantibodies and autoantibodies in healthy pregnant women and in pregnant women with rheumatoid disease. *J. Clin. Lab. Immunol.*, **11**, 143–7.

Ottaway, C. A. & Husband, A. J. 1992. Central nervous system influences on lymphocyte migration. *Brain Behav. Immunol.*, **6**, 97–116.

Ottaway, C. A. & Husband, A. J. 1994. The influence of neuroendocrine pathways on lymphocyte migration. *Iummunol. Today*, **15**, 511–7.

Ovington, K. S. 1985. Dose-dependent relationships between *Nippostrongylus brasiliensis* populations and rat food intake. *Parasitology*, **91**, 157–67.

Ownby, H. E., Frederick, J., Mortensen, R. F., Ownby, D. R., Russo, J. and the Breast Cancer Prognostic Study Association. 1986. Seasonal variation in tumor size at diagnosis and immunologic responses in human breast cancer. *Invas. Metast.*, **6**, 246–56.

Ozkan, H., Olgun, N., Sasmaz, E., Abacioglu, H., Okuyan, M. & Cevik, N. 1993. Nutrition, immunity and infections: T lymphocyte subpopulations in protein-energy malnutrition. *J. Trop. Pediatr.*, **39**, 257–60.

Paavonen, T. 1982. Enhancement of human B lymphocyte differentiation in vitro by thyroid hormone. *Scand. J. Immunol.*, **15**, 211–5.

Paavonen, T., Andersson, L. C. & Adlercreutz, H. 1981. Sex hormone regulation of in vitro immune response. Estradiol enhances human B cell maturation via inhibition of suppressor T cells in pokeweed mitogen-stimulated cultures. *J. Exp. Med.*, **154**, 1935–45.

Padgett, D. A., Loria, R. M. & Sheridan, J. F. 1997. Endocrine regulation of the immune response to influenza virus infection with a metabolite of DHEA-androstenediol. *J. Neuroimmunol.*, **78**, 203–11.

Paglieroni, T. G. & Holland, P. V. 1994. Circannual variation in lymphocyte subsets, revisited. *Transfusion*, **34**, 512–6.

Pang, C. S., Brown, G. M., Tang, P. L., Cheng, K. M. & Pang, S. F. 1993. $2[^{125}I]$iodomelatonin binding sites in the lung and heart: A link between the photoperiodic signal, melatonin, and the cardiopulmonary system. *Biol. Signals*, **2**, 228–36.

Pang, C. S. & Pang, S. F. 1992. High affinity specific binding of $2[^{125}I]$iodomelatonin by spleen membrane preparations of chicken. *J. Pineal Res.*, **12**, 167–73.

Path, G., Bornstein, S. R., Ehrhart-Bornstein, M. & Scherbaum, W. A. 1997. Interleukin-6 and the interleukin-6 receptor in the human adrenal gland: Expression and effects on steroidogenesis. *J. Clin. Endocrinol. Metab.*, **82**, 2343–9.

Pati, A. K., Florentin, I., Chung, V., De Sousa, M., Levi, F. & Mathe, G. 1987. Circannual rhythm in natural killer cell activity and mitogen responsiveness of murine splenocytes. *Cell. Immunol.*, **108**, 227–34.

Payne, L. C., Obal, F., Jr., Opp, M. R. & Krueger, J. M. 1992. Stimulation and inhibition of growth hormone secretion by interleukin-1 beta: The involvement of growth hormone-releasing hormone. *Neuroendocrinology*, **56**, 118–23.

Peisen, J. N., McDonnell, K. J., Mulroney, S. E. & Lumpkin, M. D. 1995. Endotoxin-induced suppression of the somatotropic axis is mediated by interleukin-1 beta and corticotropin-releasing factor in the juvenile rat. *Endocrinology*, **136**, 3378–90.

Pekonen, F. & Weintraub, B. D. 1978. Thyrotropin binding to cultured lymphocytes and thyroid cells. *Endocrinology*, **103**, 1668–77.

Petrovsky, N. & Harrison, L. C. 1998. The chronobiology of human cytokine production. *Int. Rev. Immunol.*, **16**, 635–49.

Pickard, R. E. 1968. Varicella pneumonia in pregnancy. *Am. J. Obstet. Gynecol.*, **101**, 504–8.

Pierpaoli, W. & Besedovsky, H. O. 1975. Role of the thymus in programming of neuroendocrine functions. *Clin. Exp. Immunol.*, **20**, 323–8.

Pierpaoli, W., Fabris, N. & Sorkin, E. 1970. Developmental hormones and immuno-logical maturation. *Ciba Found. Study Group*, **36**, 126–53.

Pierpaoli, W. & Sorkin, E. 1972. Alterations of adrenal cortex and thyroid in mice with congenital absence of the thymus. *Nature New Biol.*, **238**, 282–5.

Pietinalho, A., Hiraga, Y., Hosoda, Y., Lofroos, A. B., Yamaguchi, M. & Selroos, O. 1995. The frequency of sarcoidosis in Finland and Hokkaido, Japan. A comparative epidemiological study. *Sarcoidosis*, **12**, 61–7.

Pioli, C., Carleo, C., Nistico, G. & Doria, G. 1993. Melatonin increases antigen presentation and amplifies specific and non-specific signals for T-cell proliferation. *Int. J. Immunopharmacol.*, **15**, 463–8.

Pitson, G. A., Lugg, D. J. & Muller, H. K. 1996. Seasonal cutaneous immune responses in an Antarctic wintering group: No association with testosterone, vitamin D metabolite or anxiety score. *Arctic Med. Res.*, **55**, 118–22.

Pittendrigh, C. S. 1960. Circadian rhythms and the circadian organization of living systems. *Cold Spring Harbor Symp. Quant. Biol.*, **25**, 159–82.

Pittendrigh, C. S. & Daan, S. 1976. A functional analysis of circadian pacemakers in nocturnal rodents. I. The stability and lability of spontaneous frequency. *J. Comp. Physiol.*, **106**, 223–52.

Pittendrigh, C. S. & Minis, D. H. 1964. The entrainment of circadian oscillations by light and their role as photoperiodic clocks. *Am. Naturalist*, **98**, 261–99.

Planelles, D., Hernandez-Godoy, J., Montoro, A., Montoro, J. & Gonzalez-Molina, A. 1994. Seasonal variation in proliferative response and subpopulations of lymphocytes from ice housed in a constant environment. *Cell Prolif.*, **27**, 333–41.

Plaut, M. 1987. Lymphocyte hormone receptors. *Annu. Rev. Immunol.*, **5**, 621–69.

Plaut, S. M., Ader, R., Friedman, S. B. & Ritterson, A. L. 1969. Social factors and resistance to malaria in the mouse: Effects of group versus individual housing on resistance to *Plasmodium berghei* infection. *Psychosom. Med.*, **31**, 536–52.

Plytycz, B., Mika, J. & Bigaj, J. 1995. Age-dependent changes in thymuses in the European common frog, *Rana temporaria*. *J. Exp. Zool.*, **273**, 451–60.

Plytycz, B. & Seljelid, R. 1997. Rhythms in immunity. *Arch. Immunol. Ther. Exp.*, **45**, 157–62.

Pollettt, M., Mackenzie, J. S. & Turner, K. J. 1979. The effect of protein-deprivation on the susceptibility to influenza virus infection: A murine model system. *Aust. J. Exp. Biol. Med. Sci.*, **57**, 151–60.

Poon, A. M. & Pang, S. F. 1992. [^{125}I]Iodomelatonin binding sites in spleens of guinea pigs. *Life Sci.*, **50**, 1719–26.

Poulin, R. 1996. Sex inequalities in helminth infections: A cost of being male? *Am. Nat.*, **147**, 287–95.

Prendergast, B. J., Freeman, D. A., Zucker, I. & Nelson, R. J. 2001. The mystery of periodic arousal: Impaired immune function in hibernating, but not euthermic, ground squirrels. *Am. J. Physiol.* (In review).

Prendergast, B. J., Kriegsfeld, L. J. & Nelson, R. J. 2001. Photoperiodic polyphenisms in rodents: Neuroendocrine mechanisms, costs and functions. *Q. Rev. Biol.* (In press.)

Prendergast, B. J., Yellon, S. M., Tran, L. T. & Nelson, R. J. 2001. Inhibition of cellular immune function by *in vitro* melatonin is modulated by photoperiodic history in Siberian hamsters. *J. Biol. Rhythms*, **16**, 224–33.

Prener, A. & Carstensen, B. 1990. Month of birth and testicular cancer risk in Denmark, *Am. J. Epidemiol.*, **131**, 15–9.

Prickett, M. D., Latimer, A. M., McCusker, R. H., Hausman, G. J. & Prestwood, A. K. 1992. Alterations of serum insulin-like growth factor-I (IGF-I) and IGF-binding proteins (IGFBPs) in swine infected with the protozoan parasite *Sarcocystis miescheriana. Domest. Anim. Endocrinol.*, **9**, 285–96.

Provinciali, M., Di Stefano, G. & Fabris, N. 1992. Improvement in the proliferative capacity and natural killer cell activity of murine spleen lymphocytes by thyrotropin. *Int. J. Immunopharmacol.*, **14**, 865–70.

Puchalski, W. & Lynch, G. R. 1988. Characterization of circadian function in Djungarian hamsters insensitive to short day photoperiod. *J. Comp. Physiol. A*, **162**, 309–16.

Qazzaz, S. T., Mamattah, J. H., Ashcroft, T. & McFarlane, H. 1981. The development and nature of immune deficit in primates in response to malnutrition. *Br. J. Exp. Pathol.*, **62**, 452–60.

Raber, J., Sorg, O., Horn, T. F. W., Yu, N., Koob, G. F., Campbell, I. L. & Bloom, F. E. 1998. Inflammatory cytokines: Putative regulators of neuronal and neuro-endocrine function. *Brain Res. Rev.*, **26**, 320–6.

Råberg, L., Grahn, M., Hasselquist, D. & Svensson, E. 1998. On the adaptive significance of stress-induced immunosuppression. *Proc. R. Soc. Lond., B*, **265**, 1637–41.

Rabin, B. S., Lyte, M. & Hamill, E. 1987. The influence of mouse strain and housing on the immune response. *J. Neuroimmunol.*, **17**, 11–6.

Rabin, B. S. & Salvin, S. B. 1987. Effect of differential housing and time on immune reactivity to sheep erythrocytes and *Candida. Brain Behav. Immun.*, **1**, 267–75.

Rafii-El-Idissi, M., Calvo, J. T., Pozo, D., Hamouch, A. & Guerrero, J. M. 1995. Specific binding of $2[^{125}I]$iodomelatonin by rat splenocytes: Characterization and its role on regulation of cycle AMP production. *J. Neuroimmunol.*, **57**, 171–8.

Ralph, M. R., Foster, R. G., Davis, F. C. & Menaker, M. 1990. Transplanted suprachiasmatic nucleus determines circadian period. *Science*, **247**, 975–8.

Ramachandra, R. N., Sehon, A. H. & Berczi, I. 1992. Neuro-hormonal host defense in endotoxic shock. *Brain Behav. Immun.*, **6**, 157–69.

Rani, R. & Mittal, K. K. 1989. Immunologic and neuroendocrine interactions in pregnancy. *Indian J. Pediatr.*, **56**, 181–7.

Rao, L. V., Cleveland, R. P. & Ataya, K. M. 1993. GnRH agonist induces suppression of lymphocyte subpopulations in secondary lymphoid tissues of prepubertal female mice. *Am. J. Reprod. Immunol.*, **30**, 15–25.

Raschka, C., Schorr, W. & Koch, H. J. 1999. Is there seasonal periodicity in the prevalence of *Helicobacter pylori? Chronobiol. Int.*, **16**, 811–9.

Rasmussen, K. R., Healey, M. C., Cheng, L. & Yang, S. 1995. Effects of dehydro-epiandrosterone in immunosuppressed adult mice infected with *Cryptosporidium parvum. J. Parasitol.*, **81**, 429–33.

Ratajczak, H. V., Thomas, P. T., Sothern, R. B., Vollmuth, T. & Heck, J. D. 1993. Evidence for genetic basis of seasonal differences in antibody formation between two mouse strains. *Chronobiol. Int.*, **10**, 383–94.

Rautskis, E. 1970. Seasonal variation in parasite fauna of perch in Lake Dusia. *Acta Parasitol. Lith.*, **10**, 123–8.

Raveche, E. S., Vigersky, R. A., Rice, M. K. & Steinberg, A. D. 1980. Murine thymic androgen receptors. *J. Immunopharmacol.*, **2**, 425–34.

Redmond, H. P., Gallagher, H. J., Shou, J. & Daly, J. M. 1995. Antigen presentation in protein-energy malnutrition. *Cell. Immunol.*, **163**, 80–7.

Redmond, H. P., Shou, J., Kelly, C. J., Leon, P. & Daly, J. M. 1991. Protein-calorie malnutrition impairs host defense against *Candida albicans*. *J. Surg. Resid.*, **50**, 552–9.

Reichlin, S. 1993. Neuroendocrine-immune interactions. *N. Engl. J. Med.*, **329**, 1246–53.

Reinberg, A., Schuller, E., Delasnerie, N., Clench, J. & Helary, M. 1977. Rythmes circadiens et circannuels des leucocytes, proteines totales, immunoglobulines A, G et M; etude chez 9 adultes jeunes et sains. *Nouv. Presse Med.*, **6**, 3819–23.

Reiter, R. J. 1992. The ageing pineal gland and its physiological consequences. *Bioessays*, **14**, 169–75.

Reiter, R. J. 1993. The melatonin rhythm: Both a clock and a calendar. *Experientia*, **49**, 654–64.

Reiter, R. J. 1998. Melatonin and human reproduction. *Ann. Med.*, **30**, 103–8.

Reiter, R. J., Poeggeler, B., Tan, D. X., Chen, L. D., Manchester, L. C. & Guerrero, J. M. 1993. Antioxidant capacity of melatonin: A novel action not requiring a receptor. *Neuroendocrinol. Lett.*, **15**, 103–11.

Reppert, S. M. & Weaver, D. R. 1995. Melatonin madness. *Cell*, **83**, 1059–62.

Rettori, V., Gimeno, M. F., Karara, A., Gonzalez, M. C. & McCann, S. M. 1991. Interleukin 1 alpha inhibits prostaglandin E2 release to suppress pulsatile release of luteinizing hormone but not follicle-stimulating hormone. *Proc. Natl. Acad. Sci. U.S.A.*, **88**, 2763–7.

Rettori, V., Jurcovicova, J. & McCann, S. M. 1987. Central action of interleukin-1 in altering the release of TSH, growth hormone, and prolactin in the male rat. *J. Neurosci. Res.*, **18**, 179–83.

Rhynes, W. E. & Ewing, L. L. 1973. Testicular endocrine function in Hereford bulls exposed to high ambient temperature. *Endocrinology*, **92**, 509–15.

Richner, H., Christe, P. & Oppliger, A. 1995. Paternal investment affects prevalence of malaria. *Proc. Natl. Acad. Sci. U.S.A.*, **92**, 1192–4.

Riddle, O. 1924. Studies on the physiology of reproduction in birds. XIX. A hitherto unknown function of the thymus. *Am. J. Physiol.*, **68**, 557–60.

Riddle, O. 1928. Sex and seasonal differences in weight of liver and spleen. *Proc. Soc. Exp. Biol. Med.*, **25**, 474–6.

Riddle, O. & Krizenecky, N. 1931. Studies on the physiology of reproduction in birds. XXVIII. Extirpation of thymus and bursa in pigeons with a consideration of the failure of thymectomy to reveal thymus function. *Am. J. Physiol.*, **97**, 343–52.

Rifkind, D. & Frey, J. A. 1972. Sex differences in antibody response of CFW mice to *Candida albicans*. *Infect. Immun.*, **5**, 695–8.

Riley, E. M., Morris-Jones, S., Blackman, M. J., Greenwood, B. M. & Holder, A. A. 1993a. A longitudinal study of naturally acquired cellular and humoral immune responses to a merozoite surface protein (MSP1) of *Plasmodium falciparum* in an area of seasonal malaria transmission. *Parasite Immunol.*, **15**, 513–24.

Riley, E. M., Morris-Jones, S., Taylor-Robinson, A. W. & Holder, A. A. 1993b. Lymphoproliferative responses to a merozoite surface antigen of *Plasmodium falciparum*: Preliminary evidence for seasonal activation of CD8[+]/HLA-DQ-restricted suppressor cells. *Clin. Exp. Immunol.*, **94**, 64–7.

Rivier, C. & Rivest, S. 1991. Effects of stress on the activity of the hypothalamic-pituitary-gonadal axis: Peripheral and central mechanisms. *Biol. Reprod.*, **45**, 523–32.

Roberts, C. W., Satoskar, A. & Alexander, J. 1996. Sex steroids, pregnancy-associated hormones and immunity to parasitic infection. *Parasitol. Today*, **12**, 382–8.

Roberts-Thomson, P. J., Lugg, D., Vallverdu, R. & Bradley, J. 1985. Assessment of immunological responsiveness in members of the International Biomedical Expedition to the Antarctic 1980/81. *J. Clin. Lab. Immunol.*, **17**, 115–8.

Robertson, O. H. & Wexler, B. C. 1959. Hyperplasia of the adrenal cortical tissue in pacific salmon and rainbow trout accompanying sexual maturation and spawning. *Endocrinology*, **65**, 225–38.

Roby, K. F. & Terranova, P. F. 1990. Effects of tumor necrosis factor-alpha in vitro on steroidogenesis of healthy and atretic follicles of the rat: Theca as a target. *Endocrinology*, **126**, 2711–818.

Roenneberg, T. & Aschoff, J. 1990. Annual rhythm of human reproduction. I. Biology, sociology, or both? *J. Biol. Rhythms*, **5**, 195–216.

Rogot, E. & Padgett, S. J. 1976. Associations of coronary and stroke mortality with temperature and snowfall in selected areas of the United States. *Am. J. Epidemiol.*, **103**, 565–75.

Romero, L. M., Ramenofsky, M. & Wingfield, J. C. 1997. Season and migration alters the corticosterone response to capture and handling in an arctic migrant, the white-crowned sparrow (*Zonotrichia leucophrys gambelii*). *Comp. Biochem. Physiol. C*, **116**, 171–7.

Romero, L. M., Soma, K. K., O'Reilly, K. M., Suydam, R. & Wingfield, J. C. 1998a. Hormones and territorial behavior during breeding in snow buntings (*Plectrophenax nivalis*): An arctic-breeding songbird. *Horm. Behav.*, **33**, 40–7.

Romero, L. M., Soma, K. K. & Wingfield, J. C. 1998b. Hypothalamic-pituitary-adrenal axis changes allow seasonal modulation of corticosterone in a bird. *Am. J. Physiol.*, **274**, R1338–44.

Romero, L. M., Soma, K. K. & Wingfield, J. C. 1998c. The hypothalamus and adrenal regulate modulation of corticosterone release in redpolls (*Carduelis flammea*): An arctic-breeding song bird. *Gen. Comp. Endocrinol.*, **109**, 347–55.

Romero, L. M., Soma, K. K. & Wingfield, J. C. 1998d. Changes in pituitary and adrenal sensitivities allow the snow bunting (*Plectrophenax nivalis*), an arctic-breeding song bird, to modulate corticosterone release seasonally. *J. Comp. Physiol. B*, **168**, 353–8.

Ropstad, E., Larsen, H. J. & Refsdal, A. O. 1989. Immune function in dairy cows related to energy balance and metabolic status in early lactation. *Acta Vet. Scand.*, **30**, 209–19.

Rose, R. 1985. *Psychoendocrinology*. In: *Williams Textbook of Endocrinology*, 7th ed. (eds. J. D. Wilson & D. Foster). pp. 653–81. Philadelphia: W. B. Saunders.

Rosen, L. N., Livingstone, I. R. & Rosenthal, N. E. 1991. Multiple sclerosis and latitude: A new perspective on an old association. *Med. Hypoth.*, **36**, 376–8.

Rosenbauer, J., Herzig, P., von Kries, R., Neu, A. & Giani, G. 1999. Temporal, seasonal, and geographical incidence patterns of type I diabetes mellitus in children under 5 years of age in Germany. *Diabetologia*, **42**, 1055–9.

Rosenberg, A. M. 1988. The clinical associations of antinuclear antibodies in juvenile rheumatoid arthritis. *Clin. Immunol. Immunopathol.*, **49**, 19–27.

Rosenfeld, R. S., Hellman, L., Roffwarg, H., Weitzman, E. D., Fukushima, D. K. & Gallagher, T. F. 1971. Dehydroisoandrosterone is secreted episodically and synchronously with cortisol by normal man. *J. Clin. Endocrinol. Metab.*, **33**, 87–92.

Ross, J. A., Severson, R. K., Swensen, A. R., Pollock, B. H., Gurney, J. G. & Robison, L. L. 1999. Seasonal variations in the diagnosis of childhood cancer in the United States. *Br. J. Cancer*, **81**, 549–53.

Roubinian, J. R., Papoian, R. & Talal, N. 1977. Androgenic hormones modulate autoantibody responses and improve survival in murine lupus. *J. Clin. Invest.*, **59**, 1066–70.

Roubinian, J. R., Talal, N., Greenspan, J. S., Goodman, J. R. & Siiteri, P. K. 1978. Effect of castration and sex hormone treatment on survival, anti-nucleic acid antibodies, and glomerulonephritis in NZB/NZW F1 mice. *J. Exp. Med.*, **147**, 1568–83.

Rowan, W. 1925. Relation of light to bird migration and developmental changes. *Nature*, **115**, 494–6.

Ruby, N. F., Nelson, R. J., Licht, P. & Zucker, I. 1993. Prolactin and testosterone inhibit torpor in Siberian hamsters. *Am. J. Physiol.*, **264**, R123–8.

Ruis, J. F., Buys, J. P., Cambras, T. & Rietveld, W. J. 1990. Effects of T cycles of light/darkness and periodic forced activity on methamphetamine-induced rhythms in intact and SCN-lesioned rats: Explanation by an hourglass-clock model. *Physiol. Behav.*, **47**, 917–29.

Russell, D. H., Kibler, R., Matrisian, L., Larson, D. F., Poulos, B. & Magun, B. E. 1985. Prolactin receptors on human T and B lymphocytes: Antagonism of prolactin binding by cyclosporine. *J. Immunol.*, **134**, 3027–31.

Saad, A. H. & El Ridi, R. 1988. Endogenous corticosteroids mediate seasonal cyclic changes in immunity of lizards. *Immunobiol.*, **177**, 390–403.

Saad, A. H., el Ridi, R., el Deeb, S. & Soliman, M. A. 1987. Corticosteroids and immune system in the lizard *Chalcides ocellatus*. *Prog. Clin. Biol. Res.*, **233**, 141–51.

Saafela, S. & Reiter, R. J. 1994. Function of melatonin in thermoregulatory processes. *Life Sci.*, **54**, 295–311.

Sadleir, R. M. F. S. 1969. *The Ecology of Reproduction in Wild and Domestic Mammals*. London: Methuen.

Saino, N., Canova, L., Fasola, M. & Martinelli, R. 2000. Reproduction and population density affect humoral immunity in bank voles under field experimental conditions. *Oecologia*, **124**, 358–66.

Saino, N. & Moller, A. P. 1997. Song and immunological condition in male barn swallows (*Hirundo rustica*). *Behav. Ecol.*, **8**, 364–71.

Sakabe, K., Seiki, K. & Fujii-Hanamoto, H. 1986. Histochemical localization of progestin receptor cells in the rat thymus. *Thymus*, **8**, 97–107.

Sakamoto, M., Yazaki, N., Katsushima, N., Mizuta, K., Suzuki, H. & Numazaki, Y. 1995. Longitudinal investigation of epidemiologic feature of adenovirus infections in acute respiratory illnesses among children in Yamagata, Japan (1986–1991). *Tohoku J. Exp. Med.*, **175**, 185–93.

Sakamoto-Momiyama, M. 1977. *Seasonality in Human Mortality*. Tokyo: University of Tokyo Press.

Salas, M. A., Evans, S. W., Levell, M. J. & Whicher, J. T. 1990. Interleukin-6 and ACTH act synergistically to stimulate the release of corticosterone from adrenal gland cells. *Clin. Exp. Immunol.*, **79**, 470–3.

Salemi, G., Ragonese, P., Aridon, P., Reggio, A., Nicoletti, A., Buffa, D., Conte, S. & Savettieri, G. 2000. Is season of birth associated with multiple sclerosis? *Acta Neurol. Scand.*, **101**, 381–3.

Salimonu, L. S., Ojo-Amaize, E., Johnson, A. O., Laditan, A. A., Akinwolere, O. A. & Wigzell, H. 1983. Depressed natural killer cell activity in children with protein-calorie malnutrition. II. Correction of the impaired activity after nutritional recovery. *Cell. Immunol.*, **82**, 210–5.

Sanders, E. H., Gardner, P. D., Berger, P. J. & Negus, N. C. 1981. 6-Methoxy-benzoxazolinone: A plant derivative that stimulates reproduction in *Microtus montanus*. *Science*, **214**, 67–9.

Sandyk, R. & Awerbuch, G. I. 1993. Multiple sclerosis: Relationship between seasonal variations of relapse and age of onset. *Int. J. Neurosci.*, **71**, 147–57.

Sankila, R., Joensuu, H., Pukkala, E. & Toikkanen, S. 1993. Does the month of diagnosis affect survival of cancer patients? *Br. J. Cancer*, **67**, 838–41.

Sano, A., Miyaji, M. & Nishimura, K. 1992. Studies on the relationship between the estrous cycle of BALB/c mice and their resistance to *Paracoccidioides brasiliensis* infection. *Mycopathologia*, **119**, 141–5.

Saphier, D. 1989. Neurophysiological and endocrine consequences of immune activity. *Psychoneuroendocrinology*, **14**, 63–87.

Sapolsky, R., Rivier, C., Yamamoto, G., Plotsky, P. & Vale, W. 1987. Interleukin-1 stimulates the secretion of hypothalamic corticotropin-releasing factor. *Science*, **238**, 522–4.

Sapolsky, R. M. 1992. *Stress, the Aging Brain, and the Mechanisms of Neuron Death.* Cambridge, MA: MIT Press.

Sapolsky, R. M. 1998. *Why Zebras Don't Get Ulcers: An Updated Guide to Stress, Stress-Related Diseases, and Coping.* San Francisco: W. H. Freeman & Sons.

Saunders, D. S. 1977. *An Introduction to Biological Rhythms.* New York: Wiley.

Sawhney, R. C., Malhotra, A. S., Prasad, R., Pal, K., Kumar, R. & Bajaj, A. C. 1998. Pituitary-gonadal hormones during prolonged residency in Antarctica. *Int. J. Biometeorol.*, **42**, 51–4.

Scaccianoce, S., Alema, S., Cigliana, G., Navarra, D., Ramacci, M. T. & Angelucci, L. 1991. Pituitary-adrenocortical and pineal activities in the aged rat. Effects of long-term treatment with acetyl-L-carnitine. *Ann. N. Y. Acad. Sci.*, **621**, 256–61.

Scanga, C. B., Verde, T. J., Paolone, A. M., Andersen, R. E. & Wadden, T. A. 1998. Effects of weight loss and exercise training on natural killer cell activity in obese women. *Med. Sci. Sports Exer.*, **30**, 1666–71.

Scarborough, D. E. 1990. Cytokine modulation of pituitary hormone secretion. *Ann. N. Y. Acad. Sci.*, **594**, 169–87.

Scarselli, E., Tolle, R., Koita, Olk Diallo, M., Müller, H., Früh, K., Doumbo, O., Crisanti, A. & Bujard, H. 1993. Analysis of the human antibody response to thrombospondin-related anonymous protein of *Plasmodium falciparum*. *Infect. Immun.*, **61**, 3490–5.

Schadler, M., Butterstein, G. N., Faulkner, B. J., Rice, S. C. & Weisinger, L. A. 1988. The plant metabolite, 6-methoxybenzoxazolinone, stimulates an increase in secretion of follicle stimulating hormone and size of reproductive organs in *Microtus pinetorum*. *Biol. Reprod.*, **38**, 817–20.

Schimpff, R. M. & Repellin, A. M. 1989. In vitro effect of human growth hormone on lymphocyte transformation and lymphocyte growth factors secretion. *Acta Endocrinol. (Copenh)*, **120**, 745–52.

Schmitt, D. A. & Schaffer, L. 1993a. Isolation and confinement as a model for spaceflight immune changes. *J. Leukoc. Biol.*, **54**, 209–13.

Schmitt, D. A. & Schaffer, L. 1993b. Confinement and immune function. *Adv. Space Biol. Med.*, **3**, 229–35.

Schneider, J. E. & Wade, G. N. 1989. Availability of metabolic fuels controls estrous cyclicity of Syrian hamsters. *Science*, **244**, 1326–8.

Schneider, J. E. & Wade, G. N. 1999. Inhibition of reproduction in service of energy balance. In: *Reproduction in Context* (eds. K. Wallen & J. E. Schneider), pp. 35–82. Cambridge: MIT Press.

Schobitz, B., Holsboer, F., Kikkert, R., Sutanto, W. & DeKloet, E. R. 1992. Peripheral and central regulation of IL-6 gene expression in endotoxin-treated rats. *Endocr. Regul.*, **26**, 103–9.

Schuurs, A. H. W. M. & Verheul, H. A. M. 1990. Effects of gender and sex steroids on the immune response. *J. Steroid Biochem.*, 35, 157–72.

Schwartz, W. J. & Gainer, H. 1977. Suprachiasmatic nucleus: Use of ^{14}C-labeled deoxyglucose uptake as a functional marker. *Science*, **197**, 1089–91.

Sealander, J. A. & Bickerstaff, L. K. 1967. Seasonal changes in reticulocyte number and in relative weights of the spleen, thymus, and kidneys in the northern re-backed mouse. *Can. J. Zool.*, **45**, 253–60.

Seaman, W. E. & Gindhart, T. D. 1979. Effect of estrogen on natural killer cells. *Arthritis Rheum.*, **22**, 1234–40.

Seaman, W. E., Merigan, T. C. & Talal, N. 1979. Natural killing in estrogen-treated mice responds poorly to polyIC despite normal stimulation of circulating interferon. *J. Immunol.*, **123**, 2903–5.

Seed, J. R., Pinter, A. J., Ashman, P. U., Ackerman, S. & King, L. 1978. Comparison of organ weights of wild and laboratory *Microtus montanus* infected with *Trypanosoma brucei gambiense*. *Am. Midland Naturalist*, **100**, 126–34.

Sellmeyer, D. E. & Grunfeld, C. 1996. Endocrine and metabolic disturbances in human immunodeficiency virus infection and the acquired immune deficiency syndrome. *Endocr. Rev.*, **17**, 518–32.

Selye, H. 1936. Thymus and the adrenals in the responses of the organism to injuries and intoxications. *Br. J. Exp. Pathol.*, **17**, 234–48.

Selye, H. 1956. *The Stress of Life*. New York: McGraw-Hill.

Selye, H. 1979. Correlating stress and cancer. *Am. J. Proctol. Gastroenterol. Colon Rectal Surg.*, **30**, 25–8.

Serrano, C. J., Ramires, J. A., Venturinelli, M., Arie, S., D'Amico, E., Zweier, J. L., Pileggi, F. & da Luz, P. L. 1997. Coronary angioplasty results in leukocyte and platelet activation with adhesion molecule expression. Evidence of inflammatory responses in coronary angioplasty. *J. Am. Coll. Cardiol.*, **29**, 1276–83.

Shadrin, A. S., Marinich, I. G. & Taros, L. Yu. 1977. Experimental and epidemiological estimation of seasonal and climato-geographical features of non-specific resistance of the organism to influenza. *J. Hyg. Epidemiol. Microbiol. Immunol.*, **21**, 155–61.

Sharples, C. E., Shaw, M., Castes, M., Convit, J. & Blackwell, J. M. 1994. Immune response in healthy volunteers vaccinated with BCG plus killed leishmanial promastigotes: Antibody responses to mycobacterial and leishmanial antigens. *Vaccine*, **12**, 1402–12.

Shealy, C. N. 1995. A review of dehydroepiandrosterone (DHEA). *Integrative Physiol. Behav. Sci.*, **30**, 308–13.

Shek, P. N. & Sabiston, B. H. 1983. Neuroendocrine regulation of immune processes: Changes in circulating corticosterone levels induced by the primary antibody response in mice. *Int. J. Immunopharmacol.*, **5**, 23–33.

Shek, P. N., Sabiston, B. H., Buguet, A. & Shephard, R. J. 1995. Strenuous exercise and immunological changes: A multiple-time-point analysis of leukocyte subsets, CD4/CD8 ratio, immunoglobulin production and NK cell response. *Int. J. Sports Med.*, **16**, 466–74.

Sheldon, B. C. & Verhulst, S. 1996. Ecological immunology: Costly parasite defences and trade-offs in evolutionary ecology. *Trends Ecol. Evol.*, **11**, 317–21.

Shephard, R. J., Castellani, J. W. & Shek, P. N. 1998. Immune deficits induced by strenuous exertion under adverse environmental conditions: Manifestations and countermeasures. *Crit. Rev. Immunol.*, **18**, 545–68.

Shephard, R. J. & Shek, P. N. 1994. Infectious disease in athletes: New interest for an old problem. *J. Sports Med. Phys. Fitness.*, **34**, 11–22.

Shephard, R. J. & Shek, P. N. 1995. Heavy exercise, nutrition and immune function: Is there a connection? *Int. J. Sports Med.*, **16**, 491–7.

Shephard, R. J. & Shek, P. N. 1998. Immunological hazards from nutritional imbalance in athletes. *Exer. Immunol. Rev.*, **4**, 22–48.

Shi, H. N., Koski, K. G., Stevenson, M. M. & Scott, M. E. 1997. Zinc deficiency and energy restriction modify immune responses in mice during both primary and challenge infection with *Heligmosomoides polygyrus* (Nematoda). *Parasite Immunol.*, **19**, 363–73.

Shi, H. N., Scott, M. E., Stevenson, M. M. & Koski, K. G. 1998. Energy restriction and zinc deficiency impair the functions of murine T cells and antigen-presenting cells during gastrointestinal nematode infection. *J. Nutr.*, **128**, 20–7.

Shifrine, M., Rosenblatt, L. S., Taylor, N., Hetherington, N. W., Mathews, V. J. & Wilson, F. D. 1980a. Seasonal variations in lectin-induced lymphocyte transformation in beagle dogs. *J. Interdiscip. Cycle Res.*, **11**, 219–31.

Shifrine, M., Taylor, N., Rosenblatt, L. S. & Wilson, F. D. 1980b. Seasonal variation in cell mediated immunity of clinically normal dogs. *Exp. Hematol.*, **8**, 318–26.

Shimura, F., Shimura, J. & Hosoya, N. 1983. Biliary immunoglobulins in protein-energy malnourished rats. *J. Nutr. Sci. Vitaminol. (Tokyo)*, **29**, 429–38.

Shivatcheva, T. M. & Alexandrov, I. I. 1986. Seasonal decrease in natural cytotoxic activity of ground-squirrel spleen lymphocytes. *Dokladi Na Bolgarskata Akademiy a Na Naukite*, **39**, 121–4.

Shivatcheva, T. M. & Hadjioloff, A. I. 1987a. Seasonal involution of gut-associated lymphoid tissue of the European ground squirrel. *Dev. Comp. Immunol.*, **11**, 791–9.

Shivatcheva, T. M. & Hadjioloff, A. I. 1987b. Seasonal involution of the splenic lymphoid tissue of the European ground squirrel. *Arkhiv Anaomii, Gistologii, I Embriologii*, **92**, 48–53.

Sibley, W. A. & Paty, D. W. 1980. A comparison of multiple-sclerosis in Arizona (U.S.A.) and Ontario (CANADA); Preliminary report. *Acta. Neurol. Scand.*, **64**, 60–65.

Sidky, Y. A., Hayward, J. & Ruth, R. F. 1972. Seasonal variations of the immune response of ground squirrels kept at 22–24°C. *Can. J. Physiol. Pharmacol.*, **50**, 203–6.

Siegel, M. & Greenberg, M. 1955. Incidence of poliomyelitis in pregnancy: Its relation to maternal age, parity and gestational period. *N. Engl. J. Med.*, **253**, 841.

Silverin, B. 1981. Reproductive effort, as expressed in body and organ weights, in the pied fly catcher. *Ornis Scand.*, **12**, 133–9.

Silverin, B. 1997. The stress response and autumn dispersal behaviour in willow tits. *Anim. Behav.*, **53**, 451–9.

Silverin, B. 1998. Stress responses in birds. *Poult. Avian Biol. Rev.*, **9**, 153–68.

Silverin, B. 1999. Seasonal changes in mass and histology of the spleen in willow tits, *Parus montanus. J. Avian Biol.*, **30**, 255–62.

Silverman, A. J., Livne, I. & Witkin, J. W. 1994. The gonadotropin releasing hormone (GnRH), neuronal systems: Immunocytochemistry and in situ hybridization. In: *The Physiology of Reproduction*, 2nd ed. (eds. E. Knobil & J. D. Neill), pp. 1683–1709. New York: Raven Press.

Sinclair, J. A. & Lochmiller, R. L. 2000. The winter immunoenhancement hypothesis: Associations among immunity, density, and survival in prairie vole (*Microtus ochrogaster*) populations. *Can. J. Zool.*, **78**, 254–64.

Sinnecker, H., Sinnecker, R., Zilske, E. & Koehler, D. 1982. Detection of influenza A viruses and influenza epidemics in wild pelagic birds by sentinels and population studies. *Zentralbl. Bakteriol Mikrobiol Hyg. 1 Abt. Orig. A*, **253**, 297–304.

Sisk, C. L. & Turek, F. W. 1982. Daily melatonin injections mimic the short day-induced increase in negative feedback effects of testosterone on gonadotropin secretion in hamsters. *Biol. Reprod.*, **27**, 602–8.

Sjogren, A., Holmes, P. V. & Hillensjo, T. 1991. Interleukin-1 alpha modulates luteinizing hormone stimulated cyclic AMP and progesterone release from human granulosa cells in vitro. *Hum. Reprod.*, **6**, 910–3.

Skwarło-Sonita, K. 1996. Functional connections between the pineal gland and immune system. *Acta Neurobiol. Exp.*, **56**, 341–57.

Skwarło-Sonita, K., Rosołowska-Huszcz, D. & Sidorkiewicz, E. 1983. Diurnal changes in certain immunity indices and plasma corticosterone concentration in White Leghorn chickens. *Acta Physiol. Poland*, **34**, 445–56.

Skwarło-Sonita, K., Sotowska-Brochocka, J., Rosołowska-Huszcz, D., Pawowska-Wojewódka, E., Gajewska, A., Sepién, D. & Kochman, K. 1987. Effect of prolactin on the diurnal changes in immune parameters and plasma corticosterone in White Leghorn chickens. *Acta Endocrinol.*, **116**, 72–7.

Smith, D., Gupta, S. & Kaski, J. C. 1999. Chronic infections and coronary heart disease. *Int. J. Clin. Pract.*, **53**, 460–6.

Smith, E. M. & Blalock, J. E. 1981. Human lymphocyte production of ACTH and endorphin-like substances. Association with leukocyte interferon. *Proc. Natl. Acad. Sci. U.S.A.*, **78**, 7530–4.

Smith, E. M., Brosnan, P., Meyer, W. J. & Blalock, J. E. 1987. An ACTH receptor on human mononuclear leukocytes. Relation to adrenal ACTH-receptor activity. *N. Engl. J. Med.*, **317**, 1266–9.

Smith, E. M., Hughes, T. K., Cadet, P. & Stefano, G. B. 1992. Corticotropin-releasing factor-induced immunosuppression in human and invertebrate immunocytes. *Cell. Mol. Neurobiol.*, **12**, 473–81.

Smith, E. M., Morrill, A. C., Meyer, W. J., 3rd & Blalock, J. E. 1986. Corticotropin releasing factor induction of leukocyte-derived immunoreactive ACTH and endorphins. *Nature*, **321**, 881–2.

Smith, G. P. & Epstein, A. N. 1969. Increased feeding in response to decreased glucose utilization in the rat and monkey. *Am. J. Physiol.*, **217**, 1083–7.

Smith, V. G., Hacker, R. R. & Brown, R. G. 1977. Effect of alterations in ambient temperature on serum prolactin concentration in steers. *J. Anim. Sci.*, **44**, 645–9.

Snyder, D. S. & Unanue, E. R. 1982. Corticoids inhibit murine macrophage Ia expression and interleukin-1 production. *J. Immunol.*, **129**, 1803–5.

Sobel, E., Zhang, Z., Alter, M., Lai, S., Davanipour, Z., Friday, G., McCoy, R., Isack, T. & Levitt, L. 1987. Stroke in the Lehigh Valley. Seasonal variation in the incidence rates. *Stroke*, **18**, 38–42.

Soler, M., Martin-Vivaldi, M., Marin, J. M. & Moller, A. P. 1999. Weight lifting and health status in the black wheatear. *Behav. Ecol.*, **10**, 281–6.

Sonnenfeld, G., Measel, J., Loken, M. R., Degioanni, J., Follini, S., Galvagno, A. & Montalbini, M. 1992. Effects of isolation on interferon production and hematological and immunological parameters. *J. Interfer. Res.*, **12**, 75–81.

Spangelo, B. L., Judd, A. M., MacLoad, R. M., Goodmna, D. W. & Isakson, P. C. 1990. Endotoxin-induced release of interluekin-6 from rat mediobasal hypothalami. *Endocrinology*, **127**, 1779–85.

Spangelo, B. L., MacLeod, R. M. & Isakson, P. C. 1991. Production of interleukin-6 release from rat anterior pituitary cells *in vitro*. *Endocrinology*, **126**, 582–6.

Spangelo, B. L., Judd, A. M., Isakson, P. C. & MacLeod, R. M. 1991. Interleukin-1 stimulates interleukin-6 release from rat anterior pituitary cells *in vitro*. *Endocrinology*, **128**, 2685–93.

Spinedi, E., Giacomini, M., Jacquier, M. C. & Gaillard, R. C. 1991. Changes in the hypothalamo-corticotrope axis after bilateral adrenalectomy: Evidence for a median eminence site of glucocorticoid action. *Neuroendocrinology*, **53**, 160–70.

Spinedi, E., Suescun, M. O., Hadid, R., Daneva, T. & Gaillard, R. C. 1992. Effects of gonadectomy and sex hormone therapy on the endotoxin-stimulated hypothalamo-pituitary-adrenal axis: Evidence for a neuroendocrine-immunological sexual dimorphism. *Endocrinology*, **131**, 2430–6.

Spurlock, M. E. 1997. Regulation of metabolism and growth during growth challenge: An overview of cytokine function. *J. Anim. Sci.*, **75**, 1773–83.

Stearns, S. C. 1976. Life-history tactics: A review of the ideas. *Q. Rev. Biol.*, **51**, 3–47.

Steinman, L. 1996. Multiple sclerosis: A coordinated immunological attack against myelin in the central nervous system. *Cell*, **85**, 299–302.

Stephan, F. K. & Zucker, I. 1972. Circadian rhythms in drinking behavior and locomotor activity of rats are eliminated by hypothalamic lesions. *Proc. Natl. Acad. Sci. U.S.A.*, **69**, 1583–6.

Stephanou, A., Sarlis, N. J., Knight, R. A., Lightman, S. L. & Chowdrey, H. S. 1992. Glucocorticoid-mediated responses of plasma ACTH and anterior pituitary proopiomelanocortin, growth hormone, and prolactin mRNA during adjuvant-induced arthritis in the rat. *Neuroendocrinology*, **9**, 273–81.

Stephenson, L. S., Pond, W. G., Nesheim, M. C., Krook, L. P. & Crompton, D. W. 1980. *Ascaris suum*: Nutrient absorption, growth, and intestinal pathology in young pigs experimentally infected with 15-day-old larvae. *Exp. Parasitol.*, **49**, 15–25.

Sterky, G., Holmgren, G., Gustavson, K. H., Larsson, Y., Lundmark, L. M., Nilsson, K. O., Samuelson, G., Thalme, B. & Wall, S. 1978. The incidence of diabetes mellitus in Swedish children 1970–1975. *Acta Pædiatr Scand.*, **67**, 139–43.

Sternberg, E. M., Chrousos, G. P., Wilder, R. L. & Gold, P. W. 1992. The stress response and regulation of inflammatory disease. *Ann. Intern. Med.*, **117**, 854–66.

Sternberg, E. M., Young, W. S., 3rd, Bernardini, R., Calogero, A. E., Chrousos, G. P., Gold, P. W. & Wilder, R. L. 1989. A central nervous system defect in biosynthesis of corticotropin-releasing hormone is associated with susceptibility to streptococcal cell wall-induced arthritis in Lewis rats. *Proc. Natl. Acad. Sci. U.S.A.*, **86**, 4771–5.

Stimson, W. H. & Hunter, I. C. 1980. Oestrogen-induced immunoregulation mediated through the thymus. *J. Clin. Lab. Immunol.*, **4**, 27–33.

Stone, W. H. 1956. The J substance of cattle. III. Seasonal variation of the naturally occurring isoantibodies for the J substance. *J. Immunol.*, **77**, 369–76.

Stoop, J. W., Zegers, B. J. M., Sander, P. C. & Ballieux, R. E. 1969. Serum immunoglobulin levels in healthy children and adults. *Clin. Exp. Immunol.*, **4**, 101–12.

Styrt, B. & Sugarman, B. 1991. Estrogens and infection. *Rev. Infect. Dis.*, **13**, 1139–50.

Sullivan, D. A. & Wira, C. R. 1979. Sex hormone and glucocorticoid receptors in the bursa of Fabricius of immature chicks. *J. Immunol.*, **122**, 2617–23.

Süss, J., Schäfer, Sinnecker, H. & Webster, R. G. 1994. Influenza virus subtypes in aquatic birds of eastern Germany. *Arch. Virol.*, **135**, 101–14.

Suzuki, T., Suzuki, N., Daynes, R. A. & Engleman, E. G. 1991. Dehydroepiandrosterone enhances IL2 production and cytotoxic effector function of human T cells. *Clin. Immunol. Immunopath.*, **61**, 202–11.

Svensson, E., Råberg, L., Koch, C. & Hasselquist, D. 1998. Energetic stress, immunosuppression and the costs of an antibody response. *Funct. Ecol.*, **12**, 912–9.

Sze, S. F., Liu, W. K. & Ng, T. B. 1993. Stimulation of murine splenocytes by melatonin and methoxytryptamine. *J. Neural Transm.*, **94**, 115–26.

Szekeres-Bartho, J., Szekeres, G., Debre, P., Autran, B. & Chaouat, G. 1990. Reactivity of lymphocytes to a progesterone receptor-specific monoclonal antibody. *Cell. Immunol.*, **125**, 273–83.

Takagi, T. 1986. Longitudinal study on circadian rhythms of plasma hormone levels during Japanese Antarctic Research Expedition. *Hokkaido Igaku Zasshi*, **61**, 121–33.

Takahashi, J. S. 1996. The biological clock: It's all in the genes. *Prog. Brain Res.*, **111**, 5–9.

Takasu, N., Komiya, I., Nagasawa, Y., Asawa, T. & Yamada, T. 1990. Exacerbation of a autoimmune thyroid dysfunction after unilateral adrenalectomy in patients with Cushing's syndrome due to an adrenocortical adenoma. *N. Engl. J. Med.*, **322**, 1708–12.

Takasu, N., Ohara, N., Yamada, T. & Komiya, I. 1993. Development of autoimmune thyroid dysfunction after bilateral adrenalectomy in a patient with Carney's complex and after removal of ACTH-producing pituitary adenoma in a patient with Cushing's disease. *J. Endocrinol. Invest.*, **16**, 697–702.

Takhar, B. S. & Farrell, D. J. 1979. Energy and nitrogen-metabolism of chickens infected with either *Eimeria acervulina* or *Eimeria tenella*. *Br. Poult. Sci.*, **20**, 197–211.

Talal, N. & Ansar Ahmed, S. 1987. Sex hormones and autoimmune diseases: A short review. *Int. J. Immunother.*, **3**, 65–70.

Tälleklint, L., Jaenson, T. G. & Mather, T. N. 1993. Seasonal variation in the capacity of the bank vole to infect larval ticks (Acari: Ixodidae) with the Lyme disease spirochete, *Borrelia burgdorferi. J. Med. Entomol.*, **308**, 812–5.

Tamarkin, L., Cohen, M., Roselle, D., Reichert, C., Lippman, M. & Chabner, B. 1981. Melatonin inhibition and pinealectomy enhancement of 7,12-dimethylbenz(a)anthraceneinduced mammary tumors in the rat. *Cancer Res.*, **41**, 4432–6.

Tan, D. X., Poeggeler, B., Reiter, R. J., Chen, L. D., Hen, S., Manchester, L. C. & Barlow-Walden, L. R. 1993. Melatonin: A potent, endogenous hydroxyl radical scavenger. *Endocr. J.*, **1**, 57–67.

Tang, D., Santella, R. M., Blackwood, A. M., Young, T. L., Mayer, J., Jaretzki, A., Grantham, S., Tsai, W. Y. & Perera, F. P. 1995. A molecular epidemiological case-control study of lung cancer. *Cancer Epidemiol. Biomarkers Prev.*, **4**, 341–6.

Tang, F., Hsieh, A. C. L., Lee, C. P. & Baconshire, J. 1984. Interaction of cold and starvation in the regulation of plasma corticosterone levels in the male rat. *Horm. Metab. Res.*, **16**, 445–8.

Tartakovsky, B., De Baetselier, P., Feldman, M. & Segal, S. 1981. Sex-associated differences in the immune response against fetal major histocompatibility antigens. *Transplantation*, **32**, 395–7.

Tella, J. L., Bortolotti, G. R., Dawson, R. D. & Forenro, M. G. 2000. The T-cell mediated immune response and return rate of fledgling American kestrels are positively correlated with parental clutch size. *Proc. R. Soc. Lond., B*, **267**, 891–5.

Terrazas, L. I., Bojalil, R., Govezensky, T., & Larralde, C. 1994. A role for 17-beta-estadiol in immunoendocrine regulation of murine cysticercosis (*Taenia crassiceps*). *J. Parasitol.*, **80**, 563–8.

Teshima, S., Rokutan, K., Takahashi, M., Nikawa, T., Kido, Y. & Kishi, K. 1995. Alteration of the respiratory burst and phagocytosis of macrophages under protein malnutrition. *J. Nutr. Sci. Vitaminol.*, **41**, 127–37.

Theander, T. G., Hviid, L., Abu-Zeid, Y. A., Abdulhadi, N. H., Saeed, B. O., Jakobsen, P. H., Reimert, C. M., Jepsen, S., Bayoumi, R. A. L. & Jensen, J. B. 1990. Reduced cellular immune reactivity in healthy individuals during the malaria transmission season. *Immunol. Lett.*, **25**, 237–42.

Tian, L., Cai, Q., Bowen, R. & Wei, H. 1995. Effects of caloric restriction on age-related oxidative modifications of macromolecules and lymphocyte proliferation in rats. *Free Radic. Biol. Med.*, **19**, 859–65.

Tingate, T. R., Lugg, D. L., Muller, H. K., Stowe, R. P. & Pierson, D. L. 1997. Antarctic isolation: Immune and viral studies. *Immunol. Cell Biol.*, **75**, 275–83.

Tiuria, R., Horii, Y., Tateyama, S., Tsuchiya, K. & Nawa, Y. 1994. The Indian soft-furred rat, *Millardia meltada*, a new host for *Nippostrongylus brasiliensis*, showing androgen-dependent sex difference in intestinal mucosal defence. *Int. J. Parasitol.*, **24**, 1055–7.

Tomasi, T. E., Hellgren, E. C. & Tucker, T. J. 1998. Thyroid hormone concentrations in black bears (*Ursus americanus*): Hibernation and pregnancy effects. *Gen. Comp. Endocrinol.*, **109**, 192–9.

Tominaga, T., Fukata, J., Naito, Y., Usui, T., Murakami, N., Fukushima, M., Nakai, Y., Hirai, Y. & Imura, H. 1991. Prostaglandin-dependent *in vitro* stimulation of adrenocortical steroidogeneis by interleukins. *Endocrinology*, **128**, 526–31.

Tomkins, A. 1993. Environment, season and infection. In: *Seasonality and Human Ecology*, (ed. S. J. Ulizaszek & S. S. Strickland), pp. 123–34. London: Cambridge University Press.

Torrey, E. F., Miller, J., Rawlings, R. & Yolken, R. H. 2000. Seasonal birth patterns of neurological disorders. *Neuroepidemiology*, **19**, 177–85.

Touitou, Y., Touitou, C., Bogdan, A., Reinberg, A., Auzeby, A., Beck, H. & Guillet, P. 1986. Differences between young and elderly subjects in seasonal and circadian variations of total plasma proteins and blood volume as reflected by hemoglobin, hematocrit, and erythrocyte counts. *Clin. Chem.*, **32**, 801–4.

Trooster, W. J., Teelken, A. W., Kampinga, J., Loof, J. G., Nieuwenhuis, P. & Minderhoud, J. M. 1993. Suppression of acute experimental allergic encephalomyelitis by the synthetic sex hormone 17-alpha-ethinylestradiol: An immunological study in the Lewis rat. *Int. Arch. Allergy Immunol.*, **102**, 133–40.

Trujillo, E. B. 1993. Effects of nutritional status on wound healing. *J. Vascul. Nurs.*, **11**, 12–8.

Tsigos, C., Papanicolaou, D. A., Defensor, R., Mitsiadis, C. S., Kyrou, I. & Chrousos, G. P. 1997. Dose effects of recombinant human interleukin-6 on pituitary hormone secretion and energy expenditure. *Neuroendocrinology*, **66**, 54–62.

Turek, F. W., Alvis, J. D. & Menaker, M. 1977. Pituitary responsiveness to LRF in castrated male hamsters exposed to different photoperiodic conditions. *Neuroendocrinology*, **24**, 140–6.

Turnbull, A. V., Pitossi, F. J., Lebrun, J. J., Lee, S., Meltzer, J. C., Nance, D. M., del Rey, A., Besedovsky, H. O. & Rivier, C. 1997. Inhibition of tumor-necrosis factor-alpha action within the CNS markedly reduces the plasma adrenocorticotropin response to peripheral local inflammation in rats. *J. Neurosci.*, **17**, 3262–73.

Tyrrell, C. L. & Cree, A. 1998. Relationships between corticosterone concentration and season, time of day and confinement in a wild reptile (*Tuatara, Sphenodon punctatus*). *Gen. Comp. Endocrinol.*, **110**, 97–108.

Ueda, S. & Ibuka, N. 1995. An analysis of factors that induce hibernation in Syrian hamsters. *Physiol. Behav.*, **58**, 653–7.

U.S. Naval Observatory 1999. *Astronomical Almanac for the Year 1999*. U.S. Dept. of Defense, Washington, DC.

Valentine, G. L. & Kirkpatrick, R. L. 1970. Seasonal changes in reproductive and related organs in the pine vole, *Microtus pinetorum*, in southwestern Virginia. *J. Mammal.*, **51**, 553–60.

Van Heugten, E., Coffey, M. T. & Spears, J. W. 1996. Effects of immune challenge, dietary energy density, and source of energy on performance and immunity in weanling pigs. *J. Anim. Sci.*, **74**, 2431–40.

Vankelecom, H., Carmeliet, P., Heremans, H., Van Damme, J., Dijkmans, R., Billiau, A. & Denef, C. 1990. Interferon-gamma inhibits stimulated adrenocorticotropin, prolactin, and growth hormone secretion in normal rat anterior pituitary cell cultures. *Endocrinology*, **126**, 2919–26.

Van Loghem, J. J. 1928. Epidemiologische bijdrage tot de kennis van de ziekten der ademhalingsorganen. *Nederlands Tijdschr. Geneesk.*, **72**, 666–79.

Van Muiswinkel, W. B., Lamers, C. H. & Rombout, J. H. 1991. Structural and functional aspects of the spleen in bony fish. *Res. Immunol.*, **142**, 362–6.

Van Rood, Y., Goulmy, E., Blokland, E., Pool, J., Van Rood, J. & Van Houwelingen, H.

1991. Month-related variability in immunological test results: Implications for immunological follow-up studies. *Clin. Exp. Immunol.*, **86**, 349–54.

Varas, A., Torroba, M. & Zapata, A. G. 1992. Changes in the thymus and spleen of the turtle *Mauremys caspica* after testosterone injection: A morphometric study. *Dev. Comp. Immunol.*, **16**, 165–74.

Vasil'ev, S. V. 1979. Effect of melatonin on mitosis of the mouse asciteshpatoma Gelstein 22A. *Bioll. eksp Biol. Med.*, **88**, 595–6.

Vaughan, M. K., Hubbard, G. B., Champney, T. H., Vaughn, G. M., Little, J. C. & Reiter, R. J. 1987. Splenic hypertrophy and extramedullary hematopoiesis induced in male Syrian hamsters by short photoperiod or melatonin injections and reversed by melatonin pellets or pinealectomy. *Am. J. Anat.*, **179**, 131–6.

Vaz Nunes, M. & Saunders, D. 1999. Photoperiodic time measurement in insects: A review of clock models. *J. Biol. Rhythms*, **14**, 84–104.

Verde, T., Thomas, S. & Shephard, R. J. 1992b. Potential markers of heavy training in highly trained distance runners. *Br. J. Sports Medi.*, **26**, 167–75.

Verde, T. J., Thomas, S. G., Moore, R. W., Shek, P. & Shephard, R. J. 1992a. Immune responses and increased training of the elite athlete. *J. Appl. Physiol.*, **73**, 1494–9.

Verheul, H. A., Stimson, W. H., den Hollander, F. C. & Schuurs, A. H. 1981. The effects of nandrolone, testosterone and their decanoate esters on murine lupus. *Clin. Exp. Immunol.*, **44**, 11–7.

Vermeulen, M., Palermo, M. & Giordano, M. 1993. Neonatal pinealectomy impairs murine antibody-dependent cellular cytotoxicity. *J. Neuroimmunol.*, **43**, 97–102.

Villaseca, P., Llanos-Cuentas, A., Perez, E. & Davies, C. R. 1993. A comparative field study of the relative importance of *Lutzomyia peruensis* and *Lutzomyia verrucarum* as vectors of cutaneous leishmaniasis in the Peruvian Andes. *Am. J. Trop. Med. Hyg.*, **49**, 260–9.

Villa-Verde, D. M., Defresne, M. P., Vannier-dos-Santos, M. A., Dussault, J. H., Boniver, J. & Savino, W. 1992. Identification of nuclear triiodothyronine receptors in the thymic epithelium. *Endocrinology*, **131**, 1313–20.

Vindevogel, H., Debruyne, H. & Pastoret, P. P. 1985. Observation of *pigeon herpesvirus 1* re-excretion during the reproduction period in conventionally reared homing pigeons. *J. Comp. Pathol.*, **95**, 105–12.

Vriend, J. & Lauber, J. K. 1973. Effects of light intensity, wavelength and quanta on gonads and spleen of the deer mouse. *Nature*, **244**, 37–8.

Waage, A., Halstensen, A., Shalaby, R., Brandtzaeg, P., Kierulf, P. & Espevik, T. 1989. Local production of tumor necrosis factor alpha, interleukin 1, and interleukin 6 in meningococcal meningitis. Relation to the inflammatory response. *J. Exp. Med.*, **170**, 1859–67.

Wade, G. N. & Schneider, J. E. 1992. Metabolic fuels and reproduction in female mammals. *Neurosci. Biobehav. Rev.*, **16**, 235–72.

Wagnerová, M., Wagner, V., Mádlo, Z., Zavázal, V., Wokounová, D., Kříž, J. & Mohyla, O. 1986. Seasonal variations in the level of immunoglobulins and serum proteins of children differing by exposure to air-borne lead. *J. Hyg. Epidemiol. Microbiol. Immunol.*, **30**, 127–38.

Waid, D. D., Pence, D. B. & Warren, R. J. 1985. Effects of season and physical condition on the gastrointestinal helminth community of white-tailed deer from the Texas Edwards Plateau. *J. Wildlife Dis.*, **21**, 264–73.

Waldrop, R. D., Saydjari, R., Rubin, N. H., Rayford, P. L., Townsend, C. M. & Thompson, J. C. 1989. Photoperiod influences the growth of colon cancer in mice. *Life Sci.*, **45**, 737–44.

Warren, D. W., Pasupuleti, V., Lu, Y., Platler, B. W. & Horton, R. 1990. Tumor necrosis factor and interleukin-1 stimulate testosterone secretion in adult male rat Leydig cells in vitro. *J. Androl.*, **11**, 353–60.

Watson, H. R., Robb, R., Belcher, G. & Belch, J. J. F. 1999. Seasonal variation of Raynaud's phenomenon secondary to systemic sclerosis. *J. Rheumatol.*, **26**, 1734–7.

Weaver, D. R., Carlson, L. L. & Reppert, S. M. 1990. Melatonin receptors and signal transduction in melatonin-sensitive and melatonin-insensitive populations of white-footed mice (*Peromyscus leucopus*). *Brain Res.*, **506**, 353–7.

Weaver, D. R., Keohan, J. T. & Reppert, S. M. 1987. Definition of a prenatal sensitive period for maternal-fetal communication of day length. *Am. J. Physiol.*, **253**, E701–4.

Weaver, D. R., Liu, C. & Reppert, S. M. 1996. Nature's knockout: The Mel 1b receptor is not necessary for reproductive and circadian responses to melatonin in Siberian hamsters. *Mol. Endocrinol.*, **10**, 1478–87.

Weigent, D. A., Baxter, J. B., Wear, W. E., Smith, L. R., Bost, K. L. & Blalock, J. E. 1988. Production of immunoreactive growth hormone by mononuclear leukocytes. *FASEB J.*, **2**, 2812–8.

Weigent, D. A. & Blalock, J. E. 1995. Associations between the neuroendocrine and immune systems. *J. Leukoc. Biol.*, **58**, 137–50.

Weigent, D. A., Blalock, J. E. & LeBoeuf, R. D. 1991. An antisense oligodeoxynucleotide to growth hormone messenger ribonucleic acid inhibits lymphocyte proliferation. *Endocrinology*, **128**, 2053–7.

Weingarten, H. P. 1996. Cytokines and food intake: The relevance of the immune system to the student of ingestive behavior. *Neurosci. Biobehav. Rev.*, **20**, 163–70.

Weiss, J. M. 1968. Effects of coping response on stress. *J. Comp. Physiol. Psychol.*, **65**, 251–60.

Weksler, M. E. 1993. Immune senescence and adrenal steroids: Immune dysregulation and the action of dehydroepiandrosterone (DHEA) in old animals. *Eur. J. Clin. Pharmacol.*, **45**(Suppl. 1):S21–3; discussion S43–4.

Wells, P. W., Burrells, C. & Martin, W. B. 1977. Reduced mitogenic responses in cultures of lymphocytes from newly calved cows. *Clin. Exp. Immunol.*, **29**, 159–61.

Werb, Z., Foley, R. & Munck, A. 1978. Interaction of glucocorticoids with macrophages: Identification of glucocorticoid receptors in monocytes and macrophages. *J. Exp. Med.*, **147**, 1684–94.

Widy-Wirski, R., Berkley, S., Downing, R., Okware, S., Recine, U., Mugerwa, R., Lwegaba, A. & Sempala, S. 1988. Evaluation of the WHO clinical case definition for AIDS in Uganda. *J. Am. Med. Assoc.*, **260**, 3286–9.

Wiegers, G. J., Reul, J. M. H. M., Holsboer, F. & DeKloet, E. R. 1994. Enhancement of rat splenic lymphocyte mitogenesis after short term preexposure to corticosteroids in vitro. *Endocrinology*, **135**, 2351–7.

Wiesinger, H., Heldmaier, G. & Buchberger, A. 1989. Effect of photoperiod and acclimation temperature on nonshivering thermogenesis and GDP-binding of brown fat mitochondria in the Djungarian hamster. *Phodopus s. sungorus. Pflugers Arch.*, **413**, 667–72.

Wilder, R. L. 1995. Neuroendocrine-immune system interactions and autoimmunity. *Annu. Rev. Immunol.*, **13**, 307–38.

Williams, D. L., Climie, A., Muller, H. K. & Lugg, D. J. 1986. Cell-mediated immunity in healthy adults in Antarctica and the sub-Antarctic. *J. Clin. Immunol.*, **20**, 43–9.

Wingfield, J. C. 1994. Hormone-behavior interactions and mating systems in male and female birds. In: *The Differences Between the Sexes* (eds. R. V. Short & E. Balaban). pp. 303–30. New York: Cambridge University Press.

Wingfield, J. C., Deviche, P., Sharbaugh, S., Astheimer, L. B., Holberton, R., Suydam, R. & Hunt, K. 1994. Seasonal changes of the adrenocortical responses to stress in redpolls, *Acanthis flammea*, in Alaska. *J. Exp. Zool.*, **270**, 372–80.

Wingfield, J. C. & Farner, D. S. 1993. Endocrinology of reproduction in wild species. *Avian Biol.*, **9**, 163–327.

Wingfield, J. C. & Kenagy, G. J. 1991. Natural regulation of reproductive cycles. In: *Vertebrate Endocrinology: Fundamentals and Biomedical Implications* (ed. M. Schreibman & R. E. Jones), pp. 181–241. New York: Academic Press.

Withyachumnarnkul, B., Nonaka, K. O., Santana, C., Attia, A. M. & Reiter, R. J. 1990. Interferon-gamma modulates melatonin production in rat pineal gland in organ culture. *J. Interfer. Res.*, **10**, 403–11.

Withyachumnarnkul, B., Reiter, R., Lerchl, A., Nonaka, K. O. & Stokan, K. A. 1991. Evidence that interferon-γ alters pineal metabolism both indirectly via sympathetic nerves and directly on the pinealocytes. *Int. J. Biochem.*, **23**, 1397–1401.

Wood, D. L., Sheps, S. G., Elveback, L. R. & Schirder, A. 1984. Cold pressor test as a predictor of hypertension. *Hypertension*, **6**, 301–6.

Woodfill, C. J., Robinson, J. E., Malpaux, B. & Karsch, F. J. 1991. Synchronization of the circannual reproductive rhythm of the ewe by discrete photoperiodic signals. *Biol. Reprod.*, **45**, 110–21.

Woodfill, C. J., Wayne, N. L., Moenter, S. M. & Karsch, F. J. 1994. Photoperiodic synchronization of a circannual reproductive rhythm in sheep: Identification of season-specific time cues. *Biol. Reprod.*, **50**, 965–76.

Wu, F. C., Wallace, E. M., Howe, D. C., Kacser, E. M. & Sellar, R. E. 1991. Effects of decreasing exogenous GnRH pulse frequency on FSH and inhibin secretion in men with idiopathic hypogonadotropic hypogonadism. *Acta Endocrinol.*, **125**, 138–45.

Wunder, B. A. 1992. Morphophysiological indicators of the energy state of small mammals. In: *Mammalian Energetics: Interdisciplinary Views of Metabolism and Reproduction* (eds. T. E. Tonasi & T. H. Horton), pp. 86–104. New York: Cornell University Press.

Wurtman, R. J. & Weisel, J. 1969. Environmental lighting and neuroendocrine function: Relationship between spectrum of light source and gonadal growth. *Endocrinology*, **85**, 1218–21.

Wuthrich, R. & Rieder, H. P. 1970. The seasonal incidence of multiple sclerosis in Switzerland. *Eur. Neurol.*, **3**, 257–64.

Wybran, J. 1985. Enkephalins and endorphin as modifiers of the immune system: Present and future. *Fed. Proc.*, **44**, 92–4.

Wysocki, A., Kaminski, W. & Krzywon, J. 1999. Seasonal periodicity of perforated peptic ulcers. *Przegl. Lek*, **56**, 189–91.

Xia, Y., Hu, H. Z., Liu, S., Ren, J., Zafirov, D. H. & Wood, J. D. 1999. IL-1beta and IL-6 excite neurons and suppress nicotinic and noradrenergic neurotransmission in guinea pig enteric nervous system. *J. Clin. Invest.*, **103**, 1309–16.

Xiong, J. J., Karsch, F. J. & Lehman, M. N. 1997. Evidence for seasonal plasticity in the gonadotropin-releasing hormone (GnRH) system of the ewe: Changes in synaptic inputs onto GnRH neurons. *Endocrinology*, **138**, 1240–50.

Yamaguchi, M., Koike, K., Yoshimoto, Y., Ikegami, H., Miyake, A. & Tanizawa, O. 1991. Effect of TNF-alpha on prolactin secretion from rat anterior pituitary and dopamine release from the hypothalamus: Comparison with the effect of interleukin-1 beta. *Endocrinol. Jpn.*, **38**, 357–61.

Yamamoto, T., Hirata, H., Taniguchi, H., Kawai, Y., Uematsu, A. & Sugiyama, Y. 1980. Lymphocyte transformation during pregnancy: an analysis using whole-blood culture. *Obstet Gynecol.*, **55**, 215–9.

Yeh, S. S. & Schuster, M. W. 1999. Geriatric cachexia: the role of cytokines. *Am. J. Clin. Nutr.*, **70**, 183–97.

Yellon, S. M., Fagoaga, O. R. & Nehlsen-Cannarella, S. L. 1999. Influence of photoperiod on immune cell functions in the male Siberian hamster. *Am. J. Physiol.*, **276**, R97–R102.

Yuen, J., Ekbom, A., Trichopoulos, D., Hsieh, C. C. & Adami, H. O. 1994. Season of birth and breast cancer risk in Sweden. *Br. J. Cancer*, **70**, 564–8.

Zaman, K., Baqui, A. H., Yunnus, M., Sack, R. B., Chowdhury, H. R. & Black, R. E. 1997. Malnutrition, cell-mediated immune deficiency and acute upper respiratory infections in rural Bangladeshi children. *Acta Paediatr.*, **86**, 923–7.

Zapata, A. G., Varas, A. & Torroba, M. 1992. Seasonal variations in the immune system of lower vertebrates. *Immunol. Today*, **13**, 142–7.

Zaugg, J. L. 1990. Seasonality of natural transmission of bovine anaplasmosis under desert mountain range conditions. *J. Am. Vet. Med. Assoc.*, **196**, 1106–9.

Zeitzer, J. M., Daniels, J. E., Duffy, J. F., Klerman, E. B., Shanahan, T. L., Dijk, D. J. & Czeisler, C. A. 1999. Do plasma melatonin concentrations decline with age? *Am. J. Med.*, **109**, 343–5.

Zelazowski, P., Dohler, K. D., Stepien, H. & Pawlikowski, M. 1989. Effect of growth hormone-releasing hormone on human peripheral blood leukocyte chemotaxis and migration in normal subjects. *Neuroendocrinology*, **50**, 236–9.

Zucker, I. 1985. Pineal gland influences period of circannual rhythms of ground squirrels. *Am. J. Physiol.*, **249**, R111–5.

Zucker, I. 2001. Circannual clocks. In: *Handbook of Behavioral Neurobiology* (eds. J. Takahashi, F. W. Turek & R. Y. Moore), pp. 509–28. New York: Kluwer/Plenum.

Zuk, M. 1990. Reproductive strategies and sex differences in disease susceptibility: An evolutionary viewpoint. *Parasitol. Today*, **6**, 231–3.

Zuk, M. & Johnsen, T. S. 1998. Seasonal changes in the relationship between ornamentation and immune response in red jungle fowl. *Proc. R. Soc. Lond., B*, **265**, 1631–5.

Zuk, M., Johnsen, T. S. & MacLarty, T. 1995. Endocrine-immune interactions, ornaments and mate choice in red jungle fowl. *Proc. R. Soc. Lond., B*, **260**, 205–10.

Zuk, M. & McKean, K. A. 1996. Sex differences in parasite infections: Patterns and processes. *Int. J. Parasitol.*, **26**, 1009–24.

Zylinska, K., Komorowski, J., Robk, T., Mucha, S. & Stepien, H. 1995. Effect of granulocyte-macrophage colony stimulating factor and granulocyte colony stimulating factor on melatonin secretion in rats. In vivo and in vitro studies. *J. Neuroimmunol.*, **56**, 187–90.

Index